Fundamentals of Fire Protection

Arthur E. Cote, P.E., Editor-in-Chief

Executive Vice President and Chief Engineer
National Fire Protection Association

National Fire Protection Association
Quincy, Massachusetts

Senior Product Manager: Pam Powell
Developmental Editor: Khela Thorne
Editorial Assistant: Josiane Domenici
Editorial-Production Services: Publishers' Design and Production Services, Inc.
Composition: Publishers' Design and Production Services, Inc.
Interior Design: Cheryl Langway
Cover Design: Cameron Kelly
Manufacturing Manager: Ellen Glisker
Printer: Courier/Westford

Copyright ® 2004
National Fire Protection Association, Inc.
One Batterymarch Park
Quincy, Massachusetts 02169

NFPA No.: PFP04
ISBN: 0-87765-595-2
Library of Congress Card Catalog No.: 2003112680

Printed in the United States of America
04 05 06 07 08 5 4 3 2 1

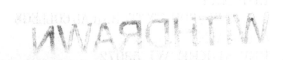

Contents

Preface

Since NFPA published the second edition of *Principles of Fire Protection* in 1988, the field of fire protection and training for careers in fire protection have both undergone substantial changes. This first edition of *Fundamentals of Fire Protection* is the successor to *Principles* and addresses those changes. For example, our understanding of the chemistry and physics of fire—and especially of the toxic effects of fire and smoke—is much more refined than in the late 1980s. The role and techniques of fire prevention, including fire and life safety education, have expanded. The technology of fire detection, fire suppression, and smoke control has become progressively more sophisticated. In addition, the role of the fire service as emergency first responders has grown from infancy to established fact.

As fire protection has changed, so too has training for fire protection careers. One influence on fire service training has been the development of a model curriculum by the Fire and Emergency Services Higher Education (FESHE) Conference organized by the National Fire Academy. A group of faculty from more than 100 institutions of higher learning with a fire science curriculum, FESHE created educational objectives and course outlines for several core courses. *Fundamentals of Fire Protection* is organized around the FESHE outline for a "Fundamentals of Fire Protection" course. The 12 chapters are designed for a 12- or 13-week semester of study, whether in the classroom or self-study. Each chapter includes specific learning objectives based on FESHE and student activities. As the following chapter list demonstrates, this book is useful to students and to anyone seeking a broad-based introduction to the principles of fire protection.

Fundamentals of Fire Protection opens with "Fire in History and Contemporary Life" and presents additional background information in Chapter 2, "Understanding America's Fire Problem," Chapter 3, "Fire Behavior," and Chapter 4, "Building Design and Construction."

The next several chapters cover fire department organization and operations. These chapters include Chapter 5, "Fire Department Structure and Management," Chapter 6, "Fire Department Facilities and Equipment," Chapter 7, "Preventing Fire Loss," Chapter 8, "Controlling Fire Loss through Active Fire Protection Systems," Chapter 9, "Fire Investigation," and Chapter 10, "Planning for Emergency Response."

The book concludes with Chapter 11, "Public and Private Support Organizations," and Chapter 12, "Careers in Fire Protection."

Any book is the work of many people, and this book is no exception. We at NFPA are grateful to Clinton Smoke and Richard Bennett for their review of the predecessor book *Principles of Fire Protection* and for the development of educational objectives for *Fundamentals of Fire Protection*. Professor Bennett also prepared the review questions and learning activities that are prominent features of each chapter. Authors Clinton Smoke, Marty Ahrens, Richard Pehrson, Jon Nisja, Richard Bennett, Robert Tutterow, Chuck Kime, Richard Custer, Bill Neville, Robert Fleming, and Doug Forsman ably translated their practical knowledge and classroom experience into chapters.

Here at NFPA, Assistant Vice President Gary Tokle, Senior Product Manager Pam Powell, and Developmental Editor Khela Thorne helped shape the individual chapters into a book that functions as a whole.

Fire protection will certainly continue to evolve with time, and NFPA codes and standards will keep pace with changes in our field. For that reason, readers should always consult the most current editions of NFPA codes and standards referenced in *Fundamentals of Fire Protection* to stay on top of new developments.

Arthur E. Cote, P.E., Editor-in-Chief
Executive Vice President and Chief Engineer, NFPA

About the Contributors

Marty Ahrens has been a fire analysis specialist at NFPA since 1997. Prior to joining NFPA, Ahrens spent 11 years as a research analyst and coordinator of the Massachusetts Fire Incident Reporting System in the office of the State Fire Marshal of Massachusetts.

Richard Bennett is an assistant professor of fire protection at the University of Akron and has presented at the NFPA's World Safety Conference and Exposition. He is a member of the National Fire Science Curriculum Committee and is a reviewer for the American Council on Education. He has reviewed management courses at the National Fire Academy for academic credit.

Richard Custer is associate principal and technical director of Arup Fire USA and has more than 35 years experience in fire protection and fire investigation. He has served on the fire protection faculties of the University of Maryland and Worcester Polytechnic Institute and as research engineer and manager at the Center for Fire Research at the National Bureau of Standards (now the National Institute for Standards and Technology). He also served as chair and secretary of the NFPA Technical Committee on Fire Investigations, chair of NFPA's Fire Science and Technology Educators Section, and on the Technical Working Group on Fire and Arson Scene Evidence of the U.S. Department of Justice.

Dr. Robert S. Fleming, a professor of management and management information systems at Rowan University, is chairman of the National Fire Academy Board of Visitors, chairman of the Pennsylvania Fire Service Certification Advisory Committee, chairman of the Chester County (PA) Local Emergency Planning Committee, first vice president of the Keystone Chapter Fire Service Instructors, southern vice president of the New Jersey Society of Fire Service Instructors, and director of governmental relations for the International Society of Fire Service Instructors. Dr. Fleming is a Certified Fire Protection Specialist and certified to the Fire Officer II and Fire Instructor II levels by the National Board on Fire Service Professional Qualifications.

Douglas Forsman has served in various fire service positions since 1964. In addition to service as both a career and volunteer fire chief, he has been a staff member of the NFPA and Director of Fire Service Programs at Oklahoma State University. For many years Chief Forsman served as the Chair of the NFPA Fire Service Professional Qualifications Projects. He is currently the Chief of the Union Colony Fire/Rescue Authority in Greeley, Colorado.

Charles H. Kime is an assistant professor at Arizona State University east campus where he coordinates the Fire Service Programs in the College of

Technology and Applied Sciences. Dr. Kime served on the Phoenix, Arizona, Fire Department more than 32 years, where he served as the Phoenix fire marshal for 13 years. His research interests include fire service related topics, organizational leadership, organizational behavior, and human resource management. Dr. Kime earned the Ph.D. in Public Administration from ASU. His book, *Organizational Leadership: Fire Services in the United States,* was published in 2001.

William Neville, who has been a fire service consultant since 1991, previously served as NFPA Assistant Vice President for Fire Service Affairs, superintendent of the National Fire Academy, fire chief for Hayward California, and assistant fire chief for the Los Angeles City Fire Department. He also served on the California governor's Bay Area Emergency Planning Council, an organization that worked to plan a response to major disasters, such as earthquakes and hazardous materials spills.

Jon Nisja is a supervisor of inspections for the Minnesota State Fire Marshal Division. He has been involved in fire prevention inspections and investigations since 1982. His areas of interest include building construction, means of egress, and fire protection systems (sprinklers and fire alarms).

Richard Pehrson is a registered fire protection engineer at Futrell Fire Consult and Design located in Minnesota, where he is also a volunteer fire fighter. After earning a B.S. in fire protection from Oklahoma State University, he went on to earn the M.S. and Ph.D. in fire protection engineering from Worcester Polytechnic Institute. He is also a former deputy state fire marshal and currently is a member of the NFPA 13 Technical Committee.

Professor Clinton Smoke is the chair of the Fire Protection Technology Program at Asheville-Buncombe Technical Community College in Asheville, North Carolina. He is also the former chair of the National Fire Science Curriculum Committee and the former editor of *Fire Command,* now part of NFPA's *Fire Journal.* He has a bachelor's degree in fire service administration from John Jay College in New York and a master's degree in business administration from The College of William and Mary in Virginia. Mr. Smoke is the secretary and serves as a director of NFPA's Fire Science and Technology Educators section.

Robert Tutterow is the health and safety officer with the Charlotte, North Carolina, Fire Department and has more than 26 years of fire service experience. He served as the department's Logistics Officer for 14 years. He is a member of the following NFPA Committees: Technical Correlating Committee on Fire and Emergency Services Protective Clothing and Equipment, Technical Committee on Structural Fire Fighting Protective Clothing and Equipment, and the Technical Committee on Fire Department Apparatus.

1 Fire in History and Contemporary Life

Clinton Smoke
Asheville-Buncombe Technical Community College

Learning Objectives

After completing this chapter, the reader should be able to do the following:

- Describe the role of fire throughout history, in early America, and in America during the 20th century
- Explain the role of the contemporary fire service in the United States
- Understand the legal foundation that underlies the fire service in the United States and its ability to provide fire protection, administer fire codes, conduct fire investigations, and carry out other department functions
- Understand how fire departments are evaluated, the impact of this evaluation on the cost of fire insurance for property owners, and the values added by having a fire department accredited by an outside agency

---

The lessons of history are important to society, especially where we experience tragedy. For longer than recorded history, fire has been a source of comfort and catastrophe for the human race. From such experiences, we have made progress in learning how to control fire and in making our world a safer place. With these thoughts in mind, let us review the events that have brought us to where we are today.

 ## ROLE OF FIRE THROUGHOUT HISTORY

Early History

Lightning, no doubt, caused the first fire seen by humans. From early fires, they saw both the value and the destructive power of fire. Early humans did

1

not know how to start fires so they learned to keep natural fires burning, under controlled conditions of course. With controlled fire, humans learned of fire's many uses. Fire kept them warm and cooked their food. Fire provided light as well, allowing them to extend their day, and to venture into caves and other normally dark areas. On a larger scale, humans used fire to clear land in order to grow crops. These applications illustrate human use of controlled fire.

> *Ever since humans first discovered the warmth of fire,*
> *it has been a source of both comfort and catastrophe.*

The step from controlling a burning fire to creating a new fire took thousands of years, and we are not sure when this first occurred. We know that Neolithic humans had the ability to create fire. Archeological evidence suggests that as early as 7000 BCE, humans had fire-making techniques, starting fires by using friction-producing tools or by striking a flint against a hard stone. Yet, even with these tools, it was easier to keep an existing fire burning than to try to start one.

Fire has been a part of our civilization since the very beginning. In addition to the uses already mentioned, fire was used to make pottery as early as 3000 BCE. Fire was used to melt and combine copper and tin to produce bronze (c. 3000 BCE) and iron (c. 1000 BCE). An entire chapter of this book could be devoted to fire, the origins of fire protection, and fire prevention over the next 5000 years. Yet this book is not intended as a history book, so the reader is directed to the Suggested Readings about the history of fire and fire protection at the end of this chapter.

Having said that, one story should be told. It illustrates many of the points this chapter aims to make, and the time and place tie in with what follows.

Great London Fire of 1666

The setting is London, England. It is likely that London has had some form of firefighting from as early as the time of the Romans. However, after the Roman armies left Britain in 415 AD, any organized attempts to fight fires were abandoned. Following the Norman Conquest in 1066, King William the Conqueror insisted that all fires should be put out at night to reduce the risk of fire in houses with straw carpets and thatched roofs. William's law of *couvre-feu* (literally, cover fire) became the modern term curfew. Even with William's law, a huge fire destroyed a large part of the city in 1212 and was said to have killed some 3000 people. This fire was known as the Great Fire of London until September 2, 1666.

On Sunday morning, September 2, 1666, the destruction of medieval London began. Within 5 days the city was destroyed by fire. In all this destruction, it is amazing that only six people died.

The fire started in the house and shop of Thomas Farynor, baker to King Charles II in Pudding Lane. Farynor forgot to douse the fire in his oven on the previous night and embers set fire to the nearby-stacked firewood. By one o'clock in the morning, 3 hours after Farynor had gone to bed, the house and shop were well alight. Farynor's assistant awoke to find the house full of smoke and he roused the household. Farynor, his wife, daughter, and one servant escaped by climbing through an upstairs window and along the rooftops to safety. The maid was too frightened to climb along the roof and stayed in the house, becoming the first fire victim.

Sparks from the burning house fell on hay and straw in the yard of the Star Inn at Fish Street Hill. The London of 1666 was a city of half-timbered and pitch-covered medieval buildings, mostly with thatched roofs. These buildings were extreme fire risks and ignited very easily. In the strong winds that blew that morning, the sparks spread rapidly, setting fire to roofs and houses as they fell. From the Star Inn, the fire engulfed St. Margaret's church and then entered Thames Street, where there were warehouses and wharves packed with flammable and combustible materials such as oil, spirits, tallow, hemp, straw, and coal. By then the fire was far too fierce to be fought with the crude hand-operated devices that were all that was available. By 8:00 AM, 7 hours after the fire started, the flames were halfway across old London Bridge. Only the gap left by a previous fire in 1633 prevented the flames from crossing the bridge and starting new fires in Southwark on the south bank of the river.

The fires burned all that day and on through the next. Fleet Street, Old Bailey, Ludgate Hill, and Newgate were all reduced to ashes. The stones of St. Paul's Cathedral were reported to be exploding with the heat, and molten lead from the roof ran down the streets in a stream. The strong easterly winds kept the flames advancing. Little could be done to stop the spread of the fire. Various laws had been enacted, obliging the parishes to provide buckets, ladders, and fire hooks, but much of the equipment was rotten because of neglect, and, away from the banks of the river, water supplies were scarce. The fire raged through the city for 3 more days, finally burning out at Temple Church near Holborn Bridge [1]. See Figure 1.1.

By the end of the fire, some four-fifths of the city had been destroyed—approximately 13,200 houses, 87 churches, and 50 livery halls spread over

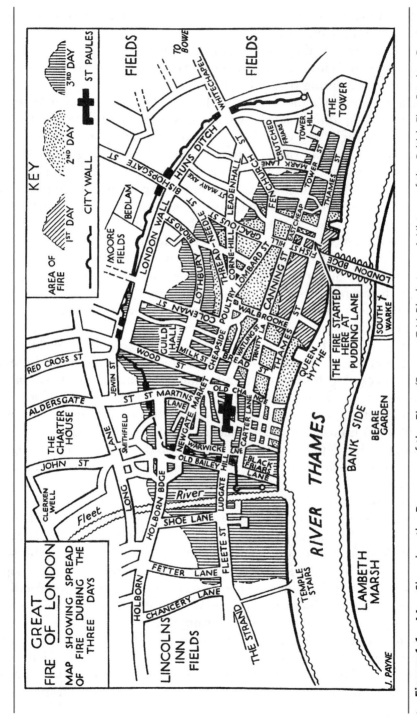

▲ **Figure 1.1** Map Showing the Progress of the Fire. *(Source: G.V. Blackstone, A History of the British Fire Service, Routledge and Kegan Paul, London, UK, 1957.)*

an area of 436 acres. Although the fire only claimed a few lives, it may actually have saved many more. In destroying the close-packed houses and other buildings, it is also likely that the fire finally put an end to the Great Plague that had devastated the city in the previous year, killing 17,440 out of the population of 93,000. Most of the rats that had helped to transmit the bubonic plague (otherwise known as Black Death) the previous year died in the fire. The number of plague victims dropped rapidly after the fire.

The Great Fire of London set in motion changes in the capital that laid the foundations for organized firefighting in the future. Wooden houses and designs dating back to the medieval period were replaced with brick and stone buildings, and owners began to insure their properties against fire damage. Christopher Wren, the great 17th-century architect, began the reconstruction of London and built 49 new churches together with the great cathedral of St. Paul's that we know today. After the fire of 1666, the face of London had changed forever.

Fire Marks

The late 17th century saw the beginnings of the insurance industry and companies soon realized that they could make money by offering fire cover policies to owners. Charters were granted allowing this business to flourish. Before long, the insurance companies realized that their losses could be minimized by employing men to put out fires that started in properties covered by their insurance. The companies introduced new fire engines with improved pumps (all manually operated) and recruited men from the group of watermen who worked on the barges and lighters on the Thames. These men would agree to be retained by the insurance company to be called out in the event of a fire.

Of course, the insurance companies were only interested in the properties that they insured, so they introduced a scheme to mark properties. Each policyholder was issued a metal badge or mark that was fixed to the outside of a building as shown in Figure 1.2. When a fire broke out, it was not unusual for firemen from several companies to arrive at the scene. If the building did not carry their mark, they would leave, often leaving the building to burn down.

The firemen arrived with a manual pump. Water could be poured into the trough from buckets and then pumped (by manually cranking the levers) through a leather hose to a metal nozzle providing a jet of water to be directed at the fire. Early water mains consisted of hollowed out tree trunks

▶ **Figure 1.2** Example of a Fire Mark.

laid beneath major roads. The fire fighters dug down into the roadway and bored a hole to open the top of the trunk. A pipe could then be inserted to take water for the pump. However, the water supply was poor so it probably was unable to support more than one pump. The first crew that arrived at a fire got water to put out the fire. Fierce competition developed—the first company to arrive and tap into the trunk would be seen to be the most effective at putting out fires and would stand to gain additional business. Fights often broke out between rival crews attempting to bore into the same trunk supply.

Throughout the 18th century, the insurance industry expanded and as it did so, it came to realize that it was in the interests of all companies for its brigades to cooperate. Eventually, in 1833, the Sun Insurance Company, together with 10 other companies, agreed to set up the first single firefighting force to cover London—the London Fire Engine Establishment [2].

ROLE OF FIRE IN AMERICA

America has been fighting fires for almost four centuries. During that time Americans have been exposed to millions of dangerous fire situations, and thousands have lost their lives. Fire has devastated our cities, our farms, and our forests. Although most of these fires were accidental, many were deliberately set for financial gain or as a savage weapon of revenge.

Fire is used to further man's most noble efforts;
it is also used to further his most ominous efforts.

At some point in our education, we learned about the brave men and women who settled in the new colonies along our eastern seaboard. They came from England, France, Spain, the Netherlands, and other countries, ready to leave their homes and traditional lifestyles for the chance of freedom and opportunity in a new land. From our history lessons we know that the early settlers faced a difficult ocean voyage, and on arrival in the New World, many were immediately challenged in establishing themselves in their new land. Harsh weather and challenging conditions added to the difficulty associated with relocation. Fire was also a problem, but this is often overlooked in history lessons.

Fire was a frequent event throughout our country's early history. Not much was in place to prevent fire tragedies from occurring again and again: there was little organized fire protection, and there was little legislation to control the design and use of buildings. Early settlers' priorities focused on obtaining food and shelter by the most expedient means available. Within their modest homes, the fireplace was an important part of life for personal warmth and for cooking food. Unfortunately, their fires often ignited unintended materials. Their houses likely contained spirits and gunpowder, both of which contributed to the rapid growth of spreading fires. Many of the roofs were covered with highly combustible thatch. In towns the houses were built close together for protection, and fire would often spread rapidly from one house to another.

We have recorded evidence of fires from as early as 1608 in Jamestown, Virginia. Just 5 years later the Dutch settlers in New Amsterdam (later New York City) lost their shelter, the ship where they were living, due to fire. In 1620, fire nearly destroyed the colony at what is now Plymouth, Massachusetts [3]. Boston had a problem with fires as early as 1630, the year the town was chartered. Two major contributors to these problems, thatch roofs and wooden chimneys, were soon outlawed, but wood continued to be the choice of materials for constructing buildings [3].

New Amsterdam took a somewhat different approach. Peter Stuyvesant, the colonial governor of New York, got legislation enacted as early as 1648 that allowed the appointment of fire wardens who, without pay, would enforce the new fire laws. Stuyvesant appointed four wardens who were empowered to inspect chimneys and fine the owners of chimneys that were not properly swept. The new wardens also had the authority to impose a fine on any citizen who negligently caused a fire that destroyed property. In 1659, he also appointed fire wardens who patrolled the streets at night looking for fires [4].

Money raised from the fines was used to buy firefighting equipment—buckets, ladders, and hooks. The hooks were attached to ropes, and if a building was threatened by fire, the fire fighters would use the hooks to peel off the roof and pull down the walls. In those days fire fighters were not effective in controlling a fully involved structure, so as an alternative they would remove the houses most exposed to the one that was on fire, and thus save a neighborhood from a greater loss.

Over the years, Boston had many fires. Conflagrations occurred in 1676 and 1679. Other cities had similar experiences. Baltimore, Maryland; Richmond, Virginia; Charleston, South Carolina; and Savannah, Georgia, all had repeated and serious fires. See Table 1.1. In nearly every case, the tragedy provoked public action. The results of these actions included efforts to organize fire protection, to regulate industrial activity, and to improve building practices through local ordinances [3].

Early Fire Protection

While the loss of property is always of concern, in our society there is even greater concern over the loss of life. A fire in Boston in 1653 claimed the lives of three children, the first recorded instance of loss of life due to fire in the new country. After this tragic occurrence, the citizens of that great city moved quickly to establish storage places for water and to provide equipment for firefighting [3].

By today's standards, early efforts to provide organized fire protection were quite modest. There were no paid fire departments, and volunteer fire companies were few and far between. In most communities, the citizens purchased buckets, ladders, and similar equipment, and called on their neighbors during time of need. Following the fire in London in 1666, English insurance companies established their own fire companies. Insurance companies in America undertook similar action. Some volunteer fire companies competed against each other to provide fire protection, starting several traditions that remain in place in some communities even today.

▶**Table 1.1**   Famous U.S. Conflagrations

Year	Location
1861	Charleston, South Carolina
1866	Portland, Maine
1871	Chicago, Illinois
1871	Peshtigo, Wisconsin
1872	Boston, Massachusetts
1889	Boston, Massachusetts
1900	Hoboken, New Jersey
1901	Jacksonville, Florida
1902	Paterson, New Jersey
1903	Baltimore, Maryland
1906	San Francisco, California
1907	Chelsea, Massachusetts
1911	Bangor, Maine
1914	Salem, Massachusetts
1915	Paris, Texas
1916	Nashville, Tennessee
1917	Atlanta, Georgia

Although Peter Stuyvesant organized the first volunteer fire protection efforts in 1648, Benjamin Franklin, early American printer and statesman, generally gets credit for organizing the first community fire department in 1736 in Philadelphia, Pennsylvania. Franklin was also interested in fire prevention and used his newspaper as a vehicle for what may have been this nation's first fire safety education program. He urged homeowners to keep their chimneys clean and warned of the dangers of carrying live coals in bed warmers, noting, "You may be forced, as I once was, to leap out of your windows and hazard your necks to avoid being over roasted" [4].

Other communities followed suit: Savannah, Georgia, had organized fire protection in 1759, and Alexandria, Virginia, had a department in 1774. That organization had several famous members, including George Washington. All of these were volunteer organizations. It would be nearly another 100 years before these organizations started to hire full-time paid employees.

If America's early fire departments seem primitive by today's standards, the equipment they had must appear ancient. Imagine fighting fire without a fire truck or even without a fire hose. That is the way it was for centuries.

There were a few water pumps for firefighting, but they were manually pow-ered and required a large number of men to effectively operate them. These pumps required water that had to be carried in buckets from the source to the pump. Both of these activities required a great number of strong and willing participants.

Two inventions of the latter 17th century improved firefighting. The first was the invention of leather hose to allow greater maneuverability of the water stream for firefighting. The second invention was a hose that would allow the pump to take water from a standing source. In today's terminology, this suction hose allowed the pump to draft water.

Fire in North America's Wars

Revolutionary War. Fire is often used as a weapon of destruction during wartime and it was no different when America was fighting for its own inde-pendence. William Dudington, captain of the British revenue schooner *Gaspee*, was conducting a siege of sorts along the Rhode Island coast and was quoted as saying that he would eventually burn the port city of Newport "around the heads of its inhabitants." On June 9, 1772, the coastal trader *Hanna* was hailed by the *Gaspee*, but rather than stopping, the *Hanna* sailed into Newport Bay. The larger *Gaspee* followed, eventually running aground. Once word of the vessel's stranding reached shore, the citizens were quick to attack the British ship. Under the cover of darkness, the colonists surprised the British, overpowered the crew, and set fire to the ship [3]. In 1775, the British sailed into the harbor at Falmouth, Maine, and set fire to the com-munity in reprisal for the aggressive action of the colonists in the region [3].

War of 1812. A few years later, America was at war again. The War of 1812 was fought on many fronts, and settlements were routinely burned. In Au-gust 1814, near the end of the war, the British attacked Washington, D.C., setting fire to the White House, the Capitol, and other government build-ings. Leaving Washington in ruins, the British sailed further up the Chesa-peake to Baltimore, Maryland. Fort McHenry defended Baltimore, and during the night of September 13, 1814, the British shelled the fort. On an-other ship, Francis Scott Key recorded his observations with words that were eventually set to music and remain with us today as "The Star-Spangled Banner."

Civil War. Half a century later, our nation was at war once again. This time it was a great civil war wherein Americans fought against one another. Fire was used as a weapon by both the North and the South, inflicting great losses on both military and civilian personnel. These fires resulted in countless lost lives, destroyed property, and ruined American business enterprises. In many cases, civilians set fire to their own property rather than allow it to be taken and used by the opposing forces.

The events of the war took interesting if not tragic twists. Richmond, Virginia, was the Confederate capital. When forced to retreat from Richmond in 1865, Southern forces set fire to three warehouses in an effort to destroy military supplies that they could not take with them. They also set fire to a bridge across the James River. During the retreat, some of the lawless elements of the city started a rampage of looting. They may have set fires along their path. High winds fanned the fires. Eventually the fires spread over most of the city's downtown area. The citizens watched helplessly as their beautiful city was destroyed. When Union forces arrived, one of their first undertakings was to bring control to the city by rounding up the marauders and helping to extinguish the fires [3].

April of 1865 was a remarkable month in our young nation's history. On April 9, 1865, General Lee surrendered the Confederate Army to General Grant at Appomattox, Virginia. Just a few days later, President Abraham Lincoln was shot at Ford's Theater in Washington, D.C. These two events overshadowed the news of an explosion aboard the *S.S. Sultana*. The *Sultana* boarded nearly 2000 Union soldiers who were recently released from Confederate prisons and had started a trip up the Mississippi River. The vessel stopped for fuel at Memphis, Tennessee, and shortly after midnight on April 27, got underway to continue the trip. About an hour later, a boiler exploded. The resulting fire quickly burned the ship to the waterline. Approximately 2100 were on board at the time; only 700 were rescued, and many of them had severe injuries. The National Fire Protection Association (NFPA) records the loss of life at 1547, making this the greatest loss-of-life fire ever recorded in our nation's history [5].

After the Civil War, rapid expansion of the country followed. The population grew rapidly, and many people moved to the western frontier. The cities in the East grew as well. The cities in the South had to be rebuilt. The cities in the North, now free of the economic restrictions of the war, saw rapid growth as well. This rapid growth was a mixed blessing for fire protection: Many older buildings in urban areas were highly combustible. Many of the newer ones were made of brick or other noncombustible material. Unfortunately, the rush to the frontier often outran fire protection, as many towns had no organized fire department. Even in the urban areas, the spread of electrical power and growth of large industrial complexes with stockpiles of coal and petroleum products set the stage for challenges that would later confront the best of the nation's fire departments.

Gradual Improvements in Fire Protection

Paid Fire Fighters. There is some debate over which city had the first paid fire department. One of those in contention for the title is Cincinnati, Ohio. As in other cities, Cincinnati's volunteer fire fighters were dedicated to providing protection for the citizens of the city, but they were even more dedicated to protecting their fire company's reputation. While buildings burned, members of various companies would fight over access to water cisterns. In *The Serene Cincinnatians,* Alvin F. Harlow describes an event that illustrates this foolishness well. He writes, "When the big saw-and-planing mill caught fire in 1851, the first two companies to arrive begin fighting over a cistern, and other companies, as they appeared, joined in the fray until ten companies were at it in a battle royal. A fire company in Covington (Kentucky, just across the Ohio River from Cincinnati), seeing the great billows of flame and smoke rolling heavenward, hastily assembled and ferried across the river to help, but instead waded into the melee while the mill and lumber yard burned to ashes" [6].

Unfortunately, this event was not unique. As a result, insurance rates soared, and the citizens of Cincinnati demanded action from their local government. The city responded and within months, placed the first steam-powered fire engine into service and shortly thereafter, hired the first paid fire fighters.

Fire Apparatus. Steam had been used to power pumps and other industrial equipment in many countries since the early 1800s, and there had even been attempts to use steam to power fire pumps in New York as early as 1840 [4]. The challenge was to design and make a piece of equipment that was both durable and reasonably transportable. One of the efforts that proved successful was the result of work by Abel Shawk and Alex B. Latta of Cincinnati. While both Shawk and Latta were experienced in building steam engines, they ran into real problems with their steam-powered pumper. The weight and size were problems of course, but the critical problem was developing a reliable method of quickly starting a fire in the boiler. The first public demonstration took place on March 1, 1852. On that occasion, the engine got up steam in 5 minutes and pumped a stream of water 130 feet (40 meters). The engine passed all tests, but its very existence was threatened by volunteer fire fighters.

The volunteer fire fighters saw the new engine as a threat to their existence but they could not say that in public. What they said was that the new engine was unnecessary and that it was just a waste of taxpayer money. The fire fighters even threatened to destroy the new engine if they had an opportunity. Not wanting to take a chance, the city asked Miles Greenwood to give the new pumper a chance. Greenwood was a respected citizen, a volunteer fire fighter, and experienced with steam-powered equipment. Shortly thereafter the engine got its first real test. As expected, his fire fighters had to fight off other volunteer fire fighters while fighting the fire. The new engine made two additional runs that first night. On each occasion, both Greenwood's fire fighters and the new engine performed well, and they drew the support of the citizens who showed up to watch.

Still the volunteers challenged the capabilities of the steam pumper. A showdown on New Year's Day, January 1, 1853, pitted the best volunteer company against the new steam-powered pumper. The volunteer fire fighters were very proficient, getting set up and pumping water before the steam engine had pressure. With a lot of hard work, their hand-pumper threw a stream an impressive 200 feet. The assembled crowd cheered them on. Then the steam-powered pumper opened up, throwing water 225 feet. Not only could the new pumper throw water further, but also it could pump several streams at once. And it could pump as long as it had fuel and water.

The next day the city started taking application for jobs with the new fire department and on April 1, 1853, the paid fire department of Cincinnati became a reality. Miles Greenwood, the volunteer fire fighter who helped everyone see the value of the new pumper, helped organize the new fire department, and became Cincinnati's first fire chief [4].

By 1860 at least 18 companies were making steam-powered pumps for fire apparatus. In addition to the Ahearns Company of Cincinnati, Ohio, two of the more famous of them were the LaFrance Fire Engine Company of Elmira, New York, and the Amoskeag Manufacturing Company of Manchester, New Hampshire. Because of their size and weight, teams of horses were required to pull these pumpers [3].

Latta and others continued to improve the design of his steam pumpers and within just a few years, new steam-powered fire engines were in service in Boston, Philadelphia, Columbus, St. Louis, New Orleans, and many other cities. The arrival of these new engines set the stage for the start of paid fire departments. The steam-powered equipment required regular maintenance and experienced personnel to operate it at fires. And the horses required regular attention. These duties represented quite a change for from the all-volunteer organizations that characterized America's early fire service. The outbreak of the Civil War delayed many cities' efforts to organize paid fire protection. The focus of the nation was on the war and a majority of the young men who would have been likely candidates for firefighting were being drafted into the army.

Hydrants and Water Supply.   Most cities had organized fire protection and these efforts were aided by the arrival of steam-powered pumpers. However, getting water to these pumpers remained a challenge, and some have argued that the lack of a good water distribution system delayed the acceptance of the steam pumpers.

However, as steam-powered pumpers were accepted, their acceptance may have stimulated improvements in water supply systems and the establishment of a system of fire hydrants. Many cities had water distribution systems consisting of hollow logs laid underground end to end. Connecting these logs was a rather crude science and as a result, very little pressure could be applied to the system.

Early efforts to use this water supply system for firefighting were also crude. At the outbreak of a fire, fire fighters dug a hole in the street and searched for the water supply system. Once the water main was located, fire fighters bored a hole in the log and water filled the excavation. At this point the water was taken to the pump, by buckets or by the suction hose of a steam-powered pumper. Once the fire was extinguished, a wooden plug was inserted into the hole in the pipe and the street restored to its original appearance. While this approach was better than nothing, clearly it was not a good system. The excavation would collapse or wash out, the water was usually muddy, and the entire process took a great deal of time and effort.

Cast-iron pipe had been around for well over a hundred years but was not generally used for community water supplies until the early 1800s. With the arrival of cast-iron pipe, two things happened that aided fire protection. One was an increase in pressure, thus providing a greater supply of water during a time of need. The second was the installation of fittings that allowed the fire department to draw water. These fittings were really nothing more than regular openings in the pipe, plugged with a wooden plug.

As these pipes were in the ground, controlling the excavation was still a problem. To overcome this problem, a metal or wooden caisson was built to provide access to the water main from the street above. The caisson prevented the erosion problem and provided a better quality of water for the pumpers. Still, water had to be taken from the hole by using either a suction hose or buckets.

The next logical step was to find a way of connecting the pumper's intake hose directly to the water supply system. Actually there were two challenges. The first challenge was to find a way of making a connection to the water supply system. A second challenge was to find a way of turning off the water pressure to that fitting so that the connection could be made. As it was, the increasing pressure in the water supply system made the connection process a difficult and drenching experience for the fire fighters. The fire hydrant gradually evolved, starting with the simple installation of a "T" in the water main, and a short riser to the surface, with a shutoff valve and fire department connection.

By the middle of the 1800s, many cities had some sort of underground water supply system and most made some provision for firefighting. Many cities still used cisterns to store water. Others developed a system of fire hydrants in their city. Even as late as 1860, there was still some skepticism regarding the use of hydrants. In Louisville, Kentucky, there were more than 100 cisterns still in use in 1861. A report of the times noted, "There are no fire (hydrants) in use, their efficiency for supplying steam fire engines with water has heretofore been regarded as doubtful" [3].

Clearly, there was debate, and there was no easy answer to the problem of providing a reliable, easy connection to a water supply system. While cisterns were widely used, they had many disadvantages, including their cost, the fact that most leaked, and the fact that water still had to be moved to the fire. Where fire departments could connect to a pressurized hydrant, the pressure in the hydrant was often sufficient for firefighting. As a result, hydrants gradually replaced cisterns throughout the country.

Similar activities were taking place in Europe. For a while, there were two methods of making this connection. The first approach was the above-ground hydrant. The second approach was to have the connection located below grade and protected by a flush metal covering. Many European cities continue to use the inground hydrant system and have very effective methods of connecting fire department pumps to the city's water supply system. Meanwhile, America led the way in establishing the system of fixed-post fire hydrants that we know today.

Sprinkler Systems

While looking at fire protection, we should also note the progress made in the use of sprinkler systems. The use of sprinklers for fire control started in the 1700s, but it was not until 1874 that Henry Parmelee introduced the automatic sprinkler as we know it today. By the late 1800s, Parmelee's sprinklers, as shown in Figure 1.3, were installed throughout New England. But Parmelee was not the only producer of automatic sprinkler systems: As many as 40 manufacturers produced as many as 80 types of sprinkler products. However in those days, there were no industry standards, so the design, installation, testing, and maintenance of these myriad systems became quite a challenge. As a result, a committee representing the sprinkler manufactur-

▶ **Figure 1.3** Parmalee Sprinkler. *(Source: Dana Gorham, Automatic Sprinkler Protection, 3rd ed., John Wiley & Sons, New York, 1923, p. 17.)*

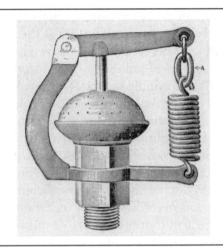

ers, the insurance industry, and the inspection bureaus met in 1895 to develop uniform rules for sprinklers. The committee was obviously quite effective, for within a year they had a new standard. But an even greater product of this committee was the creation in 1896 of an organization to develop similar standards, the NFPA.

Major 19th Century Fires

Although much progress was occurring in the area of fire protection and fire suppression, many cities had major fires during this time. Richmond, Virginia, the scene of several large fires during the Civil War, had several additional fires before the end of the century. A fire in 1882 resulted in the destruction of the Petersburg Bridge. A fire in the state penitentiary in 1888 resulted in the death of many inmates. Another fire in 1893 in a large industrial complex resulted in a large monetary loss as well as the loss of one of the city's major employers [3].

Chicago.   The fire in Chicago may be the best known of America's large fires. The Great Chicago Fire started in the evening on October 7, 1871, in a barn owned by Mr. and Mrs. Patrick O'Leary and continued for about 24 hours. The exact cause of the fire is unknown, although popular lore has it that the fire started when the O'Learys's cow knocked over a lantern.

There was some delay in notifying the fire department, and once the fire department arrived on scene, fire fighters had problems with one of their pumpers. These factors, combined with very dry conditions and a strong wind allowed the fire to spread rapidly. By the time the fire was extinguished, nearly 10 percent of the entire land area of Chicago was laid to waste. Of nearly 60,000 buildings in the city at the time, nearly 25 percent were destroyed, leaving nearly 100,000 homeless.

Considering the enormity of the event, the death toll of about 300, while tragic, was surprisingly small. The property loss was estimated at about $168 million—about $2.3 billion in today's value. Only half of the property was covered by insurance. Several insurance companies had stopped writing insurance in Chicago, being concerned about the size of the city, the number of combustible buildings, the lack of building codes, and the limitations of the city's fire department. Many of the insurance companies that remained in the city went bankrupt following the fire. As a result, less than half of the insurance claims were paid [3].

Peshtigo, Wisconsin.   On the same day as the Great Chicago Fire, a fire in a small lumber town in Wisconsin resulted in a far greater loss. The fire started in the woods south of town, possibly from lightning. The weather conditions were similar to those in Chicago: Everything was very dry and there was a mild breeze. These are favorable conditions for the rapid spread of a wildland fire. Before long the fire was at the village's door.

As might be expected in a lumber town, the houses were made of wood. Wooden sidewalks connected many of the houses. Flying brands from the wildland fire ignited the sidewalks, and soon thereafter, the houses were on fire. The fire moved faster than the people could flee. Some sought protection from the fire in the nearby river and many of them drowned as a result. By the time the village counted its losses, more than 1000 people had died, making it one of the worst fires in America's history. Every dwelling in the town was destroyed [3]. In spite of these tremendous losses, Peshtigo recovered and remains a city today. U.S. Fire Prevention Week marks the anniversary date of these fires.

Boston.   Boston is one of America's oldest cities, and at the time of the fire of 1872, it was one of America's largest cities. Given its age and size, fire protection was a concern. The city had a modest fire department. There were 21 pumpers, 6 of them in the downtown area. The downtown pumpers were rated at a pumping capacity of 500 gallons per minute, a respectable volume for the time. Unfortunately, their potential could not be fully utilized because of the city's meager water mains. Most of the water mains had only a

4-inch diameter pipe, and because of the relatively low pressure, a single engine could take all the water from a water main. This problem limited the ability of the fire department to operate effectively at the scene of large fires.

In those days there was little in the way of automatic fire suppression or detection equipment. None was installed in any of the affected buildings. There were fire alarm boxes in the city, but they were not under the control of the fire department. Most of the fire fighters were "call men," men called when needed for firefighting. The only full-time men were the drivers of the pumpers and ladder trucks and the men who operated the pump itself.

On November 12, 1872, a fire burned approximately 1 square mile of downtown Boston, destroying more than 700 buildings at a value of $75 million—about $1 billion in today's value—and taking 13 lives, including the lives of two fire fighters [3]. This event still ranks as one of the nation's largest fire losses.

In addition to the water supply problem already mentioned, fire fighters were hampered by ladders of inadequate length, by the lack of the ability to project streams above the third-floor level, and by the large amount of combustible trim on many of the buildings. Although many neighboring fire departments came to the aid of their neighbor, their effectiveness was also limited by the fact that each department had a unique hose thread, which made the hoses of one department incompatible with the hoses of another.

Many changes resulted from that fire. More men were hired, and more companies were established, especially in the high-value part of the city. The fire alarm operation was transferred to the fire department. Training programs for the department were started. Fire prevention and code enforcement activities were implemented. Improvements were made to the city's water distribution system. Unfortunately, sometimes it takes a tragic event to stimulate needed action in the area of fire protection.

New York City.   New York City, America's largest city even then, has had more than its fair share of major fires. In the latter 19th century, Manhattan was already a big city in terms of population and building stock. A fire in February of 1876 destroyed 22 of those buildings with a loss of more than $1,750,000—or $13 million in current value. This fire also resulted in the death of three fire fighters caught in a collapse [3].

In December of that same year, a fire at the Conway Theater in Brooklyn had tragic consequences. When fire broke out during a performance, the audience panicked and many died in the rush for the exits. The investigation that followed determined that there were plenty of exits and that there was plenty of time to exit, but the crowd's panic reaction packed a stairway. When the final toll was counted, 296 men and women were dead. Most of them died on the stairway. The fire called attention to the need to inspect places of public assembly regularly to ensure that adequate exits are provided [3].

Washington, D.C.   One of Washington's early landmarks was the Patent Office, an elegant building of approximately 300 by 400 feet. In the late 1800s, it housed the nation's patent office. Later the Civil Service Commission used the building. Today, the building is the home of two museums, the American Portrait Gallery and the American Museum of Art, both of which are part of the Smithsonian Institution.

The fire in 1877 destroyed a part of the building, of course, but also destroyed models of many of America's early inventions. In those days, inventors had to provide a model of their invention to the patent office, along with plans, specifications, and other documentation. Many of these unique models and many of the documents were lost in the fire.

The fire fighters of Washington were gallant in their efforts to control the fire. However, the building was huge and it was filled with combustible material. At the height of their efforts, all Washington fire companies were working at the fire in the Patent Office Building. Realizing their vulnerability, the city telegraphed the city of Baltimore, some 40 miles north, asking for assistance. Baltimore sent four companies by train, the fastest means available. This request for aid was a wise precaution, because the Baltimore fire fighters soon went to work on a second fire in the downtown area. Before the Baltimore companies departed, they were invited to the office of Interior Secretary Carl Schultz and thanked for their efforts. As they departed, the cheering citizens of Washington saluted them for the efforts [3].

▶ MAJOR 20TH AND 21ST CENTURY FIRES

America's fire problem continued into the 20th century. Major fires hit nearly every part of the country and involved many different types of occupancies. See Tables 1.2 and 1.3. In 1901, a fire in Jacksonville, Florida, destroyed 177 buildings. A year later, a fire in Paterson, New Jersey, destroyed more than 500 buildings [3].

▶ **Table 1.2**   The 25 Largest Fire Losses in U.S. History

Date and Location of Incident	Loss in Year It Occurred (in Millions)	Adjusted Loss in 2001 Dollars (in Millions)
1. The World Trade Center New York, New York September 11, 2001	$33,400	$33,400
2. San Francisco Earthquake and Fire San Francisco, California April 18, 1906	$350	$6,867
3. Great Chicago Fire Chicago, Illinois October 8–9, 1871	$168	$2,472
4. Oakland Fire Storm (wildland/urban interface) Oakland, California October 20, 1991	$1,500	$1,949
5. Great Boston Fire Boston, Massachusetts November 9, 1872	$75	$1,104
6. Polyolefin Plant Pasadena, Texas October 23, 1989	$750	$1,071
7. Wildland Fire (wildland/urban interface) Los Alamos, New Mexico May 4, 2000	$1,000	$1,028
8. Baltimore Conflagration Baltimore, Maryland February 7, 1904	$50	$981
9. Los Angeles Civil Disturbance Los Angeles, California April 29–May 1, 1992	$567	$716
10. Power Plant (auto manufacturing complex) Dearborn, Michigan February 1, 1999	$650	$690

▶ **Table 1.2**   *(Continued)*

Date and Location of Incident	Loss in Year It Occurred (in Millions)	Adjusted Loss in 2001 Dollars (in Millions)
11. "Laguna Fire" (wildland/urban interface) Orange County, California October 27, 1993	$528	$647
12. Textile Mill Methuen, Massachusetts December 11, 1995	$500	$581
13. *U.S.S. Lafayette* (formerly *S.S. Normandie* ocean liner) New York, New York February 9, 1942	$53	$575
14. *S.S. Grandcamp* and Chemical Company Plant Texas City, Texas April 16, 1947	$67	$532
15. Petroleum Refinery Norco Louisiana May 5, 1988	$330	$494
16. Cargo plane in-flight fire Near Newburgh, New York September 5, 1996	$395	$446
17. Great Fire of New York New York, New York December 16, 1835	$26	$444
18. Wildland Fire Florida May–June, 1998	$395	$428
19. One Meridian Plaza (high-rise office building) Philadelphia, Pennsylvania February 23, 1991	$325	$422
20. Forest Fire Cloquet, Minnesota October 12, 1918	$35	$410

(continues)

▶ **Table 1.2** *(Continued)*

Date and Location of Incident	Loss in Year It Occurred (in Millions)	Adjusted Loss in 2001 Dollars (in Millions)
21. Apollo Spacecraft Cabin Cape Kennedy, Florida January 27, 1967	$75	$397
22. Chemical Company Plant Pampas, Texas November 14, 1987	$215	$335
23. Nuclear Power Plant near Decatur, Alabama March 22, 1975	$100	$329
24. "Paint Fire" Goletta (wildland/urban interface) Santa Barbara, California June 27, 1970	$237	$321
25. Aluminium Plant Gramercy, Louisiana July 5, 1999	$300	$319

Note: Loss estimates are from NFPA records. Adjustments to 2001 dollars done by using the Consumer Price Index, including Census Bureau estimates for historical times. The list is limited to fires with some reliable dollar-loss estimate that occur in or over the United States.
Source: NFPA, Fire Analysis and Research Division, Quincy, MA.

▶ **Table 1.3** The 20 Deadliest Single-Building or Complex Fires and Explosions in U.S. History

Date and Location of Incident	Number of Deaths
1. The World Trade Center New York, New York September 11, 2001	2,666
2. Iroquois Theater Chicago, Illinois December 30, 1903	602

▶ **Table 1.3**   (*Continued*)

Date and Location of Incident	Number of Deaths
3. Cocoanut Grove Nightclub Boston, Massachusetts November 28, 1942	492
4. Ohio State Penitentiary Columbus, Ohio April 21, 1930	320
5. Consolidated School (gas explosion) New London, Texas March 18, 1937	294
6. Conway's Theater Brooklyn, New York December 5, 1876	285
7. Rhythm Club Natchez, Mississippi April 23, 1940	207
8. Lakeview Grammar School Collinwood, Ohio March 4, 1908	175
9. Rhodes Opera House Boyertown, Pennsylvania January 12, 1908	170
10. Ringling Brothers Barnum and Bailey Circus Hartford, Connecticut July 6, 1944	168
11. Alfred P. Murrah Federal Building Oklahoma City, Oklahoma April 19, 1995	168
12. Beverly Hills Supper Club Southgate, Kentucky May 28, 1977	165
13. Richmond Theater Richmond, Virginia December 26, 1811	160

(*continues*)

▶ **Table 1.3** (Continued)

Date and Location of Incident	Number of Deaths
14. Triangle Shirtwaist Company New York, New York March 25, 1911	146
15. Eddystone Ammunition Company plant explosion Eddystone, Pennsylvania April 10, 1917	133
16. Cleveland Clinic Hospital Cleveland, Ohio May 15, 1929	125
17. Winecoff Hotel Atlanta, Georgia December 7, 1946	119
18. The Station Nightclub West Warwick, Rhode Island February 20, 2003	100
19. Our Lady Of Angels School Chicago, Illinois December 1, 1958	95
20. Happy Land Social Club New York, New York March 25, 1990	87

Source: NFPA archive files, *1984 Fire Almanac,* and *The Great International Disaster Book,* by James Cornell, Pocket Books, New York, 1976.

1900 Through 1950

Iroquois Theater. In 1903, the Iroquois Theater in Chicago, Illinois, caught fire, resulting in the largest recorded loss of life in a fire involving one building. On the afternoon of December 30, 1903, the theater was filled beyond capacity. While designed to accommodate 1625, more than 2000 patrons were in the building. During the second act, a fire started on or over the stage and grew rapidly, despite efforts to extinguish it. An asbestos curtain, designed to protect the audience from such events, failed to function properly. On top of these problems, someone yelled "Fire!," which started a stampede for the exits. In all, 602 died and another 250 were injured [3].

The Iroquois fire was similar to the fire in a theater in Brooklyn a few years earlier. In both cases, the panic reaction added to the tragedy. In the Iroquois fire, while some of the patrons were caught in the stampede of those looking for exits, many were caught in the wave of hot gases that entered the auditorium from the intense fire burning on the stage.

City of Baltimore.   In February 1904, a conflagration in Baltimore destroyed 80 blocks in the downtown area, leveling more than 200 buildings and putting more than 50,000 people out of work. The loss was estimated at $50 million. The fire started in the basement of a business on what is now known as Redwood Street. The fire was quickly reported. The fire department responded promptly, but the fire quickly spread to nearby buildings. The fire burned for nearly 30 hours, destroying many of the buildings in the area to the east, the area that now surrounds Baltimore's Inner Harbor. The Falls Canal stopped the progress of the fire. Fortunately, there were no fatalities.

This fire is noteworthy for the fire service in that it identified the need for standardized fire hose threads. During the 1904 fire, the city of Baltimore requested assistance from as far away as New York City. Many fire departments from throughout the area responded, but many found that they could not connect to Baltimore's fire hydrants because of a different thread on their fire hose. This fire led to the establishment of standard hose threads for fire hose and fire hydrants [3].

S.S. General Slocum.   In that same year, a fire occurred aboard the steam-powered excursion boat *General Slocum*. On June 15, 1904, the *General Slocum* was underway in New York's East River with 1400 passengers, most of them parishioners from Saint Mark's Lutheran Church in New York City. Within a few minutes of getting underway, a fire broke out on the vessel's fore deck. The captain seemed to lose his head. Instead of stopping, he turned his vessel into the wind and ran at full speed.

After grounding the vessel, the captain and the pilot jumped overboard, leaving the passengers and remaining crew to fend for themselves. Some of the crew deployed fire hoses, but they were rotten and failed when water pressure was turned on. Others tried to lower the lifeboats, but they were neither organized nor proficient. The passengers found that the lifejackets were useless. The passengers had two choices: they could stand on deck and watch the approaching flames or jump overboard into the swift current. More than 1000 people lost their lives.

The conditions on the *General Slocum* were not unique. Following the fire other public vessels in New York were inspected. More than a third of them had defective life preservers and 25 percent had defective fire hose. As in the case of the *General Slocum*, many of the crews were untrained in dealing with emergency situations on board their vessel. The tragedy vividly illustrated the importance of fire prevention and the rules for effective fire safety when fire occurs on board a vessel and led to the establishment of regulations for life safety and fire protection on vessels that are still in force today [7].

San Francisco Earthquake Fires. The largest monetary loss fire ever to occur in America occurred in 1906 in San Francisco, California, following an earthquake. Early on April 18th a strong earthquake shook the Bay Area, resulting in many collapsed buildings. While the tragedy is often referred to as the Great San Francisco Earthquake, most of the losses that occurred were not from the earthquake but from the fires that followed. Gas lines and water mains were broken. The gas found ignition sources and fed existing fires. The broken water mains hampered firefighting efforts. The quake knocked down electrical power lines and the city's fire alarm system.

The fires continued for 3 days and were finally extinguished through the combined efforts of the city's fire department, the U.S. Army, and a drenching rain. More than 400 lives were lost. Nearly 5 square miles of the downtown area were destroyed, including some 28,000 buildings. Property loss was $350,000,000, nearly $6 billion in current dollars, the largest loss in the nation's history until the events of September 11, 2001 [8].

Chelsea, Massachusetts Fire. On April 12, 1908, a fire in Chelsea, Massachusetts, destroyed about half of the city. Chelsea lies north of Boston, across the Charles River. In those days, much of the city of Chelsea was made up of two- and three-story woodframe buildings. While recycling as we know it today had not yet been invented, it appears that a form of recycling was a major industry in Chelsea, because part of the city's economy depended on making and bundling rags for use in various industries. It was common practice to sort and store the rags in vacant yards in the residential area. Somehow these rags ignited, and aided by strong winds, the fire quickly spread to other rags, waste, and eventually to structures.

The Chelsea fire illustrates the need for fire prevention measures and effective fire protection at the highest levels of local government. Fire department and water supply resources were not keeping up with the growth of the city. The fire department had pointed out these problems and their concern about the fire risks in the "rag district" for years. The elected officials had other priorities. Both the insurance industry and local officials knew of the fire risks in the community, but they failed to work together to take appropriate action to reduce these risks. Chelsea was to have another conflagration before the century closed [3].

Triangle Shirtwaist Factory.   A March 25, 1911, fire in New York City provides another example of how codes are developed after a tragic event occurs. This fire occurred in the Asch Building. One of the tenants, the Triangle Shirtwaist Factory, occupied three floors of the building. A fire started in a bin of rags on the eighth floor of the building and spread quickly. Some employees attempted to extinguish the fire; others attempted to flee, but found that they could not escape because of locked exit doors. At the time, there were no laws regarding fire escapes, fire drills, or sprinkler protection in factory buildings. Within 30 minutes of discovering the fire, 150 employees, mostly young women, were dead, either as a direct result of the fire or from injuries incurred from jumping from windows to the sidewalk 100 feet below [4]. As a direct result of this fire, NFPA created its Committee on Safety to Life, which then developed the *Building Exits Code.* Later known as NFPA *101®*, *Life Safety Code®*, the Code's early focus was to make factories safer by identifying the hazards of stairways and fire escapes, the need for fire drills, and the ideal arrangement of exits.

S. S. Morro Castle.   On September 5, 1934, the 5-year old cruise ship *S. S. Morro Castle* sailed from Havana, Cuba, for New York with 318 passengers and a crew of 230. During the last evening of the trip, a small fire was discovered in one of the passenger rooms. The captain ordered an officer to investigate the fire, but he did not sound the fire alarm nor did he order the closing of any fire doors. The officer located and attempted to extinguish the fire. Meanwhile the ship continued sailing northward into a brisk northeast wind.

What followed will sound very familiar. The wind fanned the flames and they quickly spread. The growing fire cut off the obvious routes of escape to the lifeboats. Other routes of escape were available but were unknown to the passengers. The crew did little to fight the fire or help the passengers find safety. Those passengers who made it on deck were soon caught in the flames and smoke that were being fanned by the wind. Some jumped overboard and were lost and another 114 died on board—130 deaths in all.

The investigation that followed found that most of the problems were human errors. The investigation found evidence of lack of leadership, organization, and discipline within the crew. The officers had inadequately trained the crew in firefighting or in lowering the lifeboats. No provision was made for mustering the passengers at their lifeboat stations. The captain's decision to continue sailing into the strong northeast wind clearly showed poor judgment [7].

The public outcry that followed this tragedy moved Congress to establish new laws to improve the safety of life at sea. Several new regulations were enacted. At the time, the Bureau of Navigation and Steamboat Inspection, a part of the Department of Commerce, was responsible for marine safety in our country. In 1942, the functions of this agency were transferred to the United States Coast Guard, where they remain to this date [7].

Mississippi Dance Hall Fire. A dance hall in Natchez, Mississippi, was the scene of one of the 20th century's worst fire disasters in terms of loss of life. The dance hall consisted of a one-story building of about 40 by 120 feet. Only one entrance served the building and that was through a lobby and cloakroom at the front of the building. The doors swung inward. More than 700 people were in the building when fire broke out on the evening of April 23, 1940. The fire started in an interior hamburger stand located near the front of the building and spread rapidly to the decorative Spanish moss on the ceiling. The only means of escape was through the area involved in the fire. More than 200 died and another 200 were injured. Similar circumstances have contributed to other tragedies in public assembly occupancies [3].

Cocoanut Grove Nightclub Fire. While the events of World War II overshadowed many of the tragedies that occurred at home, the deadly fire at the Cocoanut Grove in Boston on November 28, 1942, will stand out in the memory of anyone old enough to remember the events of that era. Virtually all of the hazards that led to the deaths of 492 people in this fire were covered in NFPA's *Building Exits Code,* yet apparently the Code was not followed by the club's management and not enforced by safety officials, as shown in Figure 1.4.

The Cocoanut Grove was a nightclub, a holdover from the prohibition era. It looked like a one-story building with a dance hall. What was not obvious from the street was the large bar located in the basement. In retrospect, the place was a tragedy ready to happen. The building had limited exit capacity. Although the outside walls appeared sound with brick and stucco, the interior décor consisted of combustible decorative material over both walls and ceilings. On Saturday nights the place was usually filled to capacity. The rated capacity was around 600, but on the night of the fire it was estimated that nearly 1000 were in the building.

The exact cause of the fire will never be known, but the fire started in the basement and spread quickly. Many patrons were overcome by smoke before they knew what was happening. Many others tried to escape but found the limited exits already blocked by others trying to flee. The main exit consisted of a revolving door, which quickly became congested. Another exit had in-

▶ **Figure 1.4**   Damage after the Cocoanut Grove Fire. (*Courtesy of the New York Board of Underwriters.*)

ward-swinging doors, and here too, patrons quickly overwhelmed the exit. Several other doors were locked.

The fire department was on scene quickly and extinguished the fire in short order. But a tragic loss of human life had already occurred. Many people helped those who survived. More than 400 officers and men from the nearby Coast Guard base helped with firefighting and rescue efforts. The Red Cross mobilized a force of 500 emergency workers to help at local hospitals [3]. The fire led to more stringent codes regulating flammable interior finish.

Connecticut Circus Tent.   Another tragic fire occurred during World War II, this time in a circus tent in Hartford, Connecticut. In contrast to the events in several other assembly occupancies, this one occurred outdoors under a tent during a circus performance. On July 6, 1944, the weather was warm and clear, a good day for a circus. Unfortunately, fire turned the happy occasion into tragedy. The exact cause of the fire is unknown. Some speculate that it may have been a carelessly discarded cigarette; others speculate that it may have been the deliberate act of a disgruntled employee. Whatever the cause, the fire spread rapidly and quickly enveloped the entire tent.

Approximately 7000 people were in the big top when the fire started. Just before the fire broke out, the wild animals had been displayed in the rings at the east and west ends of the tent. The animals were being led back to their cages through a series of chutes. These chutes blocked two of the exits. During the fire, there was the usual rush for the remaining exits. Patrons fell, tripped, stepped on each other, were struck by the falling tent poles, and burned by the flaming tent cover. Within minutes 163 were dead and more than 200 were injured, many with severe burn injuries. This fire led to the development of fire codes covering tents and grandstands [4].

Hotel Fires of 1946.   Three hotel fires dominated the fire news in 1946. The first occurred at the La Salle Hotel in Chicago on June 5 with the loss of 61 lives. A delayed alarm, combustible interior furnishings, and open stairway and elevator enclosures all contributed to the loss. The second fire occurred less than a week later at the six-story Canfield Hotel in Dubuque, Iowa, with the loss of 19 lives. The aggressive use of aerial and ground ladders saved many lives. Again, a delayed alarm, open stairways, and combustible materials contributed to the tragedy.

The final event was the Winecoff Hotel fire in Atlanta, Georgia, on December 7. At least 119 died in this fire. Once again there was a delayed alarm to the fire department. The building did not have a fire alarm. An open stairway helped the spread of fire and smoke upward throughout the building. Open transoms (a once-common vent above the doorway) contributed to the spread of fire and fire gasses into guest rooms in all three of these events, especially at the Winecoff [3]. Most of the factors that contributed to the fires have been eliminated by building and life safety codes.

Texas City, Texas Fire.   On the morning of April 16, 1947, a ship in the Texas City harbor, the *Grand Camp*, bearing a cargo of ammonium nitrate fertilizer destined for war-torn Europe, caught fire. The fire department arrived on the scene and tried to extinguish the fire. A crowd of people, including many children, gathered to watch the fire fighters. The bright orange color that came out of the black smoke seemed to catch everyone's attention. The crowd must not have known that ammonium nitrate is highly explosive or they did not know what was in the cargo hold of the ship. The standard procedure for towing a dangerously burning ship from the harbor was not implemented until it was too late, and the tugboat that was needed did not arrive in time to prevent what happened next. A little after 9:00 AM, the Texas City Disaster, as it is often referred to, happened as the *Grand Camp* exploded. A great column of smoke shot up an estimated 2000 feet, followed in about 10 seconds by another, and even more violent shockwave. Within moments of the second blast, the Monsanto Chemical Plant was in flames. As entire buildings collapsed, trapping people inside, fires quickly spread to the nearby refineries that made up the Texas City industrial complex. Almost the entire fire department had been lost in the first explosion, along with plant workers, dockworkers, school children, and other bystanders.

All day long the work of caring for the injured and fighting the fires was underway. By nightfall, the town was filled with rescue workers, and ambulances were making repeated trips to area hospitals. Darkness did not stop the efforts to find those who were still trapped in the wreckage. Throughout the night, fear mounted because another freighter, the *High Flyer*, which was also loaded with ammonium nitrate as well as sulfur, had also been burning all day. Tugs had tried in vain to tow her out of the ruined harbor. At 1:00 AM on April 17, everyone was ordered away from the area. At 1:10 AM the *High Flyer* exploded in the most violent of all the blasts, taking with her another ship, the *Wilson B. Keene*. It also destroyed a concrete warehouse and a grain elevator and triggered even more fires.

The losses from the disaster were unprecedented. Nearly 600 deaths in a town of about 16,000 is a terrible toll. It is impossible to arrive at an exact figure because many bodies were never recovered. No one in the city was unaffected by the explosions and fires: Not a single family could be found that had not suffered a death, an injury, or severe damage to a house or business. A full week passed before the last of the fires was extinguished and a month went by before the last body was pulled from the tons of rubble. Dreadful as the disaster was, it brought the people of Texas City together as nothing else had ever done. Those that remained were determined to rebuild, and all of the industries that were damaged, stayed and rebuilt [4].

1950 Through 1999

School Fires.   Over the years, pubic and private schools have become remarkably safe places, considering the potential for tragic loss. The widespread use of school fire exit drills has prepared staff and students for dealing with tragedy and no doubt has saved countless lives. One exception occurred at Our Lady of Angels School in Chicago on December 1, 1958, in which 95 children and teachers died. The fire is believed to have started in trash in the basement, and the fire and smoke spread rapidly upward through an open stairway, trapping students and faculty in their classrooms. The tragic fire was a wakeup call for school administrators, building officials, and citizens across the country. Within the next few years, many jurisdictions took steps to install automatic detection and suppression equipment and improve housekeeping practices [3].

Bel Air, California, 1961. Bel Air and Brentwood are exclusive Los Angeles communities. By 1961, thousands of elegant homes had been built in the area, surrounded by highly combustible ground and tree cover. Access was difficult at best with winding roads and tortuous terrain. Many of the homes had wood-shingle roofs. On November 6, 1961, the weather forecast predicted high temperature, high winds, and low humidity. As a result, the Los Angeles Fire Department, one of the nation's largest and best, was on a High Hazard alert. Shortly after the start of the day shift, fire fighters were called to action to deal with a brush fire. Two hours later, 54 engine companies were on the scene and calls were going out for additional units. The attack included everything from aircraft to bulldozers.

By 8:00 PM, 12 hours after the fire started, the fire was under control, and it was extinguished the next day. The fire had destroyed 484 costly homes and 21 other buildings, and blackened more than 6000 acres. The cost of the fire was $25 million (approximately $150 million in current dollars), one of America's largest conflagrations. But we quickly forget: Five years later people rebuilt there, and the trees and ground cover returned [4].

Beverly Hills Supper Club. On May 28, 1977, a fire occurred at the Beverly Hills Supper Club in South Gate, Kentucky, just a few miles south of Cincinnati. The facility covered an area of over 50,000 square feet. The club had been renovated and expanded over the years and many of the alterations compromised the safety of the patrons and employees of the facility. Many interior furnishings were made of combustible materials, automatic detection and suppression systems were inadequate, and the exit system was inadequate for the capacity. The local government did not inspect the club for code compliance. The club was probably overcrowded on the night of the fire. Of the estimated 3000 patrons in the building, 164 died. The fire led to the codes requiring automatic sprinklers in large places of public assembly [9].

Pasadena, Texas Fire. On October 23, 1989, a series of events led to the accidental release of a huge cloud of highly flammable gas at the Phillips Petroleum chemical processing plant in Pasadena, Texas. The gas ignited with explosive force, producing an explosion that leveled several structures and seriously damaged many others over the 16-acre plant. The explosion and fire left 23 dead and 130 injured. Damage to the processing area was extensive. Loss was placed at $750 million—$1 billion in today's value.

Happy Land Social Club.   On March 30, 1990, a vengeful patron spread and ignited gasoline in the front entrance at the Happy Land Social Club in the Bronx—one of New York City's five boroughs. The fire quickly spread and blocked the only available exit. Smoke and heat spread quickly, overcoming the patrons in the one-story 22- by 58-foot building. Despite a fast response, rapid hoseline advancement, quick fire knockdown, and an immediate search of the building, the death toll was stunning: The fire claimed 87 lives. All of the fire department's efforts were for naught as the victims were overcome in spite of the quick and effective response.

Philadelphia's One Meridian Plaza.   A fire in a high-rise building is always a challenge. The fire in the 38-story One Meridian Plaza office building in downtown Philadelphia, Pennsylvania, was especially challenging. The fire started on February 23, 1991, on the ninth floor. Fire fighters found that control valves on the building's standpipe system had been installed incorrectly, thus limiting their water supply on the floors involved in the fire. The fire claimed the lives of three fire fighters. Property loss was placed at $325 million (with the current value at $400 million).

Food Processing Plant Fire.   On the morning of September 3, 1991, a fire started in the processing area at the Imperial Foods packing facility in Hamlet, North Carolina, taking 25 lives and injuring 54 others. At the time of the fire, the plant was in operation preparing and cooking chicken for restaurants. A fitting in a hydraulic line failed, spraying fluid near the natural gas cooking equipment resulting in immediate ignition. Inadequate exits and inoperable—and in some cases, locked—exit doors prevented employees from exiting the building and contributed to the death toll in the deadliest manufacturing facility fire in 20 years. In many ways these conditions were similar to those that occurred 80 years earlier in the Triangle Shirtwaist fire in New York City.

California Wildland Fires.   Fires in the wildland environment do not get the same attention as those that occur in cities. However, in the 1990s, a series of large-loss wildland fires in California received considerable attention. The first of these occurred in Oakland, on October 20, 1991, when a previously extinguished wildland fire rekindled as fire fighters were checking for hot spots. Fanned by high winds, the dry vegetation quickly ignited and spread to 2889 dwellings covering an area of 1780 acres over a 2-day period. Loss was 26 fatalities and $1580 million in property. (Current value of the loss is $1836 million.)

World Trade Center Bombing of 1993.   On February 26, 1993, a terrorist's bomb exploded in a van parked in an underground garage in the World Trade Center office building in New York City. The blast and resulting fires killed six civilians and injured at least 1000 more. The response by the city's fire department was the largest in the department's history. Following this event, the fire department and others took steps to better deal with fires and other emergencies at the World Trade Center. These preparations saved countless lives when terrorists struck again in 2001, causing the collapse of the twin towers of the World Trade Center.

Incident at Waco, Texas.   During a raid by law enforcement officials at the Branch Davidian complex outside Waco, Texas, on April 19, 1993, a fire started in the occupied portion of the complex. Because of the ongoing exchange of gunfire, fire fighters were not allowed to approach the buildings and extinguish the fire. The buildings were made of wood framing, and once ignited, they burned quickly. A total of 77 people died in the event, at least 47 of them due to the fire [10].

Oklahoma City Bombing.   On April 19, 1995, a rental truck loaded with 4000 pounds of explosives was detonated in front of the Alfred P. Murrah Federal Building in Oklahoma City, Oklahoma. The explosion and fire that followed caused one side of the nine-story office building to collapse and damaged 340 other structures and 2000 vehicles within a 48-block radius. These events led to 168 fatalities and 475 civilian injuries. Direct property damage was $136 million.

Worcester Cold Storage and Warehouse Building Fire.   As the 20th century closed, a fire in an empty warehouse in Worcester, Massachusetts, brought the kind of attention that no fire department would want. The fire on December 3, 1999, will not go into the history books as a large-loss fire in terms of civilian lives or property loss, but will always be remembered as one of the largest losses of fire fighters at that time. The fire apparently started after a candle was knocked over by homeless people seeking shelter in the abandoned building. Arriving fire fighters entered the building to conduct a search and to locate the fire. They did not return. Other fire fighters went in to search for them. Six fire fighters died in the event, as shown in Figure 1.5.

▶ **Figure 1.5**   Water Poured on Lingering Fire at the Worcester Cold Storage and Warehouse Building. *(Source: AP / Wide World / Paul Connors.)*

Starting the 21st Century

September 11, 2001.   The tragedy of September 11, 2001, is second to none in U.S. history. Thousands of Americans watched their televisions in disbelief as terrorists attacked the United States, destroying the World Trade Center in New York City, damaging the Pentagon Building in Arlington County, Virginia, and forcing an aircraft to crash in Pennsylvania, killing all on board. The total loss of life in these events was 3000. Of these, 404 fire fighters and police officers died at the World Trade Center. They went in to help others get out. The twin towers withstood the initial crash of jet planes into the buildings but not the fires afterward. In addition to the human loss, the property loss is reported at $33,400 million, making it the largest fire-related loss in history.

Rhode Island Nightclub Incident.   At least 100 people were killed and nearly 200 more were injured when a nightclub in West Warwick, Rhode Island, erupted in flames during a rock band's pyrotechnics display. The

Great White was playing its first song, "Desert Moon," and the fans were cheering as fireworks sprayed the stage with sparks. They kept cheering even as flames shot toward The Station nightclub's ceiling. Within 3 minutes, many of them were dead. Many concertgoers were caught off guard as they slowly realized the fire was not part of the show. Many were badly burned and others were trampled in the rush to escape, in large part through a single door. A TV cameraman doing a story on nightclub safety recorded the unfolding disaster, beginning with the fireworks, followed seconds later by bright orange flames climbing curtains and soundproofing behind the stage and the tragic events that followed. The fire in Rhode Island followed less than a week after a tragic non-fire event in Chicago in which security personnel incited a crowd of people when they used pepper spray in a crowded room to break up a fight. The ensuing rush to the exits left 21 dead. In both cases, exit egress systems played a role. The Rhode Island fire resulted in regulations requiring automatic sprinklers in virtually all nightclubs, regardless of size.

Lessons Learned

Each of these events has played an important role in establishing fire protection as we know it today. As we start our studies of fire protection, it is important that we have an understanding and an appreciation of the events that have occurred and that we use this knowledge in making our contributions for improving fire protection for generations to come. Until recently, events such as those described here caused public reaction that usually brought about changes in laws, regulations, or code that would likely prevent a reoccurrence. As a result of these changes, fortunately, today the tragedies occur much less frequently, and when they do, we find that there was usually not a lack of regulation, only that the regulation was not enforced.

History of Building Codes

In the previous sections, several references were made to fire and building codes. Indeed a large loss often is the catalyst in the development of new codes or the revision of existing codes. Let us briefly look at how these codes evolved. Building regulations date back to the beginning of recorded history. The Code of Hammurabi (2200 BC) included a simple but effective building code provision; if an architect built a house so negligently that it fell down and killed the owner's son, then the architect's son was put to death.

In early America, George Washington and Thomas Jefferson encouraged the development of building regulations to provide for minimum standards related to public health and safety. Present-day building codes have evolved into a comprehensive system of regulations that define safety requirements for the built environment. Today, most of the United States is covered by a network of modern building regulations ranging in coverage from fire and structural safety to health, security, and the conservation of energy.

Building codes are adopted by a state or local government's legislative body and then enacted to regulate building construction within a particular jurisdiction. The primary purpose of a building code is to regulate new or proposed construction. Building codes only apply to an existing building if the building undergoes reconstruction, rehabilitation, or alteration, or if the occupancy of the existing building changes to a new occupancy level as defined by the building code.

During the early 1900s, code enforcement officials of various communities wrote model building codes with assistance from all segments of the building industry. In 1915, code enforcement officials met to discuss common problems and concerns. Out of these meetings came the formation of three regional organizations of code enforcement officials. The first of these organizations, later known as Building Officials and Code Administrators (BOCA) International, Inc., was created in 1915 and represented code officials from eastern and midwestern portions of the United States. BOCA International's headquarters were located in County Club Hills, Illinois. The second organization, later known as the International Conference of Building Officials (ICBO), was formed in 1922 and represented the code officials from the western United States. ICBO's headquarters were located in Whittier, California. The third organization, formed in 1941, was the Southern Building Code Congress International (SBCCI) and represented the interests of code officials in the Southern United States. SBCCI's headquarters were located in Birmingham, Alabama.

In 2002, these three regional building code organizations consolidated into one organization—the International Code Council (ICC), renaming their codes the International Codes (I Codes). The new ICC began operations in 2003.

While legislative bodies are not obligated to adopt a model code and may write their own code or portion of a code, studies have indicated that 97 percent of all U.S. cities have adopted a model code or are covered by a statewide building code based on a model code. Model codes have now become the basis for the administration of building regulatory programs in cities, counties, and states throughout the United States.

EVOLUTION OF FIRE PROTECTION ORGANIZATIONS

National Fire Protection Association

The NFPA is an international nonprofit membership organization founded in 1896. Today, with more than 75,000 members representing nearly 100 nations, and 300 employees around the world, NFPA is the world's leading advocate of fire prevention and is an authoritative source on public safety. The volunteers and staff of NFPA are dedicated to the single mission of continually enhancing fire, building, electrical, and life safety.

NFPA encourages the broadest possible participation in code development, a process driven by more than 6000 volunteers who serve on 230 technical code- and standard-development committees. Throughout the entire process, interested parties are encouraged to provide NFPA technical committees with input. All NFPA members then have the opportunity to vote on proposed and revised codes and standards. NFPA develops over 300 separate codes and standards, including the *National Electrical Code®* (*NEC®*), the *Life Safety Code®* (NFPA *101®*), the *Uniform Fire Code™* (NFPA 1™), and the *Building Construction and Safety Code™* (*NFPA 5000™*). These documents form the core of a family of codes for the built environment known as the C3 codes.

National Board of Fire Underwriters

Founded in 1866, the National Board of Fire Underwriters (NBFU) began collecting and classifying fire statistics from individual fire insurance companies. There was no uniformity in the data collected, the way data was reported, or even in the period covered by the report. As a result it was hard to make comparisons and to accurately determine fire risk and the cost of fire losses.

NBFU reduced the number of occupancy categories used to collect data from 584 to a more manageable 26. In 1947, the number of occupancy categories was expanded to 115. Along with developing statistics on fire loss, the National Board of Fire Underwriters developed documents that outlined practices and procedures that could reduce fire losses. NBFU also adopted the standards developed by NFPA and republished them as NBFU standards as recommended by NFPA. This practice was discontinued in 1965. NBFU also developed the first model building code, the *National Building Code*, in 1905 and developed a method for evaluating cities and towns with regard to their overall fire protection. Teams of NBFU survey engineers graded cities and towns across the United States using the National Board Grading Schedule. In 1971, the Insurance Services Office or ISO was formed and the responsibility for updating and applying the grading schedule was transferred to ISO (see p. 49). In 1966, the board merged with the Association of Casualty and Surety Companies under the name of American Insurance Association or AIA, later renamed the American Insurance Services Group (AISG).

Factory Mutual Insurance Company (now called FM Global)

Factory Mutual is a mutual insurance company dedicated to property conservation and owned by its policyholders. While this concept is not so unusual today, there was a time when such an idea did not exist. In fact, the concept actually dates back to 1835, when a group of New England mill owners who were led by Zachariah Allen and had a mutual dedication to loss control made a plan to form their own insurance company.

By most industry standards, this new company's loss record was enviable; although losses insured by this group did occur, they were still less frequent and less severe than those experienced by most other nonmutual insurers. More important, however, was what these mutual insurance companies, later known as the Factory Mutual companies, did following their policyholder losses. Rather than paying out the loss, chalking it up to ill luck, and calling it a day, these early disciples of loss control examined the loss to determine not only how it was caused, but more importantly, what could have been done to prevent the loss from occurring in the first place. They even examined nonpolicyholder losses to help increase their knowledge base. This vital loss information was used to help identify specific industry hazards and was essential to developing loss control recommendations for policyholders with similar occupancies. Such information was shared among all the Factory Mutual (FM) insurance companies, and was particularly critical to the inspection teams staffed separately by each individual company.

As America entered the 20th century, many technological changes occurred and dozens more industry innovations were introduced. In addition to the changes in the industrial world around them, the Factory Mutuals also underwent an important transformation. Where once the mutual insurance companies focused primarily on the familiar business of textiles primarily within the Northeast, new mutual insurance companies began to form that sought business beyond the traditional geographical boundaries.

Over the next 75 to 80 years, the need for more comprehensive policyholder coverage grew, forcing a series of consolidations among the FM Mutuals; by 1987, 42 separate mutual insurance companies had become three: Allendale Mutual Insurance Company, Johnston, R.I.; Arkwright Mutual Insurance Company, Waltham, Mass.; and Protection Mutual Insurance Company, Park Ridge, Ill. These parent companies also were the remaining co-owners of Factory Mutual Engineering and Research. In 1999 these companies merged into one to form one insurance and loss control organization, FM Global.

Underwriters Laboratories Inc.

Underwriters Laboratories Inc. (UL) is a not-for-profit corporation having as its sole objective the promotion of public safety through conduct of "scientific investigation, study, experiments, and tests, to determine the relation of various materials, devices, products, equipment, constructions, methods, and systems to hazards appurtenant thereto or to the use thereof affecting life and property and to ascertain, define, and publish standards, classifications, and specifications for materials, devices, products, equipment, constructions, methods, and systems affecting such hazards, and other information tending to reduce or prevent bodily injury, loss of life, and property damage from such hazards." The organization was founded in 1894 by the fire insurance industry under whose sponsorship it operated until 1968 when it became an independent, public service-type corporation. It has no capital stock, nor shareholders, and exists solely for the service it renders in the fields of fire, crime, and casualty prevention.

The early histories of NFPA, National Board of Fire Underwriters, Factory Mutual and Underwriters Laboratories are intertwined and fascinating. For further information read NFPA Assistant Vice President Casey Grant's article about the development of these organizations on NFPA's Web site at http://www.nfpa.org/home/AboutNFPA.

Industrial Risk Insurers

Industrial Risk Insurers (IRI) was formed in 1975 with the merge of the Factory Insurance Association (FIA) and the Oil Insurance Association (OIA). IRI is an association made up of member stock insurance companies to essentially form an insurance pool that can collectively insure highly protected risks with greater liability than any member company could do individually. IRI is now owned by General Electric. IRI, like Factory Mutual, is dedicated to the concept of loss control to reduce insurance risk.

Society of Fire Protection Engineers

The Society of Fire Protection Engineers (SFPE) is the professional society for engineers involved in the multifaceted field of fire protection engineering. Organized in 1950, the purposes of the society are to advance the science and practice of fire protection engineering and its allied fields, to maintain a high ethical standing among its members, and to foster fire protection engineering education. Its worldwide members include engineers in private practice, in industry, and in local, regional, and national government, as well as technical members of the insurance industry. Forty-six chapters of the society are located in the United States, Canada, and throughout the world including Europe, New Zealand, and Australia. Membership in the society is open to those possessing engineering or physical science qualifications, coupled with experience in the field, and to those in associated professional fields.

The society serves as an international clearinghouse for fire protection engineering state-of-the-art advances and information. It publishes the SFPE Today, a newsletter with regular features, the *Journal of Fire Protection Engineering*, a peer-reviewed research journal, and *Fire Protection Engineering* magazine. *The SFPE Handbook of Fire Protection Engineering*, plus monographs, proceedings, and other technical books are also developed by SFPE.

▶ THE FIRE SERVICE IN AMERICA TODAY

Types of Fire Departments

Our country is based on democratic principles. Under the U.S. Constitution, the federal government provides for national security but defers to the state governments most of the responsibility for domestic public safety. States in turn pass this responsibility on to local governments. As a result, there are approximately 30,000 fire departments in the United States individually organized to meet the specific needs of the communities they serve. Although public fire protection is usually a function of local government, provincial, state, and federal property may also have organized fire departments for their protection. Large industries may have fire brigades or even fire departments to protect their properties. About 75 percent of the U.S. fire departments are staffed by volunteer fire fighters. Volunteer fire fighters are concentrated in the rural communities, especially in communities with a population less than 2500. See Tables 1.4 and 1.5.

Career fire departments, that is, those departments that rely mostly or entirely on career fire fighters, protect most U.S. cities, and therefore a majority of U.S. citizens. (A majority of our population is concentrated in cities.) In cities with populations more than 250,000, nearly all fire departments are career departments. In between the all-volunteer and the all-career fire departments is a mix of combination organizations that range from mostly volunteer with limited career staffing, to mostly career with some volunteer augmentation when needed.

▶ **Table 1.4**   Number of Departments and Percent of U.S. Population Protected by
                   Type of Department

Type of Department	Number	Percentage	Percentage of U.S. Population Protected
All Career	1,526	5.8	40.3
Mostly Career	1,213	4.6	18.2
Mostly Volunteer	3,671	13.9	15.6
All Volunteer	19,944	75.7	25.9
Total	26,354	100.0	100.0

Note: The foregoing is based on 8027 departments reporting.

Numbers may not add to totals due to rounding.

Source: Table 1, FEMA, USFA, and NFPA, "A Needs Assessment of the U.S. Fire Service," 2002.

Public fire departments are usually a part of local government. The head of the fire department (fire chief) reports to the chief executive officer (city manager) or to the senior elected officer (mayor). As such they are responsible for all things related to the fire department's operations. The most common type of public fire protection in larger cities is the public fire department. A less common form of municipal fire protection is a fire bureau, a part of the department of public safety. A variation of the fire bureau occurs when employees of the public safety department provide both police and fire protection. In some jurisdictions these personnel are called public safety officers.

County fire departments are a little more complicated. County fire departments are becoming more common as suburban communities grow and the demands for emergency service response outpace the capabilities of the volunteer fire companies. A model that works well is to have a county fire chief responsible for countywide fire protection, with a responsibility for coordinating the efforts of the independent volunteer fire companies. As the need arises for hiring paid staff for the volunteers' fire stations, the paid fire chief is involved with selecting, training, and assigning these people, and supervising them once they are hired. As the organization grows, the county fire chief will likely add staff assistants to deal with training and communications. As the county continues to grow, the demand for service increases proportionally. Gradually more functions, including emergency response activities, are shifted from the volunteers to the paid staff as the volume of calls and the demand for additional service increases.

▲ **Table 1.5** Department Type, by Community Size

Population of Community	All Career		Mostly Career		Mostly Volunteer		All Volunteer		Total	
	Number Depts	Percent	Number Depts	Percent	Number Depts	Percent	Number Depts	Percent	Number Depts	Percent
1,000,000 or more	12	92.3	1	7.7	0	0.0	0	0.0	13	100.0
500,000 to 999,999	24	63.2	12	31.6	1	2.6	1	2.6	38	100.0
250,000 to 499,999	42	65.6	18	28.1	2	3.1	2	3.1	64	100.0
100,000 to 249,999	158	73.5	40	18.6	12	5.7	5	2.3	215	100.0
50,000 to 99,999	305	62.6	99	20.4	56	11.5	27	5.5	487	100.0
25,000 to 49,999	389	36.9	288	27.4	226	21.5	149	14.2	1,053	100.0
10,000 to 24,999	438	15.4	493	17.3	1,094	38.5	817	28.7	2,843	100.0
5,000 to 9,999	89	2.4	174	4.8	1,194	32.9	2,172	59.9	3,629	100.0
2,500 to 4,999	30	0.7	54	1.2	629	13.8	3,858	84.4	4,572	100.0
Under 2,500	43	0.3	32	0.2	454	3.4	12,911	96.1	13,440	100.0
Total	1,526	5.8	1,213	4.6	3,671	13.9	19,944	75.7	26,354	100.0

Source: Table 2, FEMA, USFA, and NFPA, "A Needs Assessment of the U.S. Fire Service," 2002.

Fire Districts

In some states, the counties are divided into fire districts as provided by state law. These fire districts become essentially separate independent units of government, administered by some form of commission. The citizens of a fire district can elect the type of fire protection they want, and in some cases, they can elect the level of protection afforded. Usually, these citizens pay a special tax for the level of service they desire.

Another form of fire protection is established with the creation of a fire protection district. A fire protection district is a legally established tax-supported unit that contracts for fire protection with a neighboring fire department.

However, with three-quarters of the nation's fire departments being all volunteer organizations, they are found in cities, towns, fire districts, and counties. There are all-volunteer departments in cites with over 100,000 population and all-career departments in communities of less than 2500. Volunteer organizations may be supported by taxes, subscriptions, or fund-raising activities. Some volunteer companies own stations that include meeting rooms that can be rented for public and private use, thus generating a revenue source for the organization.

Federal Fire Fighters

Federal fire fighters are employees of the federal government. They work at government installations including military bases around the world. Most federal fire departments are relatively small, but collectively, federal fire fighters comprise the largest fire department in the United States. In addition to their traditional duties of fire suppression and emergency medical response, most of the federal fire fighters are very actively involved with fire prevention activities. As a result of these and other efforts, fire loss on government property is usually lower than in the surrounding community.

Both the U.S. Forest Service and the U.S. Park Service have fire personnel, including dispatchers, fire fighters, fire managers, safety specialists, researchers, and others who work together for the common goal of fire management, fire use, fire prevention, and fire suppression. Many states have similar opportunities.

Several branches of the military also have fire fighter specialties among their enlisted personnel. The Army, Air Force, and Marine Corps all have fire fighters. They are primarily involved with providing fire and rescue support for aircraft operations both at home and while deployed. The Navy and Coast Guard have a specialty rating that designates personnel who maintain the firefighting equipment on ships and smaller shore units, but actual firefighting is generally an all-hands event, especially on ships at sea.

Work Schedules

Until World War I, paid members of fire department worked a continuous–duty system with limited time off. By World War II, most departments had adopted a two-platoon system that allowed additional time off and reduced the amount of time worked per week. Since then the number of hours has continued to decline, but career fire fighters still work more hours per week than other workers in municipal employment or private industry. Today the most common arrangement is to have three platoons each working 24-hour shifts in some sort of rotation. This works out to an average of approximately 56 hours per week or 10 days on per month.

Staffing

Historically America was protected entirely by volunteer fire departments. As previously mentioned, the arrival of steam-powered equipment gradually set the stage for the arrival of paid fire fighters. As cities grew, so did the need for fire protection. The demands in terms of call volume and work hours were, in many cases, beyond the capabilities of the volunteer fire departments.

Evolving Role of Fire Departments

Where has all of this taken us? Clearly, many fire departments are doing more than just fighting fires. Total responses may be increasing but in most jurisdictions the number of fire responses is down, and down significantly. What is making up the difference? See Figures 1.6 and 1.7. More information on fire department activities is provided in Chapter 2.

Response Activities.   Across the country, calls for emergency medical service continue to increase. In communities where the fire department can provide such service, calls for hazardous materials incidents and technical rescue are also increasing.

Wildland firefighting has also become more obvious. Over the past few years, the nation has experienced an increasing number of wildland fires. In fact, as we were gathering material for this book, we noted that the 2002 wildfire season was the second worst in 50 years with more than 7 million acres burned, about twice the normal loss. The number of wildland fires in 2002 was down. The size of some of these fires, however, put them into the record books. Fires in Arizona, Colorado, and Oregon all were recorded as among the largest in a century. And with the

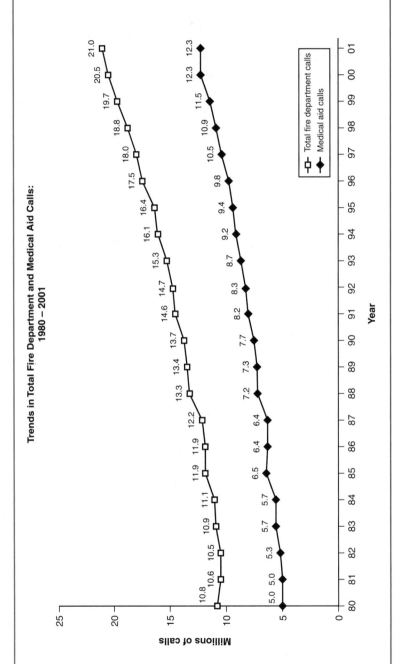

**Trends in Total Fire Department and Medical Aid Calls:
1980 – 2001**

▲ **Figure 1.6** Fire Department Emergency Calls, 1980–1999. *(Source: NFPA National Fire Experience Survey.)*

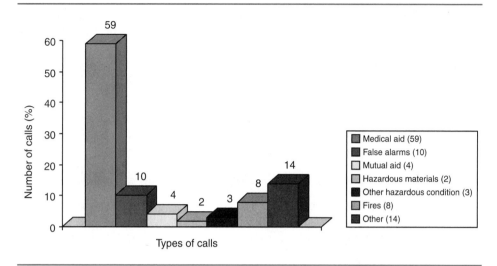

▶ **Figure 1.7** Fire Department Calls, 2001. *(Source: Michael J. Karter, Jr. "Fire Loss in the United States," NFPA, multiple years.)*

continued growth of communities into the wildland/urban interface, many of these fires were dangerously close to communities. About 3000 homes were lost. What is often overlooked—in 2002, 20 fire fighters died in wildland fire-fighting operations.

Prevention Activities.   Fire departments are also involved in a variety of prevention-related activities with an increasing emphasis on fire prevention and life safety issues. Traditional fire prevention activities include code en-forcement inspections and public education programs to remind people of the proper actions to take to prevent fires and proper actions to take when fire does occur. In addition to the traditional messages, fire departments in many areas are teaching homeowners how to protect their property from wildfire. *Firewise*, a prevention program developed by state and federal wild-land fire agencies, NFPA, and others, is gaining popularity as citizens take in-terest in the problems associated with wildfires.

In many communities, fire departments are looking beyond even these ac-tivities. In some cases, the fire department reacts to the situation and does the best it can to serve the needs of the community. Fire departments are among the first to respond to natural disasters—floods, earthquakes, tornadoes, and similar events—that often tax the resources of the organization. Fire depart-ments are also among the first on the scene at man-made disasters—bombings and threats of bombings. Planning fire department response activities for such events is now at the forefront. Much of the effort of the National Fire Academy is focused on planning for response to chemical and biological agents and weapons of mass destruction. Clearly, the fire department has taken on roles that even Ben Franklin could never have imagined.

Closely associated with this activity is a continuing concern for fire fighter safety. Responding to these events takes its toll. The most vivid example of this occurred on September 11, 2001, when police and fire fighters responded to aid their fellow citizens at the World Trade Center in New York City. Before the day was over 343 fire fighters and 61 police officers would be dead. While this dramatic example will always be on our minds, we should also remember that we lose fire fighters nearly every day. Some are lost while firefighting. Some are lost while responding to the emergency and never get to the scene. And some are lost in a variety of situations, all while trying to help someone.

▶ EVALUATING FIRE PROTECTION

Because the safety of people, property, and the environment is so vital, questions that focus on the organization and deployment of firefighting resources are especially important to communities. Although fire prevention and fire protection engineering efforts are less costly than fire suppression, many do not see the value in these preventive activities. As a result we rely on fire suppression resources to deal with emergencies that do occur. Many communities are faced with significant budget issues, and public safety, comprising both fire and police protection, is a relatively expensive undertaking. Many citizens would rather see the communities' limited financial resources spent on schools, parks, and better streets.

In evaluating the effectiveness of fire protection there are several issues. First is the already-mentioned high cost. Fire stations, fire apparatus, and on-duty personnel are indeed expensive. A second problem is determining the level of protection required. Issues include determining the number of on-duty personnel, the maximum allowable time allowed for response, and the capabilities of these responders once they arrive on the scene.

In the United States, fire protection has always been a function of local government. But sometimes, local governments, and the citizens they serve, lack an understanding of the complex nature of fire protection and may not be qualified to determine what is best, which raises a significant question: Who can best judge? Should local communities determine their own needs, or should some sort of national standard be applied?

These are challenging questions. Determining the appropriate level of protection, the cost of such protection, and the means by which these decisions are made is a daunting task. Citizens are often presented information about the cost of fire protection, but many have no real interest. Even those who are interested often lack the expertise to answer these questions. The benefits of fire prevention, public safety education, and code enforcement are not well publicized and therefore there is often little pressure exerted on their behalf. Thus, relatively few resources are allocated to these activities. However, most fire protection experts know that investments in these very activities will pay off and eventually reduce the need for more expensive fire suppression resources.

Several organizations have developed methods for communities to evaluate their fire protection resources and identify their own strengths and weaknesses. The first that we discuss here, and the first that evolved, is the rating system developed by the Insurance Services Office or ISO.

Insurance Services Office

The conflagrations that occurred in the latter 19th century in the United States placed a huge burden on the insurance industry. These events demonstrated a need for improving fire protection and a method for evaluating fire risks and fire defenses in cities throughout the country.

Grading Schedules.   One of the first efforts in this regard was a publication developed by the National Board of Fire Underwriters (NBFU) entitled *The Standard Grading Schedule for Grading Cities and Towns of the United States with Reference to Their Fire Defenses and Physical Conditions* [11]. This document provided a method of evaluating the fire department and the structural condition of the community, the use of codes, standards, and ordinances, and the availability of fire alarms systems, and assigned a rating ranging from 1 through 10, with 1 being the best. These rating were used to determine fire insurance rates and, in turn, the cost of fire insurance.

Fire insurance is unique in that the cost of the product (payment) is unknown at the time of the sale (payment of premium). To help the insurance industry better determine the probability of future losses, several fire insurance rating organizations joined together in 1971 to form the ISO. Today, ISO publishes the *Fire Suppression Rating Schedule,* a successor to the previously mentioned *Grading Schedule.* The rating schedule evaluates three areas of a fire department:

1. Receiving and handling of fire alarms
2. The fire department itself
3. The community's water supply

Field Surveys. Each of these areas is important in the timely and effective intervention of the fire department at a typical structural fire. Field surveys are conducted by ISO representatives to determine appropriate scores in each of these categories. The survey involves a review of fire department records, an evaluation of the fire department's personnel and equipment, and an assessment of the community's water supply system. Normally, this evaluation is conducted every 5 years.

The results of these surveys provide a rating. A rating of 1 through 8 indicates some sort of recognized fire department, that the nearest engine company is less than 5 miles travel distance from any protected property, and that there is an adequate fire hydrant within 1000 feet of the water supply source. This concept is new in the rating schedule. In the past the rating schedule only gave credit for hydrants. The new schedule recognizes the availability of surface water such as lakes and ponds. In such cases, ISO will allow a fire department to demonstrate its ability to deliver an adequate water supply using tankers. A Class 9 rating is similar to the previous groups, but lacks an adequate water supply. A Class 10 rating essentially indicates that there is no fire protection. Such a situation would typically exist where the protected property is beyond the 5-mile travel distance of the nearest fire station, and there is no water supply.

Here is a breakdown of the areas evaluated. In each category, the field evaluators are guided by a detailed list of performance indicators on which to base their evaluations. Many of these indicators are tied to NFPA 1710, *Standard for the Organization and Deployment of Fire Suppression Operations, Emergency Medical Operations, and Special Operations to the Pubic by Career Fire Departments* [12], and NFPA 1720, *Standard for the Organization and Deployment of Fire Suppression Operations, Emergency Medical Operations, and Special Operations to the Public by Volunteer Fire Departments* [13]. The items that comprise each of these three components are the following:

Fire Department (50 percent of total score)
- Engine companies
- Reserve pumpers
- Pump capacity
- Ladder companies
- Reserve ladders
- Distribution of the previously listed items
- Staffing of companies
- Training of personnel

Receiving and Handling of Fire Alarms (10 percent of total score)

- Telephone service including number of incoming lines, equipment, and listings in local phone directory
- Number of personnel on duty
- Number and types of dispatch circuits

Water Supply (40 percent of total score)

- Water supply, including water main capacity and hydrant distribution
- Types and method of installation of hydrants
- Hydrant inspection and maintenance activity

These scores are converted to ratings as shown in Table 1.6 [14].

The ISO survey results provide a method of determining fire insurance classifications in order to calculate fire insurance premiums. Some think, incorrectly, that ISO mandates certain elements of fire protection. ISO clearly states in its publications that this is not the case, and notes that ISO does not provide recommendations regarding loss prevention and life safety.

How does all of this improve the fire department and the fire protection provided within a community? The fire protection grading process not only determines insurance rates for individual properties but also points out deficiencies in the community's fire defenses. A good fire department will see that

▶ **Table 1.6**   Score Conversions

Score	Class	Percentage of Departments So Rated
90 to 100 percent	Class 1	0.1
80 to 89 percent	Class 2	0.7
70 to 79 percent	Class 3	2.6
60 to 69 percent	Class 4	6.8
50 to 59 percent	Class 5	12.1
40 to 49 percent	Class 6	16.0
30 to 39 percent	Class 7	15.2
20 to 29 percent	Class 8	8.8
10 to 19 percent	Class 9	34.7
0 to 9 percent	Class 10	3.0

certain actions will improve its rating. Recognizing these needs, the department can systematically make incremental steps to improve their next evaluation.

There is a clear correlation: Better fire protection leads to lower insurance costs. Local jurisdictions can take steps to improve their capabilities and lower the cost of fire insurance for their citizens. Few, if any, other municipal services have this opportunity.

NFPA Standards for Fire Department Response

The ISO rating system provides a valuable service. However, because of ISO's focus on fire, and the relatively infrequent on-site evaluations, some feel that it is too limited to help communities face future issues. In response to this need NFPA developed two documents.

NFPA 1710. NFPA 1710, *Standard for the Organization and Deployment of Fire Suppression Operations, Emergency Medical Operations, and Special Operations to the Public by Career Fire Departments*, 2001 edition, defines response capabilities for fire and medical incidents for the first-due engine company, for the entire alarm assignment, and for the first level, basic, and advanced emergency medical responders. The number of fire fighters necessary for initial fire attack and other roles are specified in the standard as are other aspects of the organization, suppression, operations, and emergency medical and other specialized services.

NFPA 1720. NFPA 1720, *Standard for the Organization and Deployment of Fire Suppression Operations, Emergency Medical Operations, and Special Operations to the Public by Volunteer Fire Departments*, 2001 edition, deals with similar issues for organizations that are staffed predominately by volunteers. Although these standards provide a road map, it is still up to the local jurisdictions to find out where they are presently located and where they would like to go with this map.

▶ ROLE OF FIRE DEPARTMENTS IN THEIR COMMUNITIES

Accreditation

Accreditation is a process by which an agency or institution evaluates and recognizes a *course of study*, such as Fire Science Program, or an *institution*, such as a college or university, as meeting certain predetermined standards. Accreditation is awarded only to programs or institutions, not to individuals. Thus, a program may be accredited by an agency such as the International Fire Service Accreditation Congress (IFSAC), and a college may be accredited by one of the several regional accrediting agencies. For colleges and universities, accreditation is essential to their ability to award recognizable degrees.

An accreditation process is now available for fire departments through Certified Fire Accreditation, International (CFAI). CFAI is affiliated with the International Association of Fire Chiefs. In this case, accreditation is purely voluntary.

The objectives of accreditation are the following:

- To create organizational motivation and self-improvement
- To provide a voluntary activity focused on self-evaluation and education as a viable means to improve service delivery
- To provide a means to recognize quality performance
- To protect the interests of the general public

Generally, accrediting agencies set standards by which agencies, programs, and institutions are measured and define the evaluation process. Most accreditation programs involve a self-assessment by the agency, program, or institution itself, a process that can take several years. The completion of the assessment is followed by a brief (2 to 3 days) onsite visit by a team representing an independent accrediting body to validate the self-assessment. The team issues a report of their findings, commenting on the strengths and weaknesses of the organization and make recommendations for further improvement. The report is usually first presented to the applicant for validation and then to the accrediting agency.

The intent of the CFAI program is to provide an accreditation process that is credible, realistic, and achievable. It can be used by fire departments, city and county managers, and elected officials to evaluate community fire risks and improve the delivery of emergency services. The thrust of the program is to improve these services and reduce the risks associated with fire and related hazards. The self-study process provides a list of questions to be answered by the department seeking accreditation.

These performance indicators must be met in order for the applicant to be accredited. For example, in the area of Essential Resources in Table 1.7, the following questions are examples of those asked. In all cases of an affirmative response, evidence must be presented to support the claim.

- Does the fire agency establish minimum fire flow requirements and total water supply needed for exiting representative structures and other anticipated fire locations?
- Is there an adequate water supply available for firefighting purposes?
- Does the agency calculate fire flow before proposed projects?

▶ **Table 1.7** Category Ratings

Category	Number of Criteria	Performance Indicators	Core Competencies
Governance and Administration	2	11	4
Assessment and Planning	4	17	11
Goals and Objectives	4	10	5
Financial Resources	3	20	5
Programs	5	59	31
Physical Resources	6	26	8
Human Resources	6	38	14
Training and Competency	4	18	4
Essential Resources	4	27	10
External Relationships	2	7	3

- Is there a mechanism for regular contact with the managers of the water department to keep the agency informed about all issues relevant to firefighting?
- Does the agency maintain current water supply and hydrant maps of its response area?
- Are fire hydrants easily located?
- Are fire hydrants regularly maintained and tested?
- Is there a plan for an alternative water supplyr those areas without hydrants, where hydrant flows are inadequate, or where a major disruption would render the public water supply inadequate?

Professional Certification

Certification is a process whereby an individual is tested and evaluated in order to determine his or her mastery of a specific body of knowledge or some portion of a body of knowledge. In simple terms, the certification process provides a yardstick by which to measure individual competency, just as accreditation measures programs and organizations. In addition to recognizing individual competency, certification may provide some measure of protection from liability. In some jurisdictions, certification is a condition of employment.

A significant step in the certification process occurred when the Joint Council of National Fire Service Agencies established the national professional qualifications system in 1972. That system provides a means of presenting the career opportunities in the fire service and defining a specified body of knowledge representing each level of each specialty area in the fire service career ladder. Standards apply to volunteer and paid personnel alike. Volunteers, just like paid personnel, need to know what to do and how to do it at the scene of any emergency. The lives of their colleagues and the lives and property of the citizens of their own community are often at stake. The Joint Council of National Fire Service Organizations consisted of the chief executive officers of 10 national fire service organizations. It was established for the purpose of facilitating cooperation in the goals in which the member organizations shared common interests. The ten agencies were the Fire Marshals Association of North America, the International Association of Fire Investigators, the International Association of Black Firefighters, the International Association of Fire Chiefs, the International Association of Fire Fighters, the International Fire Service Training Association, the International Municipal Signal Association, the International Society of Fire Service Instructors, the Metro Chiefs of the International Association of Fire Chiefs, the National Fire Protection Association, and the National Volunteer Fire Council. Having accomplished its goals, the Council disbanded in 1992.

Several of NFPA's standards deal with the professional qualifications of those who serve in the fire service. As public organizations, fire service organizations are open to public scrutiny and are held accountable for their actions. There is considerable value in being able to demonstrate that the personnel of these agencies are certified as meeting the competency standards of an entity that has itself been evaluated by an independent, thorough, objective, and public process and approved or accredited as meeting the requirements of the process. These standards are the following:

- NFPA 472, *Standard for Professional Competence of Responders to Hazardous Materials Incidents* [15]
- NFPA 1001, *Standard for Fire Fighter Professional Qualifications* [16]
- NFPA 1002, *Standard for Fire Apparatus Driver/Operator Professional Qualifications* [17]
- NFPA 1003, *Standard for Airport Fire Fighter Professional Qualifications* [18]
- NFPA 1006, *Standard for Rescue Technician Professional Qualifications* [19]
- NFPA 1021, *Standard for Fire Officer Professional Qualifications* [20]
- NFPA 1031, *Standard for Professional Qualifications for Fire Inspector and Plan Examiner* [21]
- NFPA 1033, *Standard for Professional Qualifications for Fire Investigator* [22]
- NFPA 1035, *Standard for Professional Qualifications for Public Fire and Life Safety Educator* [23]
- NFPA 1041, *Standard for Fire Service Instructor Fighter Professional Qualifications* [24]

- NFPA 1051, *Standard for Wildland Fire Fighter Professional Qualifications* [25]
- NFPA 1061, *Standard for Professional Qualifications for Public Safety Telecommunicator* [26]
- NFPA 1071, *Standard for Emergency Vehicle Technician Professional Qualifications* [27]
- NFPA 1081, *Standard for Industrial Fire Brigade Professional Qualifications* [28]

The standards listed here have been written in performance language so that candidates for certification can be fairly and objectively tested. The job performance requirements state what must be done, what tools and knowledge must be used, and to what standard of performance the task must be performed. Using these standards, fire departments can develop training programs and testing procedures that prepare members for certification. In most states, a state-level government agency administers the certification process. Several have added their own requirements to these standards.

The following are other good reasons for providing a structured program of training and education:

- Meets mandated training requirements
- Introduces new skills
- Introduces new technology
- Introduces policy changes
- Prevents skills degradation
- Expands service
- Provides higher levels of customer service
- Develops teamwork
- Promotes safety

Most of the standards address safety requirements appropriate for the position. Safety is a critical element in any training program. During training sessions, fire departments should reinforce safe practices in a controlled setting. Training is the time and place to develop good, safe habits that will become standard practice at the scene of an emergency.

Accreditation is also available for fire-service-related training and certification programs. A training organization can have its program accredited by meeting the standards outlined by the accrediting boards. The National Board of Fire Service Professional Qualifications (NBFSPQ), which was initially established in 1972 by the Joint Council of Fire Service Organizations, and the

International Fire Service Accreditation Congress (IFSAC), both evaluate and accredit fire service training programs. The NBFSPQ maintains a national registry of individuals certified by an NBFSPQ-accredited program.

An important aspect of any emergency response organization is the need for in-service or continuing training and education. Some departments make training a part of every workday's activity and nearly all volunteer departments provide training at least one night a month. The training ensures that members will be able to perform the duties that are assigned in a safe and efficient manner so that they do not expose themselves or other members of the department to unnecessary hazards.

Customer Service

Emergency service agencies are unique organizations. For the most part, they enjoy a competition-free environment. But then again who would want to work in the middle of a busy highway or a collapsed building? Who would run into a building that was on fire or filled with deadly chemicals? Who would want to take care of someone who obviously had little interest in caring for themselves? And who would want to do it in such a manner that the recipient was not just satisfied, but pleased! Customer service is very much a part of the core value of any good emergency response organization. We explore this topic further in Chapter 5.

These activities—accreditation, certification, and customer service—all provide added value to our organization and to the community we serve. With citizens asking for more from their government, with governments facing tight budgets, and with concerns for safety and health shared by providers and citizens alike, we must focus on being the best that we can be.

Let us never take ourselves, our job, or our organization for granted.
Everyday, we should take actions that will improve ourselves,
our organization, and our community.

▶ SUMMARY

This first chapter presented fire-related history since recorded time, focusing on 400 years of fire-related history in the United States. Our history is filled with both success and failure; we should learn from both. Fire departments and emergency service organizations are facing new challenges while dealing with some of the same issues that fire fighters have faced for centuries. We should be encouraged with our accomplishments, and excited about the opportunities that lie before us. To be ready for these challenges, we should take every advantage of every learning opportunity to improve ourselves and ultimately, improve our organizations and our communities. They deserve our very best.

Group Activity

In groups of three to five students, see how many fire protection advances and important events each group can come up with for each of the last four centuries: the 1600s, the 1700s, the 1800s, and the 1900s.

Review Questions

1. The London of 1666 was a city of _____ - _____ and _____ covered medieval buildings, mostly with thatched roofs.

2. What changes in fire protection resulted from the Great Fire of London?

3. In the 17th century, insurance industry and companies made what startling revelation regarding property owners?

4. How did insurance companies assist with the problem of personnel to respond to emergencies?

5. How were insured policy holders identified?

6. What were the contents of the homes of early settlers that contributed to rapid fire growth?

7. What was the role of fire wardens appointed in 1648?

8. Money raised from the fines was used for what?

9. What invention of the latter 17th century improved firefighting?

10. Once the water main was located, what did the fire fighters then do?

11. What fire protection system did Parmalee introduce?

12. What organization was created in 1896 after a committee representing sprinkler manufacturers, insurers, and inspection bureaus met a year earlier?

13. Why were there so many fires resulting from the Great Earthquake of San Francisco?

14. What was the point of origin of the Triangle Shirtwaist Fire?

15. What three activities provide added value to the contemporary fire service and the community?

Suggested Readings

A Needs Assessment of the U.S. Fire Service. Published jointly by the Federal Emergency Management Agency, the U.S. Fire Administration, and the National Fire Protection Association, Quincy, MA: 2002.

Bales, Richard F. *The Great Chicago Fire and the Myth of Mrs. O'Leary's Cow.* Jefferson, NC: McFarland & Co., 2002.

Boucher, David. *Ride the Devil Wind: A History of the Los Angeles County Forester and Fire Warden Department and Fire Protection Districts.* Fort Collins, CO: Fire Publications, Inc., 1991.

Brandt, Nat. *Chicago Deathtrap: The Iroquois Theatre Fire of 1903.* Carbondale, IL: Southern Illinois University Press, 2003.

Bugbee, Percy. *Men Against Fire.* Boston: National Fire Protection Association, 1971.

Burgess-Wise, David. *Fire Engines and Fire-Fighting.* Norwalk, CT: Longmeadow Press, 1977.

Calderone, John A. *The History of Fire Engines.* New York: Barnes & Noble, 1997.

Capron, Walter. *The U.S. Coast Guard.* New York: Franklin Watts, Inc., 1965.

Carle, David. *Burning Questions: America's Fight with Nature's Fire.* New York: Praeger, 2002.

Commission of Fire Accreditation International. *Fire and Emergency Assessment Manual.* Fairfax, Virginia: Commission of Fire Accreditation International, 2002.

Conway, W. Fred. *Firefighting Lore: Strange but True Stories from Firefighting History.* Hudson, MA: Fire Buff House Publishers, 1993.

Conway, W. Fred. *More Firefighting Lore: Forty More Strange but True Stories from Firefighting History.* Hudson, MA: Fire Buff House, 1998.

Cote, Arthur, ed. *Fire Protection Handbook.* 19th ed. Quincy, MA: National Fire Protection Association, 2003.

Cote, Arthur and Percy Bugbee. *Principles of Fire Protection.* Quincy, MA: National Fire Protection Association, 1988.

Cowan, David. *To Sleep with the Angels: The Story of a Fire.* Chicago, IL: Ivan R. Dee (Elephant Paperbacks), 1998.

Cromie, Robert. *The Great Chicago Fire*. Nashville, TN: Rutledge Hill Press, 1994.

Cytron, Barry D. *1942—Fire!: The Library is Burning*. Minneapolis, MN: Lerner Publications, 1988.

DeAngelis, Gina. *The Triangle Shirtwaist Company Fire of 1911*. Bromwell, PA: Chelsea House Publishers, 2001.

Gess, Denise and William Lutz. *Firestorm at Peshtigo: A Town, Its People, and the Deadliest Fire in American History*. New York: Holt, 2002.

Hanson, Neil. *The Great Fire of London in that Apocalyptic Year, 1666*. Hoboken, NJ: John Wiley & Sons, 2002.

Hashagen, Paul. *Fire Department City of New York*. Paducah, KY: Turner Publishing Co., 2002.

Hatch, Anthony P. *Tinderbox: The Iroquois Theatre Disaster, 1903*. Chicago: Academy Chicago Publishers, 2003.

Heys, Sam. *The Winecoff Fire: The Untold Story of America's Deadliest Hotel Fire*. Atlanta: Longstreet Press, 1993.

Hickey, Harry. *Fire Suppression Rating Handbook*. Louisville, KY: Chicago Spectrum Press, 2002.

"I'll never fight fire with my bare hands again": Recollections of the First Forest Rangers of the Inland Northwest. Lawrence, KS: University Press of Kansas, 1994.

Kurzman, Dan. *Disaster: The Great San Francisco Earthquake and Fire of 1906*. New York: William Morrow, 2001.

Leschak, Peter M. *Ghosts of the Fireground: Echoes of the Great Peshtigo Fire and the Calling of a Wildland Firefighter*. New York: Harper, 2002.

Lyons, Paul. *Fire in America!* Boston: National Fire Protection Association, 1976.

Maclean, Norman. *Young Men & Fire*. Chicago: University of Chicago Press, 1992.

Minutaglio, Bill. *City on Fire: The Forgotten Disaster that Devastated a Town and Ignited a Landmark Legal Battle*. New York: Harper Collins, 2003.

National Fire Protection Association. *Reconstruction of A Tragedy: The Beverly Hills Supper Club Fire*. Quincy, MA, 1977.

O'Nan, Stewart. *The Circus Fire: A True Story*. New York: Doubleday, 2000.

Pernin, Peter. *The Great Peshtigo Fire: An Eyewitness Account*. Madison, WI: State Historical Society of Wisconsin, 1999.

Pyne, Stephen J. *Fire in America: A Cultural History of Wildland and Rural Fire*. Seattle, WA: University of Washington Press, 1997.

Sammarco, Anthony Mitchell. *The Great Boston Fire of 1872*. Arcadia, 1997.

The San Francisco Calamity by Earthquake and Fire. University of Illinois Press, 2002.

Sawislak, Karen. *Smoldering City: Chicagoans and the Great Fire, 1871–1874.* Chicago: University of Chicago Press, 1995.

Smith, Dennis. *History of Firefighting in America.* New York: The Dial Press, 1978.

Souter, Gerry. *The American Fire Station.* New York: MBI Publishing Co., 2000.

VonDrehle, Dave. *Triangle: The Fire That Changed America.* New York: Atlantic Monthly Press, 2003.

Wallace, Deborah. *In the Mouth of the Dragon.* Brea, CA: Avery Publishing Group, 1990.

Yoder, Curt and Karen. *The Heart Behind the Hero.* Trabuco Canyon, CA: Stoney Creek Press, 2000.

2 Understanding America's Fire Problem

Marty Ahrens
National Fire Protection Association

 Learning Objectives

After completing this chapter, the reader should be able to do the following:

- Provide an overview of the U.S. fire problem, beginning with the big picture statistics for 2001 (the most recent data available at the time this chapter was written)

- Provide detailed data about fire causes and occupancies for 1999 (the most recent available data from the National Fire Incident Reporting System when this chapter was written)

- Explain how fire statistics are obtained and the role of fire service in these statistics

- Provide information on regional differences in fire experience and the role of demographic factors, such as age and gender, in injury and death from home fires, and of socioeconomic factors, such as education and poverty, in fire death rates

- Compare the U.S. fire experience to other selected countries

- Explain how today's fire service does more than fight fires as the breakdown of total fire department calls shows

- Discuss the role of local fire departments, state agencies, the National Fire Protection Association, and the United States Fire Administration in collecting and analyzing fire loss data

Local fire officials use national statistics as an important tool in evaluating their own department. Are local fire problems the same as those in the rest of the country? If not, why are they different? How do they compare with problems in similar communities? Where should prevention activities be focused? Such statistics are used to justify budgets, allocate resources, and provide a context for long-term planning.

▶ U.S. FIRES AND ASSOCIATED LOSSES

General Fire Data

In 2001, U.S. municipal fire departments responded to an estimated 1,734,500 fires. These fires killed 6196 civilians and caused 21,100 reported civilian fire injuries. Direct property damage was estimated at $44 billion dollars. Four hundred and thirty-nine fire fighters died while on duty [1]. Table 2.1 compares the fire experience in 2001 to the problem seen one year earlier in 2000, ten years earlier in 1991, and 20 years earlier in 1981. Reported fires fell 42 percent from 2,988,000 in 1980 to 1,734,500 in 2001, the second lowest point since data collection began in 1977. Reported fires rose 2 percent from the record low of 1,708,000 in 2000 to 1,734,500 in 2001.

Structure Fires

Any fire in or on a structure is considered a structure fire, even if only the contents were involved and there was no structural damage. Structure fires include building fires; fires in mobile property used as fixed structures, such as manufactured housing or portable buildings; fires in nonbuilding structures such as piers, fences, transformers, and tents; and fires in structures confined to the object of origin.

During 2001, the 521,500 structure fires accounted for 30 percent of all reported fires. Excluding the terrorist attacks at the World Trade Center in New York City and the Pentagon in Arlington, Virginia, on September 11, these structure fires caused 3220 civilian fire deaths, 17,225 civilian fire injuries, and $8.9 billion in direct property damage. The events of September 11 caused 2451 civilian fire deaths, 800 civilian fire injuries, 340 fire fighter deaths, and $33.44 billion in direct property damage.

Excluding the events of September 11, structure fires accounted for 86 percent of the civilian fire deaths, 85 percent of the civilian fire injuries, and 84 percent of the direct property loss excluding the losses of September 11, 2001. Reported structure fires fell 51 percent from 1,065,000 in 1980 to 521,500 in 2001. From 2000 (505,500 structure fires) to 2001, they rose 3 percent. Table 2.2 shows the fire experience in 2001 by incident type. Figure 2.1 shows the trends in reported fires by incident type for 1980–2001.

Seventy-six percent (396,500) of the 521,500 structure fires occurred in residential properties, including homes, hotels, motels, rooming houses, and dormitories; 74 percent (383,500) occurred in homes. (Homes include one- and two-family dwellings, apartments, and manufactured housing.) Home fires fell 48 percent from the 734,000 reported in 1980. Home structure fires rose 4 percent from the 368,000 reported in 2000.

▶ **Table 2.1** U.S. Fire Problem, 1981–2001

Reported to Fire Departments	2001	Compared To 2000 %	1991 %	1981 %
Fire incidents	1,734,500	Up 2	Down 15	Down 40
Civilian deaths				
Including 9/11/01	6,196	Up 53	Up 39	Down 8
Excluding 9/11/01	3,745	Down 7	Down 16	Down 44
Firefighter Deaths				
Including 9/11/01	439	Up 326	Up 306	Up 223
Excluding 9/11/01	99	Down 4	Down 5	Down 24
Civilian injuries				
Including 9/11/01	21,100	Down 6	Down 28	Down 31
Excluding 9/11/01	20,300	Down 7	Down 29	Down 32
Fire fighter injuries	82,250	Down 3	Down 20	Down 20
Direct property damage				
Including 9/11/01	$44,023,000,000	Up 293	Up 365	Up 559
Excluding 9/11/01	$10,583,000,000	Down 6	Up 12	Up 59
Adjusted for inflation*				
Including 9/11/01		Up 282	Up 258	Up 239
Excluding 9/11/01		Down 8	Down 14	Down 18
Civilian deaths per thousand fires				
Including 9/11/01	3.6	Up 35	Up 39	Up 34
Excluding 9/11/01	2.2	Down 7	Up 1	Up 10
Civilian deaths per million population				
Including 9/11/01	22.1	Up 51	Up 63	Down 54
Excluding 9/11/01	13.4	Down 9	Down 1	Down 7
Property damage per fire				
Including 9/11/01	$25,381	Up 287	Up 447	Up 1000
Excluding 9/11/01	$6,101	Down 7	Up 32	Up 164
Adjusted for inflation*				
Including 9/11/01		Up 276	Up 321	Up 466
Excluding 9/11/01		Down 10	Up 1	Up 36

*Inflation calculations derived from a custom table created from purchasing power of the dollar for all urban consumers at www.bls.gov/cpi on April 24, 2003.

Sources: Michael J. Karter, Jr., "Fire Loss in the United States," NFPA, multiple years. Fire Incident Data Organization (FIDO); and U.S. Census Bureau.

▲ **Table 2.2**   Fires by Incident Type, 2001 (Excluding the Events of September 11, 2001)

Incident Type	Fires		Civilian Deaths		Civilian Injuries		Direct Property Damage	
Structure fires	521,500	(30.1%)	3,220	(86.0%)	17,225	(84.9%)	$8,874,000,000	(83.9%)
Vehicle fires	351,500	(20.3%)	485	(13.0%)	1,925	(9.5%)	$1,512,000,000	(14.3%)
Outside and other fires	861,500	(49.7%)	40	(1.1%)	1,150	(5.7%)	$197,000,000	(1.9%)
Total	1,734,500	(100.0%)	3,745	(100.0%)	20,300	(100.0%)	$10,583,000,000	(100.0%)

Source: Michael J. Karter, Jr., "Fire Loss in the United States During 2001," NFPA, 2002.

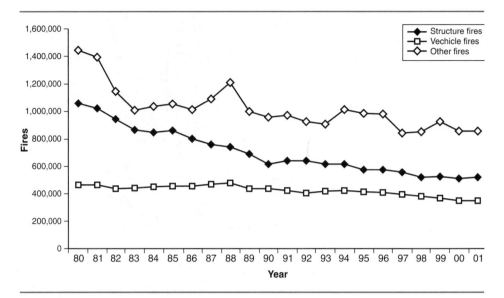

▶ **Figure 2.1** Reported Fires in the United States by Year and Incident Type.
(Source: Michael J. Karter, Jr., "Fire Loss in the United States," NFPA, multiple years.)

Fifty-seven percent (295,500) of all reported structure fires occurred in one- and two-family homes, and 17 percent (88,000) occurred in apartments.

Eighty-three percent (83%) of the 3745 non-September 11 civilian deaths occurred in home structure fires. Although only 17 percent of all reported fires occurred in one- and two-family structures, these fires caused 71 percent (2650) of the fire deaths. Apartment fires accounted for 5 percent of all reported fires, but resulted in 12 percent (460) of the deaths. Home fires are covered in more detail in the section on causal factors found later in this chapter.

With the events of September 11, total fire deaths for 2001 jumped 53 percent from 4045 in 2000 to 6196, the highest point since 1988. When those events are excluded, total fire deaths fell 7 percent to 3745, the second lowest point since tracking began. Home fire deaths fell 9 percent from 3420 in 2000 to 3110 in 2001. Total civilian fire deaths and home fire deaths specifically fell 42 percent and 40 percent, respectively, from 1980 to 2001, again excluding the events of September 11. Fire deaths followed a generally downward trend over the past two decades. Figure 2.2 shows that the lines for home fire deaths and overall fire deaths closely resemble one another in most years.

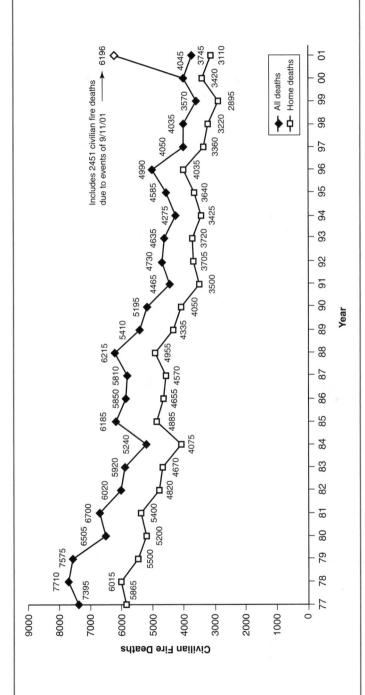

▲ **Figure 2.2**   All Civilian Fire Deaths and Home Fire Deaths, 1977–2001. *(Source: Michael J. Karter, Jr., "Fire Loss in the United States," NFPA, multiple years.)*

Vehicle Fires

During 2001, the 351,500 reported vehicle fires caused 485 civilian deaths, 1925 civilian injuries, and $1.5 billion in direct property loss. These fires accounted for 20 percent of the reported fires, 13 percent of the civilian fire deaths, 9 percent of the civilian fire injuries, and 14 percent of the total direct property damage. (These statistics exclude the events of September 11, 2001.)

The 485 civilian fire deaths accounted for six times the 80 deaths reported in nonresidential structure fires (excluding the events of September 11, 2001). During 2001, 13 percent of all (non-9/11/2001) civilian fire deaths were in highway vehicles. Less than 1 percent occurred in nonhighway vehicles such as planes, trains, ships, boats, or construction equipment. Vehicle fires fell 25 percent from 471,500 in 1980. From 2000 to 2001, highway vehicle fires rose 1 percent, while nonhighway vehicle fires rose 4 percent.

Outside or Other Fires

Fifty percent (861,500) of the 1,734,500 total reported fires were considered outside or other. These fires caused 40, or 1 percent, of the civilian deaths; 1150, or 5 percent, of the civilian injuries; and $197 million, or 2 percent, of the direct property damage (excluding the events of September 11, 2001). These fires rose 1 percent from 2000 to 2001, but have fallen 41 percent from 1980 to 2001. These outside and other fires include the following:

- 75,000 outside fires involving property of value
- 414,000 brush, grass, or wildland fires
- 208,500 outside rubbish fires
- 164,000 other fires including outside spills or leaks with ensuing fires, explosions with no after-fire, and unclassified or unknown-type fires

Fire Fighter Fatalities

Four hundred thirty-nine fire fighters died as a result of injuries or illnesses that occurred while the victims were on duty (not just at fire scenes) during 2001. Three hundred forty of these deaths resulted from the attacks on the World Trade Center on September 11, 2001. More fire fighters were killed at the World Trade Center than in any of the past 25 years. In addition, two fire department paramedics and a chaplain also died in that attack. The second highest number of on-duty fire fighter deaths since 1977 occurred in 1978. One hundred seventy-two on-duty fire fighter deaths occurred that year. In recent years, about 100 on-duty fire fighter deaths occurred per year. When the World Trade Center fatalities are excluded, the 99 on-duty deaths in 2001 are consistent with recent years. Forty of the 99 deaths were caused by heart attacks. Vehicle crashes killed 17, and four of the 99 died after being struck by vehicles [2].

Fire fighter fatalities are tracked separately from other fire statistics. The National Fire Protection Association (NFPA) actively seeks reports for every on-duty death from local and/or state fire officials and updates yearly totals should an injury subsequently lead to death.

▶ STRUCTURE FIRES BY OCCUPANCY TYPE

Residential Structures

Certain terms are commonly seen in descriptions of the fire problem. The definitions given here are for U.S. data derived from or related to the National Fire Incident Reporting System (NFIRS). Homes include one- and two-family dwellings, apartments, and manufactured housing. Residential structures include homes; rooming, boarding, or lodging houses; hotels or motels; dormitories, fraternities, and sorority houses; and residential board and care facilities. The nonhome residential properties are more stringently regulated than the home properties.

As mentioned earlier, home structure fires generally account for the largest share of fire deaths, injuries, and direct property damage. Even in 2001, more civilians were killed by or injured in fires in one- and two-family dwellings alone than in the events of September 11, 2001. (These events dwarfed all other property damage in 2001.) The construction of the homes themselves and the behavior of the people in these homes are less regulated than that of almost any other occupancy.

Because more detailed information about occupancy is captured in NFIRS, many of the following statistics are based on 1999 data. More information about how these national estimates of specific fire problems were derived is provided at the end of this chapter. During 1999, the 370,200 home structure fires accounted for 70.7 percent of the 523,600 structure fires, 94.5 percent of the 3041 civilian structure fire deaths, 85.7 percent of the 18,519 civilian structure fire injuries, and 57.2 percent of the $8.5 billion in direct property damage [3].

Nonresidential Properties

Nonresidential properties include the following types of structures.

Public Assembly Properties.   This type of structure includes amusement or recreation places, places of worship or funeral properties, clubs, libraries, museums, art galleries, courthouses, eating and drinking establishments such as restaurants, bars, and night clubs, passenger terminals, and theaters or studios. The leading cause of fires in eating and drinking establishments was cooking equipment, while intentional fire setting was the leading cause of structure fires in other public assembly properties.

Educational Properties.   This category includes day care, preschool, kindergarten through grade 12, college classrooms, and adult education properties. Dormitories are considered residential and are not included here. About half of the structure fires in educational properties were intentionally set.

Institutional Properties.   Institutional properties include nursing homes, hospitals and other health care facilities, and prisons and jails. Dryers and cooking equipment were the leading causes of fires in nursing homes, cooking equipment was the leading cause of fires in other health care facilities, and three-fifths of the fires in prisons and jails were intentionally set.

Mercantile and Business Properties.   This category includes different types of stores, facilities providing personal services such as barber and beauty shops, laundry or dry cleaning shops, service stations, vehicle or other repair shops, businesses selling professional supplies or services, and office properties such as general business offices; banks; veterinary or research offices, engineering, mailing firms, and post offices. Intentional firesetting and electrical distribution equipment were the leading causes of fires in store and mercantile properties, whereas electrical distribution equipment was the leading cause of office property fires.

Utility, Defense, Agriculture, and Mining.   This type of structure includes facilities engaged in energy production or distribution, laboratories, defense or military installation, computer or communications centers, utilities, agricultural properties, forest or timberland, and mines or quarries.

Manufacturing or Processing Facilities.   Manufacturing and processing properties are often grouped with the preceding category. Industrial equipment was the leading cause of fires in these two occupancy classes.

Storage Properties.   Storage properties include properties used for the storage of general items, agricultural products such as grain and livestock, vehicles, refrigerated products, and products requiring tank storage. The leading cause of fires in these properties was intentional firesetting.

Outside or Special Properties.   This category includes special structures such as bridges, tunnels, and toll booths. Outside, railroad, water, and aircraft areas are captured here. A new field, building status, is now used to identify buildings that are vacant, idle, or undergoing renovation or demolition. In the past, these were identified by occupancy class and all considered nonresidential.

The occupancy type breakdown of structure fires reported in 1999 is shown in Table 2.3.

▶ **Table 2.3**   Structure Fires During 1999 by Property Use

Fixed Property Use	Fires	Civilian Deaths	Civilian Injuries	Property Damage (in $ Millions)
Public Assembly	**16,800**	**11**	**194**	**381.0**
Fixed use amusement or recreation	700	0	11	13.0
Variable use amusement or recreation	900	2	4	5.5
Place of worship or funeral property	2000	3	26	110.8
Club	1200	0	17	28.1
Library, museum, courthouse, or other public property	400	0	0	11.8
Eating or drinking place	10,800	6	121	199.7
Passenger terminal	200	0	9	1.0
Studio or theater	300	0	4	3.4
Unclassified or unknown-type public assembly property	300	0	2	7.7
Educational	**7700**	**0**	**136**	**84.2**
Preschool through grade 12	5800	0	111	54.5
Adult education or college classroom	1200	0	21	26.3
Day care	400	0	0	2.8
Unclassified or unknown-type public educational property	300	0	4	0.6
Institutional	**8000**	**3**	**241**	**28.8**
Nursing home	2500	3	111	8.5
Mental retardation or substance abuse	200	0	4	0.9
Hospital or hospice	3000	0	117	8.7
Clinic or doctor's office	200	0	2	3.7
Prison, jail, or police station	1900	0	6	5.4
Unclassified institutional property	200	0	0	1.6

(continues)

▶ **Table 2.3** *(Continued)*

Fixed Property Use	Fires	Civilian Deaths	Civilian Injuries	Property Damage (in $ Millions)
Residential	**383,600**	**2921**	**16,419**	**5084.4**
Home	*370,200*	*2873*	*15,879*	*4855.2*
One- or two-family dwelling	275,000	2359	10,688	3861.5
Multi-family building	95,100	514	5191	993.7
Other Residential	*13,400*	*48*	*540*	*229.2*
Boarding or rooming house, residential hotel, or shelter	1500	9	64	13.5
Hotel or motel	4600	24	249	115.0
Residential board and care	700	0	39	2.3
Dormitory, fraternity, or barracks	2100	0	92	35.1
Unclassified or unknown-type residential	4400	15	95	63.3
Mercantile and Office	**23,300**	**15**	**315**	**1006.0**
Grocery or convenience store	4300	0	58	63.2
Textile or apparel sales	700	0	6	12.0
Household goods sales or repairs	1200	0	13	546.1
Specialty shop	1600	0	13	31.6
Personal service, recreation, or home repair	1200	0	21	23.0
Laundry, dry cleaning, or professional supplies, or services	2300	2	36	30.8
Service station or vehicle sales, service, or repair	3800	3	66	88.5
Department store or unclassified general retail	2100	5	36	46.0
Office, bank, or mail facility	5500	5	62	147.1
Unclassified or unknown-type mercantile or business	700	0	4	17.6
Utility, Defense, Agriculture, or Mining	**3800**	**23**	**162**	**690.2**
Energy production plant	200	11	107	626.8
Laboratory	300	0	11	2.0
Defense, computer, or communications center	100	0	0	1.4
Utility or distribution system	400	2	21	4.0
Agriculture	2200	10	19	47.7

▶ **Table 2.3** *(Continued)*

Fixed Property Use	Fires	Civilian Deaths	Civilian Injuries	Property Damage (in $ Millions)
Forest, timberland or woodland	100	0	2	0.1
Mine or quarry	100	0	2	2.6
Unclassified or unknown-type utility, defense, agriculture, or mining	300	0	0	5.6
Manufacturing or Processing	**13,300**	**6**	**447**	**420.3**
Storage	**38,200**	**29**	**411**	**555.7**
Grain or livestock storage	7300	2	40	125.1
Refrigerated storage	100	0	4	30.1
Outside storage tank	100	0	4	1.5
Vehicle storage, garage, or fire station	16,300	19	234	120.9
(Dwelling garage)	(13,700)	(13)	(183)	(85.8)
(Other garage)	(2500)	(6)	(51)	(35.0)
Warehouse, residential, or self-storage	5700	2	64	208.5
Unclassified or unknown-type storage property	8700	6	64	69.7
Special Property	**9200**	**6**	**83**	**64.7**
Dump or sanitary landfill	100	0	0	0.3
Bridge, tunnel, or outbuilding	2900	3	26	16.6
Open land, beach, or campsite	1600	0	15	2.6
Water area	100	0	2	3.1
Railroad area	200	0	0	1.3
Highway, street, or parking area	2700	3	30	24.9
Aircraft area	0	0	0	0.0
Construction site, oil or gas field, pipeline, or industrial plant yard	200	0	0	6.0
Unclassified or unknown-type special property	1300	0	11	9.9
Unclassified or unknown-type property use	**19,700**	**26**	**111**	**173.3**
Total	**523,600**	**3041**	**18,519**	**8488.5**

Note: These are fires reported to U.S. municipal fire departments and therefore exclude fires reported only to federal or state agencies or industrial fire brigades. Fires are rounded to the nearest hundred, deaths and injuries to the nearest one, and direct property damage to the nearest hundred thousand dollars. Sums may not equal totals due to rounding errors. An entry of zero means the number of fires or losses was zero or rounded to zero.

Source: National estimates based on NFIRS and NFPA survey.

CAUSAL FACTORS

People often ask "What is the leading cause of fires?" That question sounds simple, but a simple answer could be misleading. Is the interest in all fires, structure fires, home structure fires, or some other type of fire? The leading causes of industrial fires are very different from the leading causes of home fires. Someone from the insurance industry may be more interested in the leading cause of direct property damage rather than the number of fires. A life safety educator might be more interested in the leading causes of fire deaths or injuries.

Even the term *cause* can mean different things. Version 5.0 of NFIRS was introduced in 1999. This new version included a new field—cause of ignition—that captures the very broad designations of 1-Intentional, 2-Unintentional, 3-Failure of equipment or heat source, 4-Act of nature, 5-Cause under investigation, O-Other, and U-Cause undetermined after investigation. With the exception of intentional, these categories are too broad to provide much information that could be used for prevention.

More than half of home structure fires were characterized as unintentional, about one-third were caused by the failure of equipment or heat source, and 11 percent were intentionally set. By itself, the cause of ignition field tells us very little about the causes of fires that were not intentionally set.

Several fields have traditionally been called "causal factors." These fields include heat source, equipment involved in ignition, item first ignited, and factors contributing to ignition. These factors describe what, how, and why some form of heat ignited the specific material involved. If the fuel, the heat source, or the behavior that brought them together were eliminated, the fire could have been prevented. Fires caused by smoking or candles can be identified by heat source, those caused by cooking, heating, appliances, or electrical equipment can be identified by the equipment involved in ignition, and ignitions resulting from playing with fire can be identified by data captured under factors contributing to ignition. Flammable liquid fires can be identified by the item first ignited.

Heat Source

The heat source indicates the item or energy that provided the heat of ignition, including but not limited to, operating equipment, electrical arcing, matches, lighters, cigarettes, candles, lightning, and fireworks. In 36 percent of the home structure fires in 1999, the heat source was radiated or conducted heat from operating equipment, arcing provided the heat in 13 percent, and 6 percent were started by heat from other powered equipment.

Cigarettes were the heat source in 6 percent of the home structure fires, but these fires accounted for 23 percent of the home fire deaths. A spark, ember, or flame from operating equipment provided the heat in 5 percent of the home structure fires, and candles were the heat source in 4 percent of these incidents.

The equipment involved in ignition describes any equipment that acted as the heat source. The equipment may have malfunctioned, or more commonly, it may have been operating as designed, but used improperly. The saying "space heaters need space" was coined in response to the number of fires and associated casualties caused by heaters that were too close to combustible materials such as bedding and upholstered furniture. Automatic shutoffs on coffee pots and irons were added in response to common human errors. This field is used to identify fires caused by appliances, electrical distribution equipment and cooking and heating equipment. The five most frequent entries in this field for home structure fires were the following:

1. "No equipment" was involved (40 percent)
2. Ranges, with or without ovens (17 percent)
3. Unclassified electrical wiring (which includes all of the fixed wiring converted from earlier versions of NFIRS) (4 percent)
4. Ovens or rotisseries (4 percent)
5. Dryers (4 percent)

Factors Contributing to Ignition

Factors contributing to ignition help describe how the heat source and combustible material came together to start a fire. Examples include playing with heat source, abandoned or discarded materials or products, installation deficiency, heat source too close to combustibles, equipment unattended, failure to clean, worn out, short circuit arc, and many others. This field often describes behaviors that can be addressed with educational messages. Fires caused by "playing with heat source" (and combined with the human factor of age) would be of interest to individuals interested in juvenile fire setting. The five most frequent entries in this field (for data collected in Version 5.0) for home structure fires were the following:

1. No factors contributing to the ignition (23 percent)
2. Combustibles too close to the heat source (7 percent)
3. Failure to clean (6 percent)
4. Unclassified factor (5 percent)
5. Equipment left unattended (4 percent)

Item First Ignited

Although somewhat less of an actual causal factor, the item first ignited describes what the heat source ignited, such as structural components, finishes or insulation, furniture, mattresses, bedding, clothing or other soft goods, decorations, fuel, and other products. In 1999, cooking materials, including food, were first ignited in 21 percent of the home structure fires; structural members or framing were first ignited in 8 percent; electrical wire or cable insulation were first ignited in 7 percent; rubbish, trash, or waste products were first ignited in 6 percent; and mattresses or bedding were first ignited in 6 percent of the fires, but mattress and bedding ignitions accounted for 12 percent of the home fire deaths. Upholstered furniture was first ignited in only 3 percent of the home structure fires, but these fires accounted for 16 percent of the home structure fire deaths.

Fire Cause in Confined Structure Fires

Causal data is generally not collected for certain types of confined structure fires in Version 5.0 of NFIRS. A method was needed to ensure that these fires were not forgotten when looking at fire causes. It seems reasonable to say that the equipment involved in a confined cooking fire was unspecified cooking equipment, that food or cooking materials were the item first ignited, and that the fire started in a kitchen or cooking area. Similarly, the equipment involved in a confined chimney fire is inferred as an unspecified-type chimney, the fire started in the chimney area, and the item first ignited was film or residue, including creosote. The NFPA and some other organizations are using the rules shown in Table 2.4 to assign inferred codes for analysis purposes.

Hierarchy of Major Causes

As mentioned earlier, different fire causes are identified by looking at data from different fields. Ranking causes across the multiple dimensions is difficult. The United States Fire Administration (USFA) developed a sorting hierarchy with 12 separate known cause categories. Specific data in the different causal factors, exposure number, and incident type resulted in an incident being assigned to a specific major cause category. A ranking system was used to strip off fires once they were assigned so they would not be double counted. Tables 2.5 and 2.6 show home structure fires and nonresidential structure fires in 1999 by these cause categories. This hierarchy is merely a summary. Although cooking and heating equipment were the leading cause of home structure fires, it does not mean that defective equipment was the leading cause.

▶ **Table 2.4**   Confined Structure Fires and Inferred Codes or Causes

Type of Confined Fire	Codes to Be Inferred or Assigned
113—Cooking fire, confined to container	Area of origin = 24 (kitchen) Item first ignited = 76 (cooking materials, including edible materials) Equipment involved = 640 (a new code defined as cooking equipment of unknown type); this code will also sort these fires into Cooking among the 12 major causes in the hierarchical sort.
114—Chimney or flue fire, confined to chimney	Area of origin = 57 (chimney, conversion code only) Item first ignited = 95 (Film, residue, included are paint, resin, and chimney film or residue and other films and residues produced as a by-product of an operation. Creosote was coded as rubbish, trash, or waste in older versions of NFIRS.) Equipment involved = 129 (a new code defined as chimney or flue of unknown type); this code will also sort these fires into Heating among the 12 major causes in the hierarchical sort.
115—Incinerator overload or malfunction	Area of origin = 64 (incinerator area) Item first ignited = 96 (rubbish, trash, or waste) Equipment involved = 352 (incinerator); this code will also sort these fires into Other Equipment among the 12 major causes in the hierarchical sort.
116—Fuel burner/boiler malfunction	Area of origin = 62 (Heating room or area or water heater area) Item first ignited = 69 (a new code for flammable or combustible liquid of unknown type) Equipment involved = 130 (a new code for boiler, furnace, or central heating unit of unknown type); this code will also sort these fires into Heating among the 12 major causes in the hierarchical sort.
117—Commercial compactor, confined to rubbish	Area of origin = 46 (chute or container for trash, rubbish, or waste) Item first ignited = 96 (rubbish, trash, or waste) Equipment involved = 812 (trash compactor); this code will also sort these fires into Other Equipment among the 12 major causes in the hierarchical sort.
118—Trash or rubbish fire, contained	Area of origin = UUU (unknown) Item first ignited = 96 (rubbish, trash, or waste) Equipment involved = NNN (none)

Source: P. A. Frazier, et al., "Revised Proposed Analysis Rules for Fire Incident Data in NFIRS 5.0," Memorandum dated September 25, 2000.

▲ **Table 2.5** Major Causes of 1999 Home Structure Fires (Unknown Cause Fires Allocated Proportionally, National Estimates)

Major Cause	Fires		Civilian Deaths		Civilian Injuries		Direct Property Damage (in Millions)	
Cooking equipment	102,400 #1	(27.7%)	378 #4	(13.2%)	4,440 #1	(28.0%)	$531.0 #4	(10.9%)
Heating equipment	53,100 #2	(14.3%)	383 #3	(13.3%)	1,447 #6	(9.1%)	$682.5 #3	(14.1%)
Intentional	42,400 #3	(11.4%)	533 #2	(18.6%)	1,741 #4	(11.0%)	$793.0 #2	(16.3%)
Electrical distribution	40,100 #4	(10.8%)	226 #5	(7.9%)	1,166 #7	(7.3%)	$804.7 #1	(16.6%)
Open flame, ember, or torch	31,000 #5	(8.4%)	175 #6	(6.1%)	1,991 #2	(12.5%)	$523.5 #5	(10.8%)
Appliance, tool, or air conditioning	24,300 #6	(6.6%)	71 #9	(2.5%)	594 #8	(3.7%)	$223.4 #9	(4.6%)
Smoking materials	23,500 #7	(6.4%)	739 #1	(25.7%)	1,794 #3	(11.3%)	$329.9 #6	(6.8%)
Exposure	15,200 #8	(4.1%)	26 #11	(0.9%)	91 #11	(0.6%)	$234.8 #8	(4.8%)

(continues)

▲ **Table 2.5** (Continued)

Major Cause	Fires		Civilian Deaths		Civilian Injuries		Direct Property Damage (in Millions)	
Child playing	15,100 #9	(4.1%)	159 #7	(5.5%)	1,678 #5	(10.6%)	$255.7 #7	(5.3%)
Other heat source	12,800 #10	(3.5%)	113 #8	(3.9%)	557 #9	(3.5%)	$196.6 #10	(4.0%)
Natural causes	5,300 #11	(1.4%)	11 #12	(0.4%)	69 #12	(0.4%)	$169.3 #11	(3.5%)
Other equipment	5,000 #12	(1.4%)	60 #10	(2.1%)	312 #10	(2.0%)	$110.9 #12	(2.3%)
Total	370,200	(100.0%)	2,873	(100.0%)	15,879	(100.0%)	$4,855.2	(100.0%)

Note: These are fires reported to U.S. municipal fire departments and therefore exclude fires reported only to federal or state agencies or industrial fire brigades. Major cause classes are based on a hierarchy developed by the U.S. Fire Administration. Fires are expressed to the nearest hundred, deaths and injuries to the nearest one, and property damage to the nearest hundred thousand dollars. Totals under "Total" may not equal sums because of rounding errors.

Each entry shows the estimated number, percent share of total in parentheses, and rank among the twelve major cause groups below each entry. Percentages are calculated on the actual estimates, so two figures with the same rounded-off estimates may have different percentages. Fires in which the cause was unknown have been allocated proportionally among fires of known causes. The values in this table represent the sum of the data for (a) fires in one- and two-family dwellings and manufactured housing, and (b) fires in apartments.

Source: National estimates based on NFIRS and NFPA survey.

▲ **Table 2.6**   Major Causes of 1999 Nonresidential Structure Fires (Unknown Cause Fires Allocated Proportionally, National Estimates)

Major Cause	Fires		Civilian Deaths		Direct Property Civilian Injuries		Damage (in Millions)	
Intentional	34,100 #1	(24.3%)	46 #1	(38.1%)	340 #1	(16.2%)	$1,592.1* #1	(46.8%)
Electrical distribution	16,500 #2	(11.8%)	11 #4	(9.5%)	257 #4	(12.3%)	$519.1 #2	(15.2%)
Cooking equipment	14,700 #3	(10.5%)	7 #6	(6.0%)	219 #5	(10.4%)	$86.0 #9	(2.5%)
Open flame, ember, or torch	13,700 #4	(9.8%)	16 #2	(13.1%)	288 #3	(13.7%)	$207.7 #5	(6.1%)
Other equipment	11,000 #5	(7.8%)	14 #3	(11.9%)	311 #2	(14.8%)	$283.4 #3	(8.3%)
Heating equipment	10,900 #6	(7.8%)	7 #7	(6.0%)	217 #6	(10.3%)	$228.3 #4	(6.7%)
Exposure	10,300 #7	(7.4%)	3 #9	(2.4%)	18 #11	(0.8%)	$123.2 #7	(3.6%)
Other heat source	9,600 #8	(6.9%)	4 #8	(3.6%)	130 #8	(6.2%)	$94.0 #8	(2.8%)

(continues)

▲ **Table 2.6** *(continued)*

Major Cause	Fires		Civilian Deaths		Direct Property Civilian Injuries		Damage (in Millions)	
Appliance, tool, or air conditioning	7,200 #9	(5.1%)	0 #10	(0.0%)	138 #7	(6.6%)	$35.0 #11	(1.0%)
Smoking materials	7,200 #10	(5.1%)	11 #4	(9.5%)	127 #9	(6.1%)	$134.4 #6	(3.9%)
Child playing	2,700 #11	(1.9%)	0 #10	(0.0%)	46 #10	(2.2%)	$17.2 #12	(0.5%)
Natural causes	2,200 #12	(1.6%)	0 #10	(0.0%)	8 #12	(0.4%)	$83.7 #10	(2.5%)
Total	140,100	(100.0%)	120	(100.0%)	2,100	(100.0%)	$3,404.1	(100.0%)

* The direct property damage is inflated by one fire with a miscoded property loss. On average, intentionally set fires accounted for 28.6 percent of the nonresidential fires during the 5-year period of 1994–1998.

Note: These are fires reported to U.S. municipal fire departments and therefore exclude fires reported only to federal or state agencies or industrial fire brigades. Major cause classes are based on a hierarchy developed by the U.S. Fire Administration. Fires are expressed to the nearest hundred, deaths and injuries to the nearest one, and property damage to the nearest hundred thousand dollars. Totals under Total may not equal sums because of rounding errors. Each entry shows the estimated number, percent share of total in parentheses, and rank among the twelve major cause groups below the entry.

Percentages are calculated on the actual estimates, so two figures with the same rounded-off estimates may have different percentages. Fires in which the cause was unknown have been allocated proportionally among fires of known causes.

Source: National estimates based on NFIRS and NFPA survey.

Contributing factors such as unattended equipment and combustibles too close to the equipment play an important role in these fires. Both candles and torches are captured under the category of open flame, ember, or torch. Prevention strategies and educational messages are very different for these two heat sources. The major cause category is an important tool, but it is not a substitute for the detail found in the separate fields.

The causes of home structure fires and structure fire deaths are shown in Figure 2.3. Table 2.7 shows the top five causes of structure fires, direct property damage, civilian deaths, and civilian injuries from fires in homes and in nonresidential structure fires.

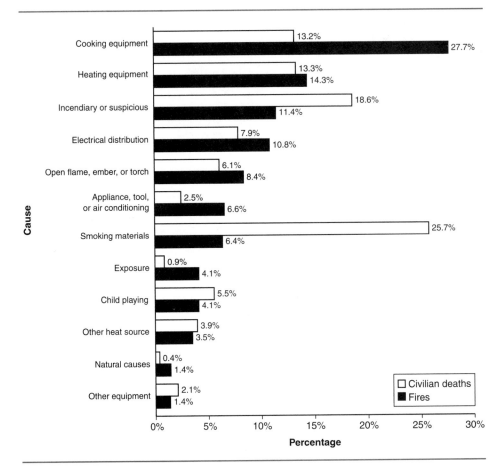

▶ **Figure 2.3** Causes of Home Fires and Home Fire Deaths, 1999. *(Source: Marty Ahrens, "The U.S. Fire Problem Overview Report: Leading Causes and Other Patterns and Trends," NFPA, 2003.)*

▶ **Table 2.7**   Top Five Major Causes for Home and Nonresidential Structure Fires in 1999

Homes		Nonresidential	
Fires			
1 Cooking equipment	27.7%	Intentional	24.3%
2 Heating equipment	14.3%	Electrical distribution	11.8%
3 Intentional	11.4%	Cooking equipment	10.5%
4 Electrical distribution	10.8%	Open flame, ember, or torch	9.8%
5 Open flame, ember, or torch	8.4%	Other equipment	7.8%
Civilian Deaths			
1 Smoking materials	25.7%	Intentional	38.1%
2 Intentional	18.6%	Open flame, ember, or torch	13.1%
3 Heating equipment	13.3%	Other equipment	11.9%
4 Cooking equipment	13.2%	Electrical distribution	9.5%
5 Electrical distribution	7.9%	Smoking materials	9.5%
Civilian Injuries			
1 Cooking equipment	28.0%	Intentional	16.2%
2 Open flame, ember or torch	12.5%	Other equipment	14.8%
3 Smoking materials	11.3%	Open flame, ember, or torch	13.7%
4 Intentional	11.0%	Electrical distribution	12.3%
5 Child playing	10.6%	Cooking equipment	10.4%
Direct Property Damage			
1 Electrical distribution	16.6%	Intentional	46.8%
2 Intentional	16.3%	Electrical distribution	15.2%
3 Heating equipment	14.1%	Other equipment	8.3%
4 Cooking equipment	10.9%	Heating equipment	6.7%
5 Open flame, ember, or torch	10.8%	Open flame, ember, or torch	6.1%

Source: NFIRS

Fires of Unknown Cause

A certain share of fires will always have information that is unknown. However, this share is not constant. When looking at the percentages of fires by cause, it is possible to look just at the percentage of fires with known cause. In most NFPA analyses, fires with unkown data are all rated proportionally among fires for which the data was known.

Careful readers of fire statistics may notice slight variations in the numbers and percentages of fires of different causes, which are often due to differences in what is considered unknown.

▶ ARSON

Definitions and Analysis

In general, most fires are not criminal acts. Arson, as defined in the American Heritage dictionary, is the "crime of willfully setting fire to buildings or other property." It must be proved that there was intent to cause harm. In an arson case, it is necessary to first establish that a crime was committed and then to prove guilt. Different states have different legal definitions and different age minimums at which a juvenile could be charged with arson.

Most statistical analyses of the arson fire problems in past years combined the older codes for incendiary, which required "a legal decision or physical evidence that a fire was set," and suspicious, which was to be used when "circumstances indicated the possibility that the fire may have been deliberately set, multiple ignitions were found, or there were suspicious circumstances and no accidental or natural ignition factor could be found" [4]. Many fire investigators rightly complained that suspicious was not an appropriate final cause determination. However, some fires are never fully investigated, and, even when investigations are completed, the causal information is often not updated. The Federal Bureau of Investigation's (FBI's) Uniform Crime Reporting (UCR) System captures arson offenses from law enforcement agencies. These statistics have tended to resemble the incendiary only fires [5].

Version 5.0 of NFIRS dropped suspicious entirely and replaced incendiary with intentional. (The wildland fire module continues to use incendiary.) Intentional would indicate that the fire was deliberately set, but there is no legal connotation attached. Some fires were reported to have been intentionally set by children under 5 years old.

Arson Data

The most current statistics about structure arson come from NFPA's annual fire department survey. In 2001, U.S. fire departments responded to an estimated 45,000 intentionally set structure fires. These fires caused 2781 civilian fire deaths and $34 billion in direct property damage. Because the events of September 11, 2001, were intentional, the losses were higher than in the past. When these events are excluded, intentionally set structure fires caused 330 civilian fire deaths and $1 billion in direct property damage. The 39,500 intentionally set vehicle fires caused an estimated $219 million in direct property damage. Beginning in 2001, NFPA no longer asked about suspicious fires. These statistics also do not include any fires of unknown cause, nor do they capture any outside arsons [1].

With NFIRS data, fires with unknown causes can be proportionally allo-cated. Based on NFIRS data, structure fires accounted for about 18–21 per-cent of the intentionally set fires during the 1990s, vehicle fires accounted for 12–15 percent of these fires, and outdoor and other fires accounted for 65–69 percent of these fires. See Figure 2.4. Fifteen percent of the structure fires in 1999 were intentional. However, this varies widely by occupancy type. Eleven percent of home fires in 1999 were intentionally set compared to 24 percent of nonresidential fires. There is wide variation within the nonresi-dential structures. Between one-half and two-thirds of the structure fires in prisons and jails, educational properties, and structures that are vacant, idle, or under construction or demolition were intentionally set, as compared to only 6 percent of the structure fires in industrial and manufacturing prop-erties.

Statistics about the fires and associated losses caused by intentionally set fires tell only part of the arson story. Only about one-third of the intention-ally set fires were filed with the police as arson offenses. Only 15–20 percent of this third resulted in arrest or were cleared by other means. (Cases are often cleared when a suspect is charged with other crimes but is believed to have been involved in others with less compelling evidence.) This means that only 5–7 percent of the intentionally set fires resulted in an arrest. About one-half of the arrests resulted in prosecution, and two-thirds of the

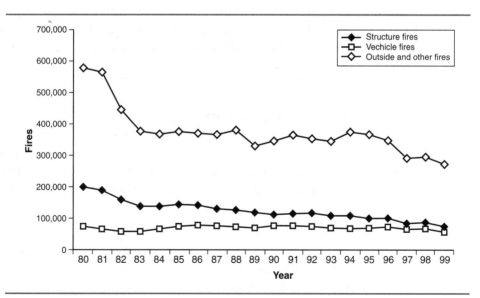

▶ **Figure 2.4**   Intentionally Set Fires by Year. *(Source: J. Hall, Intentional Fires and Arson, NFPA, 2003.)*

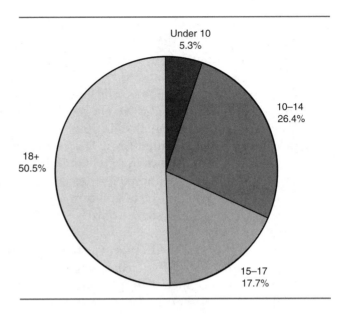

▶ **Figure 2.5** Arson Arrestees in 2001 by Age. *(Source: FBI's Crime in the U.S. series, summarized in J. Hall's Intentional Fires and Arson, NFPA, 2003.)*

prosecutions resulted in convictions. Taken together, this means that about 2 percent of all intentionally set fires actually led to convictions. About two-thirds of those convicted were sentenced to prison [6].

As shown in Figure 2.5, half of the people arrested for arson in 2001 were under the age of 18 [7]. This does not necessarily mean that half of all arsonists were under 18, because information is not available on perpetrators who were not caught.

▶ FIRE LOSS BY GEOGRAPHIC REGION

The four regions, as defined by the U.S. Bureau of the Census and used in this text, are the South, the Northeast, the West, and the North Central. The different regions show a variety of patterns of fires and assorted losses [8, 9].

South

The Southern region includes the following states: Alabama, Arkansas, Delaware, District of Columbia, Florida, Georgia, Kentucky, Louisiana, Maryland, Mississippi, North Carolina, Oklahoma, South Carolina, Tennessee, Texas, Virginia and West Virginia. As Figures 2.6 to 2.9 show, the South has traditionally led the country in per capita rates of fires and fire deaths.

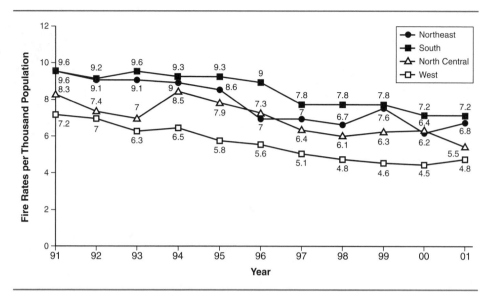

▶ **Figure 2.6**   Trend in Fire Incident Rates by Region: 1991–2001. *(Source: Michael J. Karter, Jr., "U.S. Fire Experience by Region," NFPA, multiple years, and Michael J. Karter, Jr., "Fire Loss in the United States," NFPA, multiple years.)*

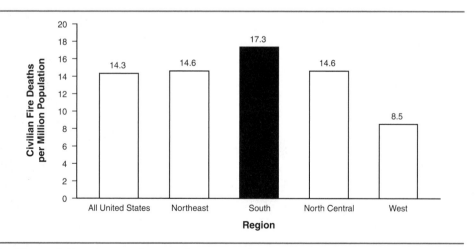

▶ **Figure 2.7**   Fire Death Rates by Region (Excluding the Events of 9/11/01): 1997–2001, Annual Averages. *(Source: Michael J. Karter, Jr., "U.S. Fire Experience by Region," NFPA, multiple years, and Michael J. Karter, Jr., "Fire Loss in the United States," NFPA, multiple years.)*

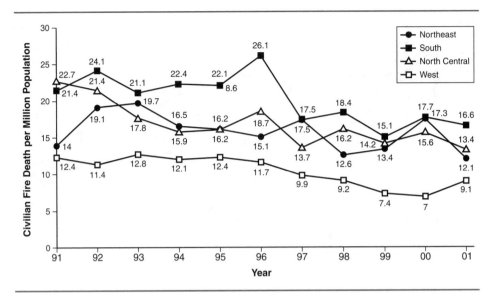

▶ **Figure 2.8** Trend in Civilian Fire Death Rates by Region: 1991–2001 (Excluding the Events of 9/11/01). *(Source: Michael J. Karter, Jr., "U.S. Fire Experience by Region," NFPA, multiple years, and Michael J. Karter, Jr., "Fire Loss in the United States," NFPA, multiple years.)*

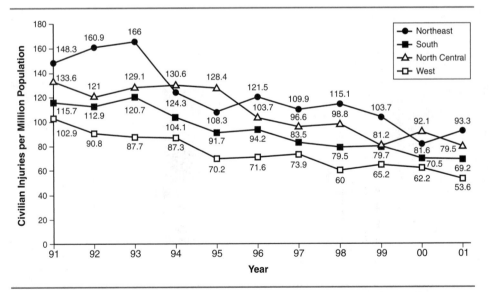

▶ **Figure 2.9** Trend in Civilian Fire Injury Rates by Region: 1991–2001 (Excluding the Events of 9/11/01). *(Source: Michael J. Karter, Jr., "U.S. Fire Experience by Region," NFPA, multiple years, and Michael J. Karter, Jr., "Fire Loss in the United States," NFPA, multiple years.)*

The South's 5-year average civilian fire death rate for 1997–2001 (excluding the events of September 11, 2001) was 21 percent higher than that of the country as a whole. People in the South are more likely to live in rural areas than are people in other parts of the country. Rural fire and fire death rates are higher than these rates for any other size of community.

Northeast

The Northeastern region includes the following states: Connecticut, Maine, Massachusetts, New Hampshire, New Jersey, New York, Pennsylvania, Rhode Island, and Vermont. With September 11, 2001, the Northeast had the highest rate with a civilian fire death rate of 58.4 per million population and the highest per capita property damage of $687 for the year 2001. When these events are excluded, the Northeast's 2001 fire death rate and per capita property damage ranked third among the four regions. The Northeast usually ranks first in civilian fire injury rates. Injuries, however, are harder to define than deaths. There may be some differences in the nature of injuries that are documented in different parts of the country or even in different fire departments.

West

The Western region includes the following states: Alaska, Arizona, California, Colorado, Hawaii, Idaho, Montana, Nevada, New Mexico, Oregon, Utah, Washington, and Wyoming. The West has consistently had the lowest per capita rates of reported fires, fire deaths, and reported civilian fire injuries. Although the West is generally below the national average in per capita direct property damage, the highest per capita direct property damage (excluding the events of September 11, 2001) ever reported was seen in that region after the Oakland fire storm in 1991.

North Central

The North Central region includes the following states: Illinois, Indiana, Iowa, Kansas, Michigan, Minnesota, Missouri, Nebraska, North Dakota, Ohio, South Dakota, and Wisconsin. In 2001, the North Central region was third in per capita rates of fires, civilian fire injuries, direct property damage, and, when the events of September 11 were included, of civilian fire deaths. When these events are excluded, the region's 2001 per capita fire death rate ranked second.

▶ DEMOGRAPHIC CHARACTERISTICS AND FIRE

Demographic data, including data collected by the U.S. Census and others, such as age, gender, education, housing density, and smoking status, can be used to describe a population. Certain characteristics are correlated with higher risk of death and injury from fires. Males, young children and older adults face higher risk of fire death and injury [9]. Children under 6 and people over 65 years of age face a risk of home fire death twice that of the general population. People over 75 have a risk of home fire death that is 3 times as high, and those over 85 have a risk that is 4.6 times that of the general population. The risk of home fire injury is spread somewhat more evenly. People over 85 have an injury risk 1.9 times that of the general population. Interestingly, young adults between 20 and 29 had the second highest injury risk, with a rate 1.3 times the general population. Figure 2.10 shows the risk of home fire death and injury by age group.

Males of all ages face a higher risk of home fire death than do females, and except for the 75–84 age group, a higher risk of home fire injuries. In two studies examining blood alcohol levels in people killed by fire, 39–51 percent of the victims over 20 years old were legally intoxicated.

Poverty, lack of education, and smoking are correlated with high state death rates [10]. In his report "U.S. Fire Death Patterns by State," John Hall examines the demographic characteristics of states in relation to their rank in fire death rates. States with higher fire death rates tend to have larger shares

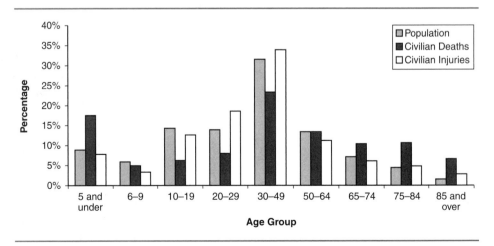

▶ **Figure 2.10** Civilian Home Fire Deaths and Injuries by Age Group, **1994–1998.** *(Source: J. Hall, Patterns of Fire Casualties in Home Fires by Age and Sex, NFPA, 2001.)*

of the population below the poverty line, of people age 25 or older who lack a high school diploma, and of people 18 or older who are current smokers. These factors are also correlated with each other. States having larger percentages of people living in rural areas also tend to have higher fire death rates.

▶ **DEATH RATES IN SELECTED INDUSTRIALIZED COUNTRIES**

The United States and Canada have had some of the highest fire death rates in any industrialized democracy outside the former Soviet bloc. Figure 2.11 shows the fire death rates per year in the United States, Canada, the United Kingdom, Japan, and Sweden.

There has been relatively little change in the fire death rates in the United Kingdom, Japan and Sweden over the past 20 years. The death rates in Canada and the United States have been fairly similar, with some seesawing back and forth. In recent years, Canada's death rate has generally been lower than that of the United States.

Different countries have different fire problems. Incendiary suicides are a particularly large share of Japan's fire deaths, accounting for about half of the fire deaths among Japanese adults between 21 and 60 years of age. Without this problem, the Japanese death rate in the 1990s would have been about half that of the United States [11].

According to a study done by Tri-Data Corporation for the USFA, many foreign cities expect a much larger fire department response time and put far more resources into preventing fires rather than extinguishing them [12].

▶ **FIRE DEPARTMENT ACTIVITIES**

Emergency Calls

Municipal fire departments responded to 20,965,500 emergency calls during 2001. This was the largest number of calls since the NFPA survey began collecting this data. Only 1,734,500, or 8 percent, were to fires; 12,331,000, or 59 percent, were to medical aid calls [1], as illustrated in Figure 2.12.

Figure 2.13 shows that medical aid responses and false alarms (including malicious reports and unwanted activations of automatic systems), and other calls, including hazardous material calls and other hazardous condition calls, all at least doubled; mutual aid responses tripled; and reported fire incidents fell 42 percent from 1980 to 2001.

Information about hazardous material calls was first collected in 1986. Since then, these calls have more than doubled, from 171,500 in 1986 to

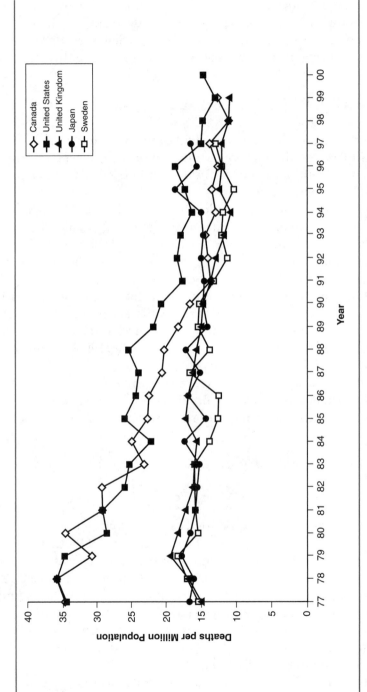

▲ **Figure 2.11** Fire Death Rates in Selected Countries. *(Source: J. Hall, International Comparison Reports: U.S.A. vs. Canada, U.S.A. vs. Japan, U.S.A. vs. Sweden, U.S.A. vs. U.K., NFPA, multiple years.)*

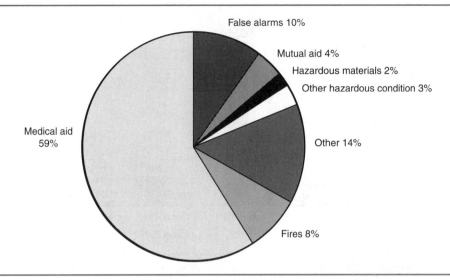

Medical aid 59%

False alarms 10%

Mutual aid 4%

Hazardous materials 2%

Other hazardous condition 3%

Other 14%

Fires 8%

▶ **Figure 2.12**   Fire Department Calls, 2001. *(Source: Michael J. Karter, Jr., "Fire Loss in the United States," NFPA, multiple years.)*

381,500 in 2001. Other hazardous condition calls increased 90 percent from 318,000 in 1986 to 605,000 in 2001. To put these numbers in perspective, the average frequency for different types of calls in 2001 is shown in Table 2.8.

False Alarms

Although the decrease in fires is encouraging, Figure 2.14 shows that the total of reported fires and false alarms has averaged about 3.6 million since 1980. Through most of the 1990s, the total was above this average. In 2001, the combined total was at its highest point since 1980. Fire departments cannot presume a call is a false alarm and must respond as they would to a fire.

Fire Prevention Activities

Although fire department statistics show that fire service responsibilities are growing and changing, they do not show the complete picture. Most fire prevention activities, such as inspections, plan review, and public fire education, are not captured in call counts. Training is generally ongoing to ensure that fire fighters can protect their communities as safely and efficiently as possible. Equipment must be checked regularly. Many time-consuming, essential activities are invisible to the general public.

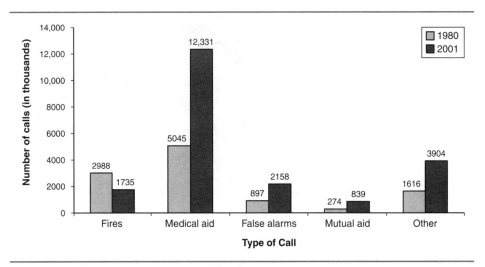

▶ **Figure 2.13** Fire Department Calls, 1980 and 2001. *(Source: Michael J. Karter, Jr., "Fire Loss in the United States," NFPA, multiple years.)*

▶ **Table 2.8** Average Frequency of Fire Calls by Type

Type of Call	Rate
All fire department calls	38.9 per minute (or 1 every 1.5 seconds)
Medical aid call	23.2 per minute (or 1 every 2.6 seconds)
Any fire call	3.2 per minute (or 1 every 18.2 seconds)
Structure fire call	1.0 per minute
False alarm call	4.1 per minute
Hazardous material call	0.7 per minute (or 1 every 1.4 minutes)

▶ **FIRE DATA SOURCES**

NFIRS—Three-Tiered System

Almost all of the details about the national fire experience come from USFA's NFIRS. The three tiered system, made up of local fire departments, state fire agencies, and the federal government's USFA, all play critical roles in collecting NFIRS data.

Local Fire Departments. Fire officers and fire fighters at the local level provide the basis of the system. Without their efforts, this system could not

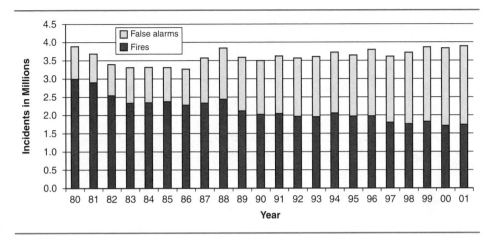

▶ **Figure 2.14**   Reported Fires and False Alarms, 1980–2001. *(Source: Michael J. Karter. Jr., "Fire Loss in the United States," NFPA, multiple years.)*

work. All the data in NFIRS comes originally from local fire officers and fire fighters. Approximately 14,000, or almost half, of the nation's public fire departments report at least one fire to NFIRS each year. Between one-third and one-half of the fires reported to local fire departments are captured by NFIRS, making it the largest and most detailed fire incident database in the world. Although not legally required to do so, most states are either currently collecting local NFIRS data or are developing the capability to do so.

State Fire Agencies.   States generally provide the forms or form layout used by local departments. Some states use the standard NFIRS forms without modification. Some have added a few additional data elements or changed the format slightly. A few states have modified the NFIRS data classification to meet their own needs. These states convert the data back to the standard format before forwarding the data.

Federal Government (USFA).   The USFA's National Fire Data Center compiles the data from participating states. There is generally a considerable delay before a full year of national data is obtained. After the data is compiled nationally, other users, such as the NFPA, the U.S. Consumer Product Safety Commission, and private entities may obtain the data for their own analyses. The USFA regularly updates its publication, *A Profile of Fire in the United States.*

The USFA's National Fire Information Council (NFIC) is comprised of state NFIRS program managers and NFIRS managers from fire departments protecting populations of at least 500,000. NFIC advises and assists the USFA with NFIRS development and training in addition to providing training to its own members.

Version 5.0 of NFIRS is designed to capture all emergency responses. Until 1999, when NFIRS 5.0 was introduced, the fire incident codes were based on the 1976 edition of NFPA 901, *Uniform Coding for Fire Protection*. The current title is *Standard Classification for Incident Reporting and Fire Protection Data*. However, NFIRS stayed with the 1976 edition. New fields were added, some were dropped, and some definitions changed. A conversion program was developed to allow data from the two systems to be analyzed together. Because the conversions are not exact, some apparent differences will actually be due to system changes. Detailed information about Version 5.0 of NFIRS can be obtained from www.nfirs.fema.gov.

NFPA Fire Department Survey

Each January, NFPA sends its survey to all fire departments protecting 100,000 or more and a random sample, stratified by population, of the smaller departments. The 3303 departments that responded protect about 37 percent of the U.S. population. The data is used to develop national estimates of the fire problem. A press release about the number of fires and fire deaths is issued each August. The full report "Fire Loss in the United States" is generally available in September [1].

NFPA's survey is not terribly detailed. It asks about fires by type and broad occupancy class, intentional structure and vehicle fires, and civilian and fire fighter deaths and injuries. It also captures nonfire calls. Because the NFPA survey is general, it is combined with the NFIRS to develop national estimates of specific problems. NFIRS has far more detail, although it usually lags substantially behind the survey in terms of timeliness.

By comparing the totals from NFPA's survey to corresponding totals in NFIRS, analysts estimate what fraction of total U.S. fires, deaths, injuries, and lost dollars were captured by the NFIRS sample. These fractions are inverted into ratios that are used to scale up the NFIRS numbers on specific fire problems to national estimates of the total size of these problems. Different ratios are used for residential structure fires, nonresidential structure fires, vehicle fires, and other fires. Within each group, analysts use different ratios for fires, injuries, deaths, and direct property damage [13].

Statistical Estimates. It is important to remember that not all fires are captured by statistics. In many cases, a fire is handled privately or by an industrial fire brigade without fire department involvement. These cases will not be counted. If no paperwork or documentation was done on a fire, in terms of statistics, the fire did not happen. If a fire occurred in a community that was not participating in NFIRS, that fire will not be part of the national database, although it might be captured in very broad terms by the NFPA survey. In some years, difficulties at the state level prevent any data from the state to be submitted to the USFA. California did not submit any NFIRS data in 1997 or 1999. The national statistics also do not capture fires handled by state or federal agencies. The process of collecting statistics is illustrated in Figure 2.15.

Occasionally, readers will see statistics that seem to disagree with the facts, a circumstance most likely to happen with unusual events. For example, five people died in a fraternity fire in Pennsylvania in 1994, and another five were killed in a fraternity fire in North Carolina in 1998. National estimates of fraternity and sorority house fire deaths were zero in both years [14]. These fires were among the half to two-thirds of reported fires that did not make it into NFIRS.

Many fire fighters have wondered why, if they faithfully complete their fire incident reports, they are also asked to complete NFPA's survey. Because different states have different reporting practices, NFIRS cannot estimate the total number of reported fires per year. Analysts need to use the survey results and NFIRS together to develop national estimates. In the older versions of NFIRS, nonfire responses (except for hazardous material incidents in 4.1) were not forwarded to the USFA. Until all fire departments around the country report using NFIRS 5.0 to states that are processing all incidents, the NFPA survey is the only national source of this data. Because the survey is completed by June each year, it can produce numbers about a year earlier than NFIRS.

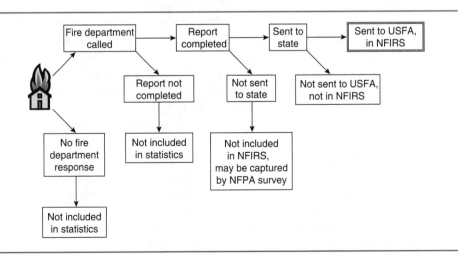

▶ **Figure 2.15**   **How a Fire Gets Counted.** *(Source: Fire Protection Handbook, NFPA, 2003, Figure 3.2.2)*

Death Certificate Data.   Death certificates have been around far longer than NFIRS and the NFPA survey. The National Safety Council (NSC) has been using a consistent estimating procedure based on state health department records of death certificates to track unintentional fire and burn deaths dating back to 1913, although their definition excludes some fire deaths, most notably those involving vehicle accidents, arson, or self-immolations. Figure 2.16 shows the steady decline in the rate of fire deaths per million population from both sources (excluding the events of September 11, 2001).

▶ SUMMARY

Incident reports from and surveys of the fire service are the source of what we know about the nation's fire experience. Without the detail provided by NFIRS, it would be impossible to answer many of the most basic questions about fire causes, use and effectiveness of smoke alarms or sprinklers, or causes of fire fighter injuries. Most of the statistics are projections, rather than an actual census documenting every single incident. It is important to pay attention to the definitions when reading statistical reports—"intentional" is not identical to the legal definition of arson, and residential occupancies include more than just homes.

The statistics show that the number of fires has been dropping, but fire department calls have increased. The largest share of fire deaths and injuries continue to result from home fires, even with the events of September 11, 2001. The fire problem varies substantially by occupancy and by type of fire. The cause profile is also different for different measures of fire loss. Cooking equipment is the leading cause of home structure fires and fire injuries, whereas smoking is the leading cause of home fire deaths. Young children, older adults, and males have a higher risk of fire deaths. Young adults and adults over 75 face the highest risk of fire injury. Males of all but one age group also have a higher risk of fire injury.

The states with the highest death rates tend to have larger shares of adult smokers, people living below the poverty line, adults without high school diplomas, and people living in rural areas. Many countries have traditionally had lower fire death rates than the United States, but the United States has been making progress in reducing the fire death toll in most of the recent years.

Reports on many different aspects of America's fire problem are updated annually by NFPA's Fire Analysis and Research Division. Many of these reports are free to NFPA members and/or the fire service. Readers are encouraged to contact NFPA's One-Stop Data-Shop for additional information.

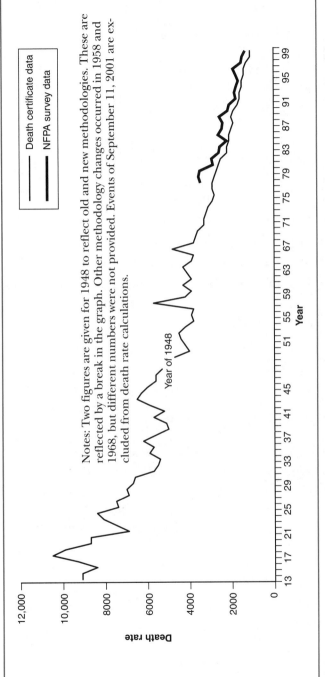

▲ **Figure 2.16**   Fire Death Rates per 100,000 Population.   *(Source: National Safety Council, Injury Facts, 2002 edition, pp. 44-45, and NFPA National Fire Experience Survey.)*

Group Activity

Discuss among yourselves the fire statistic (for example, the number of fires, fire deaths, fire fighter fatalities, or the financial costs) that best describes the greatest concern for understanding America's fire problem. Pick a spokesperson to explain the group's answer.

Review Questions

1. Where did 76 percent of structure fires occur?
2. What was the source of heat in over one-third of structure fires in 1999?
3. In an arson case, what two things is it necessary to prove?
4. About what percentage of intentionally set fires resulted in arrests?
5. What groups in our population are at highest risk of dying in home fires?
6. What are poverty and lack of education related to?
7. Which two countries have had some of the highest fire death rates?
8. What system provides most of the national fire statistical details?
9. What is the leading cause of home structure fires and fire injuries?

Suggested Readings

Ahrens, M., P. Frazier, and J. Heeschen. "Use of Fire Incident Data and Statistics." In *Fire Protection Handbook*, 19th ed. Quincy, MA: National Fire Protection Association, 2003.

Ahrens, M., S. Stewart, and P. Cooke. "Fire Data Collection and Databases." In *Fire Protection Handbook*, 19th ed. Quincy, MA: National Fire Protection Association, 2003.

NFIRS Analysis: Investigating City Characteristics and Residential Fire Rates, FA-179. Emmitsburg, MD: U.S. Fire Administration, April 1998.

A Profile of Fire in the United States, 1989–1998, 12th ed. Emmitsburg, MD: U.S. Fire Administration, August 2001.

3 Fire Behavior

Richard Pehrson*
Futrell Fire Consult and Design

Learning Objectives

After completing this chapter, the reader should be able to do the following:

- Explain the basic chemistry of fire
- Define and explain the difference between the fire triangle and the fire tetrahedron
- Describe the effects of oxygen on fire
- Describe the properties of a fuel
- Describe the three methods of heat transfer
- Illustrate the classifications of fire
- List and describe the modes of fire spread
- Describe fire extinguishment theory
- Explain the properties of combustible solids
- Define the characteristics of flammable and combustible liquids
- Define the properties of gases
- Define backdraft, flashover, and products of combustion and describe their implications for fire fighter safety

A basic understanding of the chemical and physical nature of fire is essential for anyone interested in fire protection. This is true for the architect, engineer, fire protection system designer, fire fighter, fire marshal, investigator, and anyone else who designs, builds, or inspects buildings where the potential for uncontrolled fire exists. At its simplest, fire behavior describes the

*The author thanks Thomas Woodford for providing the basic groundwork for the information in this chapter.

combination of chemistry and physics that applies to fire. However, in this chapter we also look at how fires grow in buildings and how the smoke and hot gases spread, a young branch of science known as fire dynamics. It is important to realize that fire is a complex natural phenomenon that can appear to be random, which complicates its study immensely. For more on these topics, see the Suggested Readings at the end of this chapter.

▶ CHEMISTRY OF FIRE

To understand fire, we need to define what fire is and how it occurs. The National Fire Protection Association document, NFPA 921, *Guide for Fire and Explosion Investigations*, defines fire as a rapid oxidation process with the evolution of light and heat in varying intensities [1]. Let us look at the parts of this definition and add a few refinements.

Elements of the Chemical Reaction

Oxidation. The interaction between a fuel and an oxidizer (usually oxygen from the air, but not always) is a chemical reaction called *oxidation*. The oxidation process involves the transfer of electrons from one compound (the fuel or reducing agent) to another (the oxidizer or oxidizing agent). Many forms of oxidation reaction can occur, including the rusting of steel, tarnishing of silver, the generation of energy from foods in our body, and the browning of fruit. Although we concentrate on the specific oxidation reactions known as fire or combustion, it is important to know that many others are occurring around us at all times.

Oxidation as a chemical reaction in fire breaks down the fuel and oxidizer, allowing the atoms to recombine in the form of compounds different from those with which we started. The fuel and the oxidizer are called *reactants* and the chemical compounds that result from the reaction are called *products*. The reactants undergo a chemical reaction (i.e., fire) to produce new substances different in composition, called products.

Fire is a rapid oxidation process. The process of rust forming on steel is chemically the same oxidation reaction as fire; the difference is the speed at which they occur. Rusting is a slow oxidation process. The rate at which energy is released from rust is slow, allowing the heat to dissipate without a noticeable increase in the temperature. With fire, the energy release causes temperatures to rise significantly where the reaction is occurring, often by thousands of degrees. This extra energy sometimes heats fuel that is not yet ignited, allowing a fire to spread, but to continue burning requires some mechanism to allow the reaction to continue over time.

Self-Sustaining Reaction.   Fire is a self-sustaining reaction. Once the chemical reaction starts, it can provide enough energy to continue the reaction between the fuel and oxidizer. If this were not the case, the reaction would stop and the fire would cease to exist.

Exothermic Reaction.   Fire is exothermic, meaning it gives off more energy than is consumed. The reaction releases energy because the products have lower internal energy levels than the reactants. Energy is released by the fire in the form of heat, light, and sound. Depending on the fuel and the conditions surrounding the fire, we may observe varying intensities of each of these.

Smoke-Producing Process.   In addition to light and heat, fire also produces smoke. Smoke is a broad term that encompasses all the products of combustion, including air surrounding the fire that is mixed with the smoke due to turbulence as it rises above the fire. The composition of the smoke, like the characteristics of the energy released, depends on the fuel, the oxidizer, and the conditions surrounding the fire. Smoke is defined more precisely later in this chapter.

Fire Versus Combustion

If we put these parts together, we get a more complete definition of fire. Namely, fire is a rapid, self-sustaining oxidation process that evolves heat, light, and smoke in varying quantities. This definition also fits combustion, so the terms fire and combustion are often used interchangeably. As we study fire behavior in more detail, it will be helpful to differentiate between these two terms. Combustion usually implies a level of control or design in the oxidation of fuel not associated with fire. As an example, the internal combustion engine in your car is a carefully designed machine that extracts as much energy from the fuel (gasoline or diesel) as possible. Combustion processes regulate the fuel and the oxidizer (usually the oxygen in the air) and strive to high levels of efficiency, while also reducing pollution or other undesirable by-products. Based on how the combustion system is designed, we know in advance what fuel will be burned and where it will be burned. Other examples of these "controlled" combustion processes include furnaces, stoves for cooking, gas and coal electrical generation plants, and many others.

Fire is often an unplanned or uncontrolled event ("uncontrolled" combustion), especially when one considers that the fuels are not selected or even known in advance. Consider a fire starting in the bedroom of a home; as a fire grows and spreads through the home, what is burning may not be known, or could be different from what would have burned the day before— conditions change continually and are not planned. Spontaneous ignition of a pile of coal outside a power plant may involve a fuel used for electricity generation (i.e., combustion), but because it is an unplanned or undesirable event, would best be considered a fire.

▶ FIRE TRIANGLE

The fire triangle has been used to explain the conditions necessary for combustion or fire to occur and includes fuel, oxygen, and heat. These three items are put together as shown in Figure 3.1, where each represents a different side of the triangle. When all three sides are present in sufficient quantities, the result is fire. Remove one side and the fire is extinguished. Fuel provides the stored energy, oxygen provides the key to release the energy, and the heat provides the kick to get the chemical reaction to occur at a rate we recognize as fire.

Fuel

Just like food for our bodies, fuel provides the energy stored in a chemical form that is released by the chemical reaction. Fuel molecules are generally made up of compounds containing hydrogen and carbon, known as *hydrocarbons*, although there are many others. A *molecule* is a group of atoms that are joined in a set combination, where the energy is stored in the bonds that hold the various atoms together to form the molecule. When the fuel molecules are broken up during the oxidation process, energy is released. Common fuels include the following:

▶ **Figure 3.1** Fire Triangle.

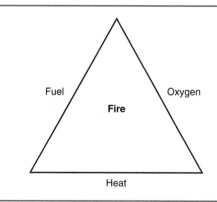

- Wood and wood products composed of cellulose. Cellulose is made up of linked hydrocarbons (combinations of hydrogen and carbon atoms that form more complex molecules).
- Simple hydrocarbons like methane, butane, and propane, which are usually gases at room temperature and pressure.
- Petroleum products, such as gasoline, oils, and plastics. These products are composed of complex hydrocarbons, especially where the hydrocarbons join up and connect to each other in long chains.

Where does a pile of fireplace wood originally get its energy? As a small plant grows, it absorbs energy from the sun in its leaves and uses that energy to build more of the plant, and eventually the plant becomes a mature tree. All of the energy the tree has absorbed from the sun has been transformed and absorbed into the wood. When you put a log on the fire, you are using the chemical reaction of fire to release the sun's energy stored by the tree in the wood.

Oxygen

Oxygen, or any oxidizer (some other examples include chlorine, nitrates, chlorates, and concentrated nitric acid), provides the agent necessary to release the stored energy in the fuel. Fuel in the presence of an oxidizer begins to break down into compounds with even lower amounts of stored energy, thus the oxidizer is the key that unlocks the stored energy. This reaction occurs whenever a fuel and an oxidizer are placed together and not otherwise prevented from reacting.

Right now you are reading this sentence written in a book made of paper (fuel) and surrounded by oxygen in the air you breathe (oxidizer), so why do we not see this oxidation reaction occurring? Because the reaction is moving so slowly that it is imperceptible. If you return in 20 years after the book has been sitting on a shelf, you may notice a yellowing of the paper, one sign of a slow oxidation of the cellulose making up the paper. Without a way to set in motion a faster rate of reaction (e.g., a match), we would not otherwise notice this oxidation reaction already occurring. The following reaction shows the oxidation of a simple hydrocarbon fuel such as methane (CH_4), which you may recognize as being one of the main components of natural gas, occurring in pure oxygen, O_2:

$$CH_4 + 2O_2 \xrightarrow[\text{Ignition Energy}]{} CO_2 + 2H_2O$$

This reaction is just a simplified way of describing the chemistry of one molecule of methane, CH_4, combining with two molecules of oxygen (O_2) to form one molecule of carbon dioxide (CO_2) and two molecules of water (H_2O). We usually call the molecules on the left side of our chemical reaction the *reactants*, and those on the right side the *products*. Unfortunately, fire is not usually so easy to describe because typically a number of different fuels are all burning at the same time. Looking at wood, for example, we may have dozens of different reactants given off at one time, each undergoing a different combustion reaction with numerous steps.

Especially when we look at fires inside buildings, as the fire grows larger one would expect that there is not always enough oxygen to burn all of the fuel that is produced, as shown by returning to our example of methane combustion:

$$CH_4 + \tfrac{7}{4}O_2 \xrightarrow[\text{Ignition Energy}]{} \tfrac{1}{2}CO + \tfrac{1}{2}CO_2 + 2H_2O$$

Now the one molecule of methane reacts with less than two molecules of oxygen ($\tfrac{7}{4}$ in this case) to give one-half of a molecule of carbon monoxide (CO), one-half of a molecule of carbon dioxide (CO_2) and finally two molecules of water (H_2O). This equation is a simple representation for a complex chemical reaction because we are talking about fractions of whole molecules ($\tfrac{7}{4}$, $\tfrac{1}{2}$, etc.). Our simple equation is just a model for introducing what occurs overall in such a fire and not for each methane and oxygen molecule, because $\tfrac{7}{4}$ of a molecule is not practical either mathematically or chemically. Finally, we should recognize an important fact that firefighters deal with everyday: When a hydrocarbon fuel burns with insufficient oxygen, other toxic *products of combustion*, like carbon monoxide, are formed and unfortunately can impact people at very low concentrations, causing incapacitation or death.

Heat

Heat is the third part of the fire triangle and is necessary to provide the energy for ignition to occur. First, heat provides energy to transform the fuel into a gaseous state (if necessary) so that it can react with the oxidizer. Both solid fuels like wood and liquid fuels like gasoline require sufficient heat energy to result in a phase change to the gaseous state. Fuels such as hydrogen, methane, and acetylene are already in the gas phase at room temperature, so no additional heating is necessary and is one of the reasons these fuels can be considered much more dangerous.

Heat also is required for the oxidation reaction to happen fast enough to become fire. This heat can come from many mechanical, electrical, chemical, or nuclear sources. For the purposes of fires, we often consider the heat coming from the ignition source being a spark or flame of sufficient size (energy) and duration (time) to begin the combustion reaction. Returning to our simplified chemical reactions for methane, we have included *ignition energy* under the arrow pointing us in the direction of the reaction to describe the energy necessary to start the reaction going. Examples of mechanical ignition sources include heat from friction, such as in the grinding of steel. We often hear of electricity starting fires through arcing, sparks, overheating in a wire or appliance, static electricity, or even lightning. Finally, an organic material (like oily rags or hay) undergoing decomposition may produce sufficient heat to finally result in an open flame (this self heating is often called *spontaneous combustion*). Even a model as simple as the fire triangle can be used to further our understanding of how fires grow and how they can be controlled.

▶ FIRE TETRAHEDRON

Fire Triangle vs. Fire Tetrahedron

The fire triangle identifies fuel, oxidizer, and heat as necessary conditions for a fire and is a good starting point, especially when describing the process to the public as part of fire prevention efforts. Fire science, however, has shown that a continuing chemical reaction is also required for sustained burning. The fire tetrahedron is a model that adds a fourth side to the fire triangle to address this continuous chemical chain reaction, as shown in Figure 3.2. The *fire tetrahedron* is therefore drawn as a pyramid

▶ **Figure 3.2** Fire Tetrahedron. *(Source: NFPA 921, 2004, Figure 5.1.2)*

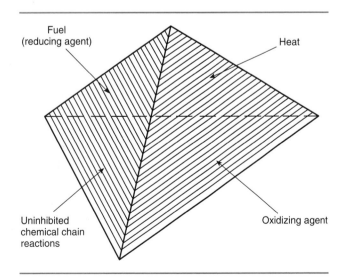

shape to represent the fact that if any one of the following four sides is removed, the structure collapses (i.e., the fire goes out):

1. Fuel
2. Oxidizer
3. Heat
4. Uninhibited (self-sustaining) chemical chain reaction

Uninhibited Chemical Chain Reaction

In this case, the chain reaction is a chemical process yielding products that are able to act as initiators for other reactants, leading to further reactions of the same kind. The previous simple chemical equations for the combustion of methane (CH_4) only give part of the story, namely the overall result of burning methane with oxygen to produce carbon dioxide and water. What actually occurs, however, is a series of intermediate steps where the (CH_4) and (O_2) molecules first break into smaller carbon (C), hydrogen (H), and oxygen (O) pieces either individually or in combination with other atoms or molecules. These *free radicals* are intermediate building blocks in the reaction that may exist for only a short time. After enough steps and combinations, the free radicals have combined in the proper number with other free radicals to form the final products, namely CO_2 and H_2O. Even for a simple fuel like methane, it would take many additional equations to represent all of the intermediate steps in the actual chemical combustion reaction.

As our reactants (CH_4 and O_2) start to break down into smaller pieces, energy is given off. This energy can heat more reactants so they begin to participate in the burning process. This chain reaction of breaking into smaller pieces, along with the evolution of energy driving more reactants to break down is what supports continued burning. If anything breaks up or interrupts this chain reaction, the fire will go out, just as when we remove the fuel, oxidizer, or heat.

Chapter 8 presents a special type of suppression agent called *dry chemical* that can control a fire by interrupting this chain reaction. When the solid particles of dry chemical are discharged on a fire from an extinguisher, their small size allows them to vaporize (change to the gas phase) very quickly. Although this action absorbs some heat, it is not enough to put out the fire. Instead, the dry chemical is effective at absorbing the intermediate free radicals, thus removing them from participation, and by doing so interrupts the chemical chain reaction.

HEAT TRANSFER AND FIRE EXTINGUISHMENT

Heat Transfer

The primary driving force for fire spread is the transfer of heat energy from the flame to other areas by three mechanisms: conduction, convection, and radiation. These mechanisms can spread fire to fuel that is located next to the fire or at some distance and they represent the movement of heat energy from an area of higher temperature to lower temperature.

Conduction.   Conduction is heat transfer through or within a material as a result of direct physical contact. As an example, consider a steel bar with one end placed in a campfire. If you hold the other end you will notice that it gets hot. This is because the heat travels from the fire to your hand by conduction through the bar. Conversely, if we hold a branch from a tree in the fire, we will not feel as much of a change in temperature at our hand, because wood is a poor conductor of heat as compared to steel. The amount of energy transferred depends on the material; because there is less resistance due to the molecular structure of the material, steel is a better conductor of heat than wood. Conduction can also occur within liquids and gases and at the interface between them and a solid.

Convection.   Convection is the transfer of energy due to the movement of a fluid (a general term used to indicate both liquids and gases). *Natural convection* relies on gravity to move the fluid from areas of higher temperature to lower temperature. When air is heated, it expands and becomes less dense than the surrounding cold air, which has a higher density. *Buoyancy* causes the less dense air to rise and cooler air to move downward and is responsible for the rising column of smoke above a campfire. *Forced convection* instead relies on mechanical means such as fans and blowers to move air or heated gases to other areas. When smoke and toxic gases cool as they move further from the fire, they have lost buoyancy by reaching the same temperature as the surrounding air. These cooled fire gases will not continue moving due to natural convection, but forced convection from a furnace or fan is able to transport the dangerous gases throughout a building, possibly endangering the occupants.

Radiation.   Radiation is the transfer of heat by electromagnetic waves. Unlike conduction or convection, radiation does not require an intervening media such as a solid or fluid through which it must pass. Solar heat from the sun is probably the most familiar example of radiation heat transfer and describes energy being transported from an area of high temperature (the sun) to a body with a lower temperature (the earth), while traveling through empty space.

Fire Extinguishment Theory

Removal of Heat. The fire tetrahedron provides a good basis for discussing a theory relating to how fires are extinguished, known as the *theory of extinguishment*. Removing any one side of the tetrahedron disrupts the combustion reaction and causes the fire to go out. It usually is not necessary to completely remove all of one side of the model for the fire to go out; a reduction below a minimum threshold value is often sufficient.

Once ignition has occurred and a fire is burning, one common way of controlling such an unwanted fire is to remove the heat necessary for vaporizing more fuel, called *smothering the fire*. If applied in sufficient quantities, water will absorb heat on the surface of a solid fuel, cooling it and reducing the production of combustible gases from the surface. At the same time, the water can soak into the solid and make it more difficult for oxygen to reach the fuel. The science of this is very complicated, even for the relatively simple problem of a single piece of wood burning in isolation.

Removal of Fuel. How a fire grows and its level of hazard are strongly influenced by what fuel is burning, suggesting that it is important to spend additional time reviewing this side of the fire tetrahedron in more detail.

▶ FLAMMABILITY PROPERTIES OF FUELS

Gas Phase Fuels

If we can remove the fuel from the reaction, we prevent one of the required inputs from participating in the chemical reaction. Obviously if all the fuel is gone, there will be no fire, but we can also achieve the same result by removing only some of the fuel. When mixed with air, gaseous methane has a lower flammability limit (LFL) of 5 percent fuel concentration (95 percent air) and an upper flammability limit (UFL) of 15 percent fuel concentration (85 percent air). In practical terms, whenever the concentration of methane is between 5 percent and 15 percent by volume, the mixture can burn. Because gas mixtures burn very rapidly or can even explode, this range may also be termed the lower explosive limit (LEL) and the upper explosive limit (UEL).

When the fire department responds to a natural gas leak in a house, the usual practice is to open windows and doors to vent the concentration of gas below the lower flammable limit (also called fuel lean), thus creating a condition in which the gas cannot ignite. A more dangerous condition exists when the concentration of fuel is above the upper flammable limit, or fuel rich, even though the mixture will not burn, the same as if there is too little fuel. To ventilate a building in the fuel rich case, however, means reducing the fuel concentration by mixing with air from a point where there is too much fuel to burn, through the flammable range, and finally below the lower flammability limit. Should an ignition source be present, the mixture may ignite during ventilation operations when the concentration is in the flammable range. Each gas has a different flammable range; some like methane are narrow, while acetylene can burn at concentrations from 2.5 percent to nearly 100 percent.

Liquid Fuels

As with solids, liquid fuels must be converted to the gas phase before they can burn. When a container of liquid is left open to the atmosphere, a small fraction of the liquid molecules have enough energy to actually break through the liquid surface and into the surrounding air, a process we are familiar with called *evaporation*. If given long enough, we would find that much, if not all, of our liquid evaporates. At the same time, molecules of the liquid that are now in the gas phase above the surface tend to diffuse or be blown away, while a small number collide back with the liquid surface again due to random motion. Because more molecules leave the liquid than return, it appears over time that the liquid is disappearing (i.e., evaporating).

If the same liquid at room temperature is placed inside a container, a similar process begins as soon as the liquid enters the container, which we assume is half full. By placing a tight fitting lid on the container, we prevent the molecules in the gas phase from escaping outside the container. Over time, the concentration of molecules in the gas phase reaches a point where they are returning to the liquid as fast as molecules are leaving the liquid through evaporation. At this point of equilibrium, molecules are entering and leaving the liquid at the same rate, thus the level of the liquid no longer appears to change.

Consider the upper part of the closed container above the liquid surface. Right when the lid was put on the container, this space contained mostly air, but over time now contains the same air, plus the evaporated liquid (now too in the gas phase like the air). Because the volume has remained basically the same, by putting more molecules of gas in the same space, we know the pressure must go up and would be higher than atmospheric pressure where we started. This increased pressure resulting from the vaporized liquid is known as the *vapor pressure* and is the reason why one hears a brief hissing sound as the pressure is relieved when a gasoline can is opened on a warm day.

One can also think of the vapor pressure as the pressure exerted by the molecules in the liquid against the surface as they try to escape. Returning to our open container of liquid, atmospheric pressure is greater than the vapor pressure, leading to conditions where some, but not all of the molecules can be kept in the liquid phase. If the liquid is heated, the molecules become more energetic, leading to an increase in the vapor pressure. With additional heating, the bulk temperature of the liquid approaches the *boiling point temperature*, and the vapor pressure increases until it is just less than atmospheric pressure. At this point, liquid is vaporizing at a high rate, but technically not yet boiling. By heating the liquid just a little more, the temperature equals the boiling point, the vapor pressure equals atmospheric pressure, and we notice a change in the behavior of the liquid. Vaporization can now occur throughout the liquid, which is recognizable by the formation of bubbles of gas below the surface, along with turbulence as they mix the liquid while rising to the surface. Prior to boiling, atmospheric pressure is greater than the vapor pressure, preventing the formation of any bubbles, while at temperatures equal to the boiling point, atmospheric pressure is no longer able to suppress the vaporization. To be precise, the boiling point is the temperature of a liquid when the vapor pressure equals atmospheric pressure. Because atmospheric pressure is lower high atop a mountain, a pan of water will boil at a lower temperature, thus cooking pasta takes longer at higher altitudes than at sea level.

The boiling point temperature is the first important measure of how volatile, or potentially dangerous, a liquid fuel may be. A related measure of liquid hazard is the *flashpoint*, that is, the liquid temperature where the vapor pressure is just sufficient to allow enough vapors to form over the liquid surface such that an external ignition source (i.e., spark or small flame) will ignite the vapors over the surface. This momentary flash is enough to burn off the vapors, but does not provide enough heat energy to allow the fire to continue burning after the initial flash. Because it represents the lowest temperature where brief ignition can occur, the flashpoint is used as a conservative measure of what temperature is necessary for hazardous conditions to be formed.

The *ignition temperature*, which is higher than the flashpoint, represents the case where the fuel not only quickly flashes, but continues to burn when ignited by the external source. When heated to an even higher temperature, the vapors over an ignitable liquid may begin to burn without an external ignition source, a process known as autoignition. For most liquid fuels, the

▶ **Figure 3.3**   Flammable
Range for Acetone and
Ethyl Alcohol. *(Source: Fire
Protection Handbook, NFPA,
2003, Figure 8.6.1)*

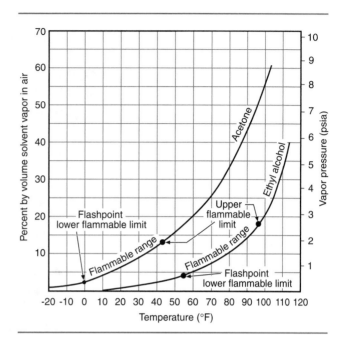

autoignition temperature is greater than the flashpoint and boiling point, thus describing a point where any liquid fuel has already vaporized and would be in the gas phase. These limits are shown in Figure 3.3 for two common flammable liquids.

Liquids are categorized based on how easily they vaporize and ignite as a function of liquid temperature. *Flammable liquids* are considered very easy to ignite, because they produce sufficient vapors above the liquid surface to burn at room temperature, thus only require an ignition source to burn. *Combustible liquids,* however, require heating above normal room temperature to burn. A strong ignition source, such as a large flame from a welding torch, for example, can heat a small area of combustible liquid sufficiently to form a combustible mixture above the liquid, leading to ignition.

Flammable liquids are ignitable liquids with flashpoints below 100°F (37.8°C) and are further broken down as follows (based on NFPA 30, *Flammable and Combustible Liquids Code*) [2]:

- Class IA = flashpoint below 73°F (22.8°C) and boiling points below 100°F (37.8°C)

- Class IB = flashpoint below 73°F (22.8°C) and boiling point at or above 100°F (37.8°C)

- Class IC = flashpoint at or above 73°F (22.8°C) and below 100°F (37.8°C)

Combustible liquids are ignitable liquids with flashpoints at or above 100°F (37.8°C) and are also further broken down into different classes:

- Class II = flashpoint at or above 100°F (37.8°C) and below 140°F (60°C)
- Class IIIA = flashpoint at or above 140°F (60°C) and below 200°F (93°C)
- Class IIIB = flashpoint at or above 200°F (93°C)

This classification system for liquids is helpful for separating them into different groups based on ease of ignition and is often used by fire prevention codes and standards to control the quantity and types of liquids stored and used inside buildings. Whereas it may be acceptable to use a hazard Class II or IIIA solvent for cleaning, many fire codes prohibit the use of Class I flammable liquids for cleaning due to the ease with which they can be ignited.

Once an unintended ignition of a flammable or combustible liquid has occurred, fire prevention has broken down and a severe fire should be expected. Based on the amount of energy given off with each pound (kilogram) of liquid consumed, most hydrocarbon liquids tend to burn with similar _heat release rates,_ a measure of the energy liberated per second, which is discussed more at the end of this chapter.

Solid Fuels

Although wood is a common fuel, wood combustion is a rather complex reaction due to the wide number of its components that can burn, variation in moisture content, and the nonuniformity of its structure due to variations in the grain depending on what side of the wood is burning. We discussed earlier that it is not the solid wood that actually burns, but instead it is necessary to convert the fuel to the gas phase before burning can occur. In fact, the process is even more complicated than simply converting a solid or liquid to a gas through vaporization, as occurs when we boil off water in a pan.

As we heat a piece of wood slowly with a small flame, we eventually notice portions that have received the most heating begin to change color, usually turning darker. We may be able to smell the wood getting hot, but if we could sample the environment just above the surface, we would measure a large number of different combustible gases that could burn. As the solid wood breaks down when exposed to heat, it undergoes a change known as _pyrolysis,_ breaking down into other chemicals (molecules), different from the original wood. Products of this decomposition in the gas phase will eventually mix with air to burn, while others will remain in liquid and solid phase and can be seen as dark tarry, char residues left over from pyrolysis.

With enough heating, wood can produce sufficient combustible decomposition products to support a region above the surface that is within its combustible range. Measuring the surface temperature at this point is often referred to as the *piloted ignition temperature*. In our example, a flame is used to heat the wood and thus can provide the necessary pilot ignition to start the fire. With even more heating, the *autoignition temperature* is reached, allowing the fire to start even without an outside source of energy such as from a flame.

Classification Based on Fuel Type

A classification system has been developed to identify different groups of fuels based on their burning behavior and can be used as a quick way to indicate firefighting techniques or suppression agents appropriate for the given class. This system uses the letters A, B C, D, and K to designate the fuel type and can be found prominently displayed on fire extinguishers, for example, to assist the operator in selecting the correct appliance to use. See NFPA 10, *Standard for Portable Fire Extinguishers*, for more detailed information on the classification of fires and the selection of suppression agents [3].

Class A.   Class A fires include ordinary combustible materials, such as wood, cloth, paper, rubber, and many plastics. Class A fuels should burn with an ember and leave an ash. Most extinguishing agents, including water, are effective on these fuels.

Class B.   Class B fires include flammable liquids, combustible liquids, petroleum greases, tars, oils, oil-based paints, solvents, lacquers, alcohols, and flammable gases. These fuels present a higher fire hazard, and, for many of these fuels, water may not be an extinguishing option or could actually make the fire worse.

Class C.   Class C fires involve energized electrical equipment and would otherwise be Class A or Class B, except for the added hazard of the electricity. Special techniques and agents are required to reduce the hazard to the user.

Class D.   Class D fires include combustible metals, such as magnesium, titanium, zirconium, sodium, lithium, and potassium. Due to extremely high flame temperatures, if water is accidentally applied in an attempt to extinguish a burning metal, it could break into its component parts of hydrogen and oxygen, resulting in enhanced burning or even an explosion.

Class K.   Class K fires are a new division and include fires in cooking appliances that involve combustible cooking media (vegetable or animal oils and fats). The primary hazard is the deep fat fryer, which had formerly fallen under the Class B designation. With the change of dietary practices in the United States, many commercial cooking operations shifted from animal fat-based cooking oils to vegetable oils, while at the same time better insulating equipment for improved energy efficiency. These vegetable oils have autoignition temperatures closer to the cooking temperatures, which results in an increased likelihood of ignition and a more likely reignition after extinguishment. These fires are more challenging and it was determined that a separate classification was appropriate.

▶ TYPES OF FIRES

One way to classify fires is by the mechanism of burning. In some cases, the fuel and oxidizer are mixed before burning, while at other times not. How the fuel burns can be described as a diffusion flame, premixed flame, or smoldering fire.

Diffusion Flames

A flame is simply a zone where a chemical reaction between fuel and oxidizer in the gas phase is carried out at elevated temperature. In a diffusion flame, the fuel and gas are both separated and first come in contact with each other when they react in the flame, as each approaches from a different side of the flame. Candle flames, as well as most natural fires, are diffusion flames.

Far from the flame, the concentration of oxygen in air is approximately 21 percent by volume. Inside the flame, oxygen is consumed by the fire, leading to a zone with lower oxygen concentration. As we saw with heat transfer, energy moves from areas of high temperature to low temperature and the same idea can be said for differences in concentration. *Diffusion* is the processes whereby fluids (liquids or gases) are transported from areas of high concentration to low concentration, thus oxygen molecules tend to move from areas outside the flame (high concentration) toward the flame (low concentration). The flame basically acts as the dividing line between the fuel and the oxidizer.

A similar diffusion process moves fuel into the reaction zone. A candle uses wax as a fuel source and oxygen in the air as the oxidizer. Heat from the flame converts the wax from a solid form to a liquid. The liquid moves up the wick by *capillary action*, the ability of the wick with its closely spaced fibers to use surface tension to draw liquid wax from the pool below, as illustrated in Figure 3.4. As the liquid wax travels up the wick, it is heated by the flame

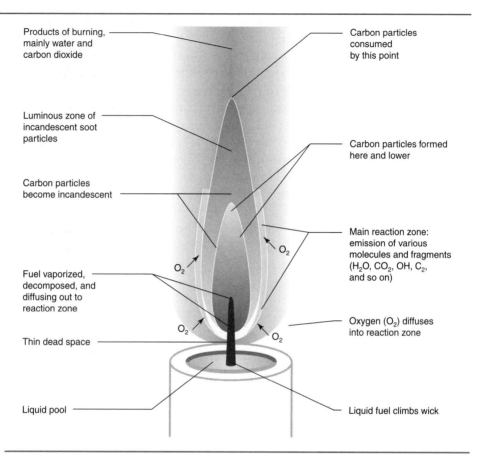

Products of burning, mainly water and carbon dioxide

Carbon particles consumed by this point

Luminous zone of incandescent soot particles

Carbon particles formed here and lower

Carbon particles become incandescent

Main reaction zone: emission of various molecules and fragments (H_2O, CO_2, OH, C_2, and so on)

Fuel vaporized, decomposed, and diffusing out to reaction zone

O_2

O_2

Oxygen (O_2) diffuses into reaction zone

Thin dead space

O_2

O_2

Liquid pool

Liquid fuel climbs wick

▶ **Figure 3.4** Capillary Action Shown on a Candle Flame. *(Candle illustration from Fire by John W. Lyons, © 1985 by Scientific American Books, Inc. Reprinted by permission of Henry Holt and Company, LLC.)*

until it vaporizes. The wax vapor rises inside the flame and moves into the flame, also by diffusion. A diffusion flame can thus be thought of as a thin sheet where fuel and oxygen react, so the flame over a candle is not filled entirely with reacting fuel and oxygen but instead is primarily vaporized fuel, with a thin sheet of fire actually at the interface between fuel and oxygen.

Premixed Flames

There are many cases where fuel and oxidizer are combined prior to burning, leading to what are called premixed flames. Although we are less likely to see premixed accidental fires, they can and do occur. As an example, a pipe containing natural gas can leak over time, allowing the fuel to spread throughout a building while mixing with air. Once the mixture is within its flammable range, all that is necessary for combustion is to introduce an ignition source. From the point of ignition, we expect the flame to spread out through the region where fuel and oxidizer have already mixed.

Combining the fuel and oxidizer before burning can improve the combustion reaction, so many of the machines and equipment we use to heat or power our buildings rely on premixed flames. An example of premixing is a Bunsen burner used in a chemistry lab. When properly adjusted, the fuel is supplied at the base of the burner, where it is allowed to mix in surrounding air at different rates depending on how large the air intake opening is made. The combustible mixture continues up the tube of the burner until it is discharged out the top where it burns. If the mixture in the tube can burn, why does the flame not rush back down into the Bunsen burner?

Each fuel has a different *flame speed*, a measure of how fast a flame can travel through a given combustible mixture of the fuel and air. By sizing the tube in the Bunsen burner small enough, we can make sure the velocity of the fuel/air mix is greater than the speed at which the flame can travel back to the source of fuel. If the supply of fuel and air is reduced too low, the flame can travel back down the burner until it reaches the unmixed (pure) gas supply, at which point the fuel concentration is too rich to burn, preventing the flame from continuing down the hose or pipe back to the source of the gas. By making it easier for the burning fuel to find oxygen, premixed combustion can be more efficient, thus using less fuel and reducing emissions of gases hazardous to our environment.

Smoldering Fires

There need not be a visible flame for fuel and air to burn. A smoldering fire has no flame and is characterized by hot glowing embers. A common example of a smoldering fire is a charcoal grill after the lighter fluid has burned off, where now the combustion is not occurring above the fuel, but very near or below the surface. This type of fire is often hot, slow, and very smoky because the oxidizer must travel down to the surface of the fuel, leading to less efficient combustion.

Flaming combustion requires higher oxygen concentrations than smoldering combustion, allowing fuel to continue burning even if the visible fire has gone out. A wood fire goes in to a smoldering phase when the atmosphere reaches about 14 percent oxygen concentration and we no longer observe flaming combustion, but it can continue to burn down to oxygen concentrations approaching 10 percent. This mechanism is important in creating a hazardous fire behavior known as *backdraft*, discussed in the next section.

▶ STAGES OF FIRE GROWTH AND SPREAD

Few fires remain the same size throughout their entire existence. Instead, they tend to start small, spread to involve other fuel, and continue growing until they run out of additional materials to burn, at which point they begin to burn out. Fire departments, fire sprinklers, and fire extinguishers are the result of human attempts to interrupt a fire's growth to reduce the damage to people, property, environment, and business continuity. The stages a typical fire goes through include preignition, ignition (incipient), growth (free burning), fully developed, and decay (burnout). Not every fire goes through this sequence because the process could be interrupted by a person using a fire extinguisher, for example. While reading this section, refer to Figure 3.5.

Preignition

To initiate a fire, fuel and oxygen need to be intimately mixed in the proper amounts at the point of ignition. This mixing usually requires that both be in the gas phase. In most cases we are dealing with oxygen in air as the oxidizer, already in a gaseous form. The fuel can be in any state: solid, liquid, or gas. Liquid fuels must be *vaporized* to burn, which can be done by heating the liquid alone or in combination with misting the liquid by a mechanical spraying process. Vaporization does not include a chemical change; it is a change

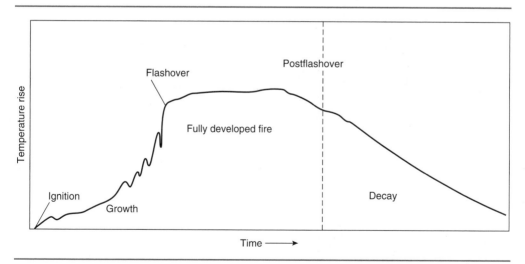

▶ **Figure 3.5**   Stages of Fire Growth When Allowed to Progress Without External Influences such as Extinguishment. *(Source: The SFPE Handbook of Fire Protection Engineering, 3rd ed., Figure 3-6.1. © Society of Fire Protection Engineers. Used with permission.)*

in physical phase. Water vapor is chemically the same as ice or snow, merely at different states (gas versus solid).

Solid fuels also break down into gaseous components when heated through a process called *pyrolysis* or *thermal decomposition*. For wood, the flammable gases that come from heating are chemically different from the solid. They are not wood vapor but simpler hydrocarbon vapors. The large numbers of molecules that come together to form a solid hydrocarbon are too long to be easily vaporized, thus the energy heating the fuel instead results in breaking of the structure to form shorter, more volatile units. When a fuel is heated in preparation for burning, solid fuel pyrolysis (a chemical change) should be thought of as different from liquid fuel vaporization (a phase change).

Both vaporization and pyrolysis require energy transfer to occur, thus a portion of the heat in the fire triangle or tetrahedron goes to preparing solid and liquid fuels to burn. Each fuel that makes up a couch, for example, has different physical properties that impact how quickly it will burn. Even before we can see a visible flame, energy is required to transform the fuel to a state that will allow fire to begin during the preignition phase.

Ignition (Incipient)

This event defines the onset of fire or combustion. The time leading up to ignition is highly variable, from nearly instantaneous (a match or an arc in a flammable gas), to several hours (cigarette discarded in an upholstered chair), to days (spontaneous combustion of oily rags or wet hay). It involves the processes that bring all sides of the fire tetrahedron together in the proper amounts, resulting in a flame. Typically this phase is controlled by the fuel properties and the heat or ignition source. If we look at a fire scenario originated in a small space such as a bedroom, we can expect that in many cases the amount of energy given off by the ignition process is tiny enough that we notice little or no increase in temperature prior to ignition.

This small fire that is established after ignition and limited to the initial burning fuel is sometimes termed an *incipient fire*, as illustrated in Figure 3.6. At this stage in a fire's life, it is still small enough that a willing occupant can use a fire extinguisher to control the fire, or close a valve on a broken natural gas pipe, thus removing the fuel. The fire is also expected to be small enough that occupants in the fire room should be able to exit the space to safety if they are awake, aware that there is a fire, and capable of self-preservation (i.e., moving). In some cases, a fire this small can cause injury, or even death, to an occupant intimate with ignition such as when a person's clothes catch on fire.

Although many fires start with a flame or other strong ignition source, this need not be the case. As discussed, a fuel can be heated to its autoignition temperature where burning starts even in the absence of an outside ig-

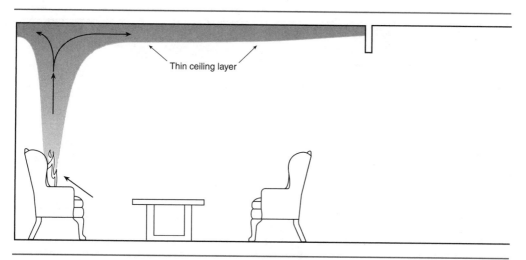

▶ **Figure 3.6**   Incipient Stage Fire Just After Ignition and at the Point Where the Fire Is Starting to Impact the Inside of the Compartment Through the Formation of a Ceiling Jet. *(Source: NFPA 921, 2004, Figure 5.5.4.2.)*

nition source. *Spontaneous combustion* can also result in open flaming, due to the buildup of heat within a fuel package resulting from the decomposition of material that is trapped within the fuel. As the heat builds up, the temperature rises, which in turn speeds up the decomposition reactions and increases the heat generation. Given the right circumstances of fuel type, outside temperature, amount and arrangement of the fuel, the temperature within the fuel pile can reach ignition temperatures.

Growth (Free Burning)

Once ignition occurs, additional fuel can ignite using the initial flame as a heat source. Flame spread across the surface of a fuel can be seen as a series of ignitions following the initial ignition of the fuel. The flame heats the fuel and releases flammable vapors that mix with the oxygen in the air. The flame acts as a pilot and heats the fuel vapor/air mixture until it ignites. At this stage in a fire's development, the only thing limiting its size is how quickly the flame can heat additional fuel, resulting in a larger burning area. Should the compartment contain a door or window, smoke will begin spreading from the space into adjacent building areas or outside the building, as shown in Figure 3.7. These same openings also allow a flow of air into the room to replace the air consumed by the fire or mixed with the fire gases and discharged out any openings. With time, the fire continues spreading, but due to the availability of oxygen, is still limited only by the amount of fuel available. If the chair in

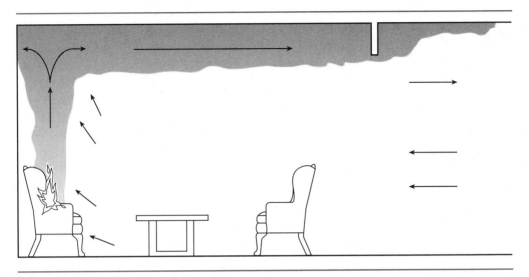

▶ **Figure 3.7** Ceiling Jet Is Transporting Hot Fire Gases from the Compartment and into the Adjacent Room. *(Source: NFPA 921, 2004, Figure 5.5.4.2.1.)*

Figure 3.7 is located far enough from other combustibles, the fire may consume the chair while spreading no further, thus eventually decaying until only noncombustible components of the chair (springs, nails, etc.) and some ash remains.

The growth stage of a fire takes us from ignition, through the period when the fire is spreading, up to the time when many of the fuels in the room are burning. To an external observer, temperatures and smoke levels in the room would go from approximately room temperature and no smoke at ignition, through a period of increasing temperature and smoke concentration. Conditions in the room are no longer safe for humans at the end of the fire growth period. Because conditions become life threatening during the fire growth period, it is the critical period when fire protection engineering steps into play through the activation of smoke detectors and alarms to warn building occupants or the discharge of water from sprinkler systems to halt the growth of the fire.

Fully Developed

If the fire has been able to spread over much, if not all of the available fuel, it has entered the *fully developed stage* where conditions such as temperature and smoke level in the room become more constant. From Figure 3.5, temperatures are expected to be at their highest during this period, resulting in thermal damage not only to occupants, but the structure of the building itself.

Decay (Burnout)

Even the largest fire must eventually run out of fuel. As a fire spreads over and consumes the available fuel, it enters the final *decay or burnout* stage where temperatures begin to decrease as the fire reduces in intensity. From a fire safety standpoint, the decay period tends to be of less interest because the damage to people, the building, and contents has already been done. Based on how fire safety professionals expect buildings to perform, if a fire reaches the decay period, fire protection and prevention would be considered a failure because of the following:

- Fire prevention was unsuccessful in preventing ignition, our first stage in the development of fire.
- Fire extinguishers or fire suppression systems (sprinklers) were either not provided, used, or did not operate properly, thus allowing the fire to continue growing.
- Fire walls and doors were either not provided or did not function properly, allowing the fire to spread during the fully developed period from the room or area of origin.
- The fire department was unable to deliver enough water to the base of the fire during the growth or fully developed stage to control or suppress the fire.

▶ SPECIAL CHALLENGES TO FIRE FIGHTERS

The fully developed period is the most dangerous time for fire fighters to be operating on the fire ground. While fire fighters are busy searching for occupants or applying water to the fire, the building around them is constantly being attacked, reducing the structural performance of the components designed to keep the building standing. Incident commanders must continually evaluate the interaction of the fire and the building to determine if the risk to fire fighters outweighs the possible benefit interior firefighting will have. Four situations are special challenges to fire fighters: preflashover, flashover, postflashover, and backdraft.

Preflashover

In some cases, different fuels are located close enough to each other that flame is able to spread by direct flame contact, as when a wood table is located directly adjacent to an upholstered chair. A fire is able to spread, however, even when the fuels are separated by longer distances, as shown in Figure 3.8.

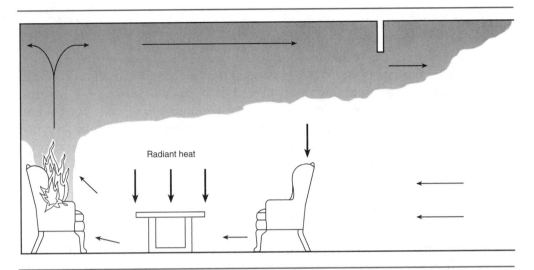

Radiant heat

▶ **Figure 3.8** The Fire and Hot Gases at the Ceiling Are Heating Everything in the Compartment, Driving the Fire Closer to Flashover. *(Source: NFPA 921, 2004, Figure 5.5.4.2.2.)*

Since ignition, the fire's flame and hot gases have heated both the compartment boundaries and other fuels that have yet to start burning, increasing their temperature. Hot gases collecting at the ceiling transport this energy through radiation heat transfer, allowing all fuels throughout the compartment to be brought closer to their ignition temperature at the same time, even if remote from the flame. The fire is about to undergo a rapid transformation called *flashover*, where it goes from being limited in area, to involving many, if not all, of the fuels in the compartment. Looking into the fire room from the outside, it would appear that the fire goes from a limited area around the point of ignition, to almost instantaneously involving the entire room.

The time period from ignition to this point depends on factors such as fuel (type, arrangement, quantity), size of the compartment (floor area and height), number of openings and where they are located (small vent at the ceiling versus a wide open door), and how well the room is insulated (concrete blocks versus gypsum drywall board with fiberglass insulation). Each arrangement is unique, so the process can take less than a minute, several minutes, or even an hour or longer if the fire starts out slowly smoldering. For furnished bedrooms or living rooms typical of residential buildings, times to flashover can be as short as a few minutes. The public often underestimates how quickly a fire can grow, with disastrous consequences.

As the fire moves closer to flashover, the chances for an unprotected occupant to survive approach zero, as shown in Figure 3.9. Even if rescued by the

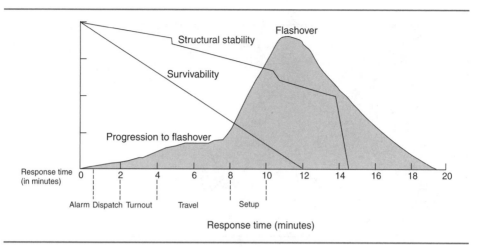

▶ **Figure 3.9**   As a Fire Approaches Flashover, the Probability for Survival Diminishes Rapidly. The times given in this figure were created for this example and should not be considered to represent reality or referred to as absolutes. *(Source: Structural Fire Fighting, NFPA, 2000, Figure 4.8.)*

fire department, an unconscious occupant may have already received a lethal dose of toxic fire gases or severe burns. Example fire department response times are shown and compared to one possible fire development curve in Figure 3.9. Looking at an average time for response including dispatch and scene setup periods, we have come to realize that there are cases where building occupants are in danger well before the fire fighters are in a position to make a difference. To improve the situation, we can install smoke detectors that will alert occupants and the fire department sooner, or build fire stations closer together to reduce the response time. Both of these strategies, however, do little to improve the outcome for occupants who may not be able to move away from danger fast enough, such as in hospitals and prisons. Building designers, along with the code enforcement community, have recognized this fact, leading to changes in building and fire codes (like NFPA 1, *Uniform Fire Code*™, and *NFPA 5000*™, *Building Construction and Safety Code*™) that require automatic fire sprinklers to be located in many new and some existing buildings [4, 5].

Flashover

Flashover marks the rapid transition from a fire of limited size and impact to a very hazardous situation where the extent of the fire grows nearly instantaneously to include the available fuels in the compartment, as shown in Figure 3.10. Few occupants, including fire fighters in full protective equipment, will survive a flashover.

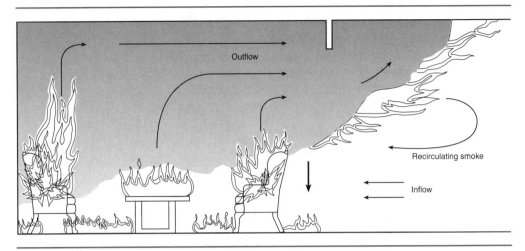

▶ **Figure 3.10**   The Fire Has Flashed Over, Creating a Condition in Which the Fire
Goes from a Small Area to Involving All of the Fuels in the Room at the Same Time.
(Source: NFPA 921, 2004, Figure 5.5.4.2.6.)

Due to the wide range of possible fuel and ventilation arrangements, the
transition to flashover creates a number of fire behaviors of interest to fire
fighters. Just before flashover, fuels in the room are being heated by the
flame and hot smoke flowing across the ceiling, causing the release of large
quantities of volatile gases from the fuels. These hot combustible gases can
be entrained into the flow at the ceiling and out the openings of the compartment, leading to conditions where gases contain fuel that is hot enough
to burn. All that is missing is oxygen. As the hot gases travel across the ceiling and out an opening, the flow eventually mixes in enough fresh air to provide the oxygen necessary for burning, resulting in a narrow region of
flaming at the interface between fuel and air. To a fire fighter entering the
fire room through the open doorway, the smoke above appears to suddenly
start burning in a way characteristic of the name given to this phenomenon,
a *rollover*. Fire fighters use the observance of rollover as a sign of impending
flashover, although it is not a reliable indicator because many flashovers
happen without creating a rollover.

During the transition to flashover, rapid heating of the compartment atmosphere leads to expansion of the fire gases, pushing the hot smoke and
unburned vaporized fuel out of the compartment. As the hot gases are
ejected from the compartment, they mix with air, leading to what appears to
be a fireball once the gases begin burning.

Returning to Figure 3.10, as smoke travels out of the compartment
through a door, it tends to mix with the fresh air entering near the floor. This

process can lead to contamination of the cleaner atmosphere near the bottom of the room, reducing visibility in the area where victims are often found. Flashover also marks the point where smoke and hot gases are pushed throughout a building from the fire room, increasing the exposure to occupants far from the fire. Before flashover, it is occupants in the fire room that are at greatest risk, while after flashover those occupants on the entire fire floor or above it are in danger. An incident commander that does not recognize this fact will not direct the fire fighting efforts to best use.

Flashover also marks the transition from a fire that is controlled by the amount of fuel available (fuel controlled), to a situation where the amount of burning is instead dependent on how much air (oxidizer) is able to enter the compartment through the openings. The fire is now said to be *ventilation controlled*. As discussed in the next section, when insufficient oxygen is available to allow complete combustion of the available fuel, the rate at which toxic gases are produced increases dramatically, leading to a greater hazard to occupants remote from the fire. For this same reason, fire fighters try to get water on a fire as soon as possible, because this may be a more effective first action than rescuing victims trapped near the fire. Rapidly reducing the production of toxic smoke from the fire can improve conditions faster than the more time-consuming task of searching a smoke-filled building for victims.

Postflashover

After flashover, enough fuel is being produced in the fire compartment that it is not all able to burn inside the space due to insufficient oxygen. This postflashover burning is characterized by ventilation-controlled conditions that lead to flames extending outside compartment openings. Unprotected occupants and fire fighters in full protective equipment will not survive exposure to postflashover conditions, as illustrated in Figure 3.11.

At this point in a fire's development, focus must instead shift to the damage the fire is doing to the building's structural support. Fire temperatures are at their highest, resulting in a weakening of steel and concrete. Due to flames extending outside windows and through openings, the fire can now be pushed to other spaces in the building or spread over the exterior of combustible buildings.

When a building is provided with fire walls and doors, it is with the intent to limit the spread of smoke and hot fire gases from a fully developed fire, providing more time for occupants remote from the fire to escape. *Structural fire protection* can also be provided for the wood, steel, or concrete skeleton of a building to reduce the impact of heat on the ability of the building's columns, beams, walls, and floors to support the weight of the structure.

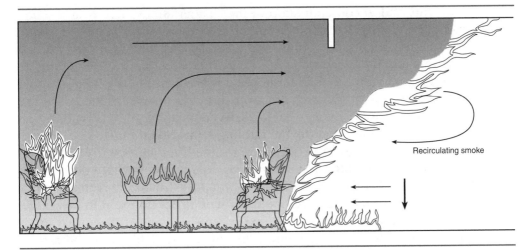

Recirculating smoke

▶ **Figure 3.11** Postflashover Fire. *(Source: NFPA 921, 2004, Figure 5.5.4.2.7.)*

Backdraft

We have already discussed that an incipient fire grows with little impact on the compartment such that oxygen concentrations in the room remain nearly normal (21 percent oxygen in air by volume). In a well-sealed room with the doors and windows closed, only minor construction imperfections such as cracks around doors, electrical outlets, lights, and so forth are able to allow air into the space. As the fire grows, it consumes the available oxygen, leading to disappearance of the visible flame. At these reduced concentrations, solid fuels can still burn in a smoldering state. The environment in the space may be hot enough to allow continued pyrolysis, resulting in the accumulation of hot flammable gas that only lacks the oxygen necessary to burn. Over time, the fire appears to go out, yet the fuel has collected to fill the room. Temperature increases lead to expansion of the gases, pressurizing the room and pushing them out any small openings by what appears to be "puffing." Due to tars and soot in the smoke, windows can appear to be dark or frosted over with dark stains. The conditions ideal for a *backdraft* have just been created, as illustrated in Figure 3.12.

As soon as an opening is made in the closed compartment, oxygen is able to mix with the hot vaporized fuel, leading to instant burning. As with some flashovers, rapid expansion of the gases that fill the compartment push them out through the opening, resulting in a characteristic rolling fireball being discharged. Backdrafts can occur so quickly, they have been called smoke explosions due to the appearance of very rapid burning through what otherwise appears to be dark smoke. Proper firefighting technique

- Low oxygen
- High heat
- Smoldering fire
- High fuel vapor concentrations

AIR
AIR

PREBACKDRAFT

- Introduction of oxygen causes fire of explosive force.

BACKDRAFT

▶ **Figure 3.12**   Backdraft. *(Source: Redrawn from Essentials of Fire Fighting, 4th ed., International Fire Service Training Association (IFSTA) 1998.)*

involves ventilating the compartment as high as possible to allow the fuel vapors to escape the space before a lower opening is made for entry.

▶ EFFECTS OF FIRE

Products of Combustion

One of the fundamental tenets of fire protection is to minimize or prevent building occupants from being exposed to the products of combustion, especially for vulnerable populations such as the sick or elderly. Although smoke can be important for the damage it can do to merchandise in a store or to the environment as the result of a large forest fire, here we concentrate on how fires can injure or kill human beings.

To this point, we have been using the terms *products of combustion, smoke, fire gases,* and *fuel vapors* quite interchangeably. As already stated, fire produces heat, light, and smoke in varying quantities. You should have a good understanding of heat and light. Smoke is a different matter. What really makes up smoke and are there different types of smoke that we should be concerned about in the study of fire behavior?

We will consider *smoke* to be a general term used to describe gaseous volumes composed of different combinations of the following. Unburned fuel is what was not consumed by a fire, especially if conditions were not ideal for complete combustion such as insufficient oxygen, low temperatures, or the fuel concentration near the limits of flammability (too lean or too rich).

Products of complete combustion represent the final materials produced by burning a fuel with sufficient oxygen. Returning to the first chemical reaction displayed in this chapter, these products are the molecules on the right-hand side of the equation. For common hydrocarbon fuels, we expect the *products of complete combustion* to include CO_2, H_2O vapor, plus others.

Products of incomplete combustion include fuel, oxygen, or other reactants that were partially consumed by the fire but have not been entirely converted to products that we would expect to represent complete combustion. Remember we identified fire as representing a number of intermediate steps needed to convert reactants to products, so if the reaction is interrupted or not allowed to go to completion, remnants of the partially complete reaction will be left over. Especially when there is insufficient oxygen for combustion, carbon monoxide (CO) can end up being produced instead of carbon dioxide (CO_2), one of the components of smoke that causes many deaths because only a small concentration is necessary to cause incapacitation (to pass out) or death.

Other solid or liquid products of combustion known as *soot* or *tar* can be formed from unburned or partially consumed fuel and other chemicals that liquefy or solidify, especially as the smoke cools. If observed under a microscope, soot appears to be a chaotic assembly of mainly carbon atoms that attach to each other. It is one of the main components giving smoke its characteristic dark color.

Ambient air that has mixed in with the other products due to entrainment of surrounding air into the plume or ceiling jet is a part of smoke. Although air mixed into the fire gases is not technically a product of combustion, it is worthwhile to include it in the list because it increases the quantity of smoke while it cools it.

Fire Fighter Safety

Fires cause harm in a number of ways to occupants and fire fighters who come into contact with products of combustion, causing dangerous conditions including thermal injury, toxicity, asphyxiation, and obscuration.

Thermal Injury.   Thermal injury to skin or respiratory tract (nose, trachea, lungs) can be the result of heat transferred from a hot material to the human body. Short term exposure, such as direct flame contact, can lead to damage at the point where the heat is transferred. Slow heating over a longer period is also possible and produces a dangerous bodywide elevation in temperature known as *hyperthermia*. When addressing thermal injury, it is necessary to consider the following modes for heat transfer and the possible damage each will inflict:

- *Conduction*—Contact of skin with burning fuel or hot steel resulting in burns at the point of contact.
- *Convection*—Hot smoke traveling down a hallway can surround a person crawling to safety with high temperatures, leading to overheating.
- *Radiation*—The hot smoke layer gathered at the ceiling in a room near flashover can result in fatal burns to an occupant lying unconscious due to radiation heat transfer from the smoke layer to the floor.
- *Mixed modes*—Many cases involve multiple modes of heat transfer. If a person walks into a cloud of hot smoke and takes a breath, the gases will be transported deep into the lungs where both conduction and convection can lead to extensive tissue damage. Water vapor (steam) in the smoke can increase the damage done due to the capacity for the humid air to inflict additional burns.

Toxicity.   Toxicity, and not burn injury, is the primary cause of most fire fatalities. Components of smoke, especially products of incomplete combustion, are able to induce harm due to chemical change or reaction within the victim's body. Carbon monoxide (CO) is a dangerous toxic fire gas because it is able to bind with the hemoglobin in blood when absorbed in the lungs, thus preventing the blood from transporting oxygen to vital organs. The burning of wool can produce hydrogen cyanide (HCN), another toxic gas that is able to travel through the blood throughout the entire body and do damage at the cellular level. Many products of combustion could be classified as cancer causing when exposed to over long periods (i.e., cigarette smoking). From a fire protection standpoint, it is usually the short-term exposures leading to immediate injury or death that are of the most concern, although exposure to smoke from power plants or forest fires through the environment, for example, are also a concern.

Asphyxiation.   Asphyxiation, or lack of oxygen, can result in a reduction in reasoning or muscle coordination (impairment) when oxygen concentrations drop to between 14 percent and 17 percent, which is less than the 21 percent normally available in clean air. Oxygen concentrations below 10 percent can lead to unconsciousness and eventually death. Fires obviously consume oxygen to continue burning, thus smoke should be considered to contain reduced levels of oxygen and be treated as hazardous. Unfortunately, as the body experiences lower oxygen levels (or higher CO_2 levels), the rate of breathing increases, thereby also leading to higher intake of toxic gases.

Obscuration.   Obscuration represents the ability of smoke to block or reduce the visual input to a person necessary to move away from danger. Especially in buildings like hotels or movie theaters occupied by people who are unfamiliar with the premises, escape during a fire depends on the ability to visualize the route and then travel to a point of safety. Smoke can also contain corrosive or irritating components, leading to difficulty breathing or causing an inability to open the eyes.

▶ SUMMARY

A solid understanding of fire behavior forms the basis for the study of fire protection because it provides the background critical to explain why fire and building codes contain certain provisions regulating how buildings are built and used; provides insight on how buildings fail during a fire, a topic important to fire fighters and their families; and illustrates the need to properly design, install, and test fire protection equipment such as alarm and sprinkler systems.

This chapter is only able to introduce the rapidly growing field of enclosure fire dynamics and to cover some of the basics concepts of fire behavior, such as the fire triangle and fire tetrahedron, heat transfer, and fire extinguishment theory. Just as the world is becoming more technically based, the field of fire increasingly involves science to allow discovery of how fires start, spread, and ultimately do damage. This information can be used to further fire protection efforts and protect the fire fighters who risk their lives fighting fires.

Group Activity

Draw a figure explaining the fire tetrahedron and extinguishment theory.

Review Questions

1. What three things are required for a fire?
2. What two elements provide the components for the chemical reaction to occur?
3. How must the fuel and oxidizer be mixed?
4. When does piloted ignition occur?
5. What are the three methods of heat transfer?
6. List the five classifications of fire based on the type of fuel burning.

Suggested Readings

Drysdale, D. *An Introduction to Fire Dynamics*, 2nd ed. John Wiley & Sons, 1998. This text is commonly used in Fire Protection Engineering classes and is considered the standard reference on the subject of fire behavior.

Karlsson, B. and Quintiere, J. *Enclosure Fire Dynamics*. CRC Press, 2000. Requires some math and science background, but includes plenty of very good examples.

Lyons, J. W. *Fire*. Scientific American Library, 1985. Although out of print, Lyons's book is a great introduction to all aspects of fire that is intended for people with no scientific or fire background.

Quintiere, J. *Principles of Fire Behavior*. Delmar Publishers, 1998. A very good book for beginning to learn more about fire behavior.

4 Building Design and Construction

Jon Nisja
Minnesota State Fire Marshal Division

Learning Objectives

After completing this chapter, the reader should be able to do the following:

- List some objectives of fire safe building design
- Describe several types of building construction
- Explain modern construction techniques
- List common materials used in building construction
- Explain the impact of fire in buildings
- Describe building systems and services
- Describe possible building failures and concerns

By far, the majority of fire deaths occur in buildings. There have been significant improvements in certain aspects of fire safe building design over the past several decades in the way of automatic fire suppression systems, automatic smoke detection, and improved egress systems. In other aspects, fire safe building design has been eroding; particularly with the advent of inexpensive, lightweight construction techniques and materials. The type of construction and the materials used in a building can greatly influence how a fire will behave and also determines how well the building can withstand the effects of a fire.

Clearly, the advent of building codes and regulations has had a positive impact on life loss from fires. But even into the 21st century, America was shocked to witness the collapse of the World Trade Center towers in New York City from a fire following a terrorist attack when fuel-laden jet airplanes flew into the buildings. None of the codes and regulations could ever contemplate or protect against such an extreme event. Eighteen months later, 100 people perished in a bar and nightclub fire in Rhode Island. Many people thought that the codes and construction technology had evolved to a point where these types of incidents could not happen. Yet these disasters reminded people that the possibility of losing hundreds of people in a building fire-related incident still existed.

Many fire safety experts, and perhaps society as a whole, had come to believe that the days of a significant conflagration and loss of life from fire had passed. A hundred years ago, major fires involving multiple buildings were common occurrences in communities. In the late 1800s and early 1900s, many of America's cities experienced significant conflagrations that destroyed large portions of the town. These fires occurred because of combustible construction and poor municipal planning and community layout.

▶ BUILDING AND SITE PLANNING FOR FIRE SAFETY

Decisions need to be made early in the design or planning process for a new building to provide an effective and safe design from a fire safety standpoint. Considerations need to be given to both active fire protection (sprinklers) and passive fire protection (fire walls, fire doors, fire extinguishers, etc.). If the building is not sprinklered, there is heavy reliance on manual firefighting techniques.

Table 4.1 lists many of the factors that should be addressed when designing or reviewing the design for a new building. In the case of some of these factors, the building and fire codes have incorporated these considerations into the code requirements.

Traffic and Transportation Patterns

Response time for fire departments and other emergency vehicles (such as ambulances) is a vital consideration in designing a building or site. Major roads, thoroughfares, traffic congestion patterns, and private roads all should be reviewed and considered for new site projects so that emergency vehicles have limited response times and distances.

Fire Department Access

Fire apparatus access is a major consideration in new construction. The fire department must be able to get hoselines to all portions of the building. Multistory buildings may require that fire department aerial platforms or ladder trucks have ready access to the building. Fire apparatus access roads need to be wide enough to support the equipment used by the fire depart-

▶ **Table 4.1**   Elements of Building Fire Safety

Building Design and Construction Features Influencing Fire Safety	
Fire Propagation	• Fuel load and distribution • Finish materials and their location • Construction details influencing fire and products of combustion movement • Architectural design features
Smoke and Fire Gas Movement	• Generation • Movement —Natural air movement —Mechanical air movement • Control —Ventilation —Heating, ventilating, air conditioning —Barriers —Pressurization • Occupant protection —Egress —Temporary refuge spaces —Life support systems
Detection, Alarm, and Communication	• Activation • Signal • Communication systems —To and from occupants —To and from fire department —Type (automatic or manual) —Signal (audio or visual)
People Movement	• Occupant —Horizontal —Vertical —Control —Life support • Fire Fighters —Horizontal —Vertical —Control
Suppression Systems	• Automatic • Manual (self-help; standpipes) • Special

▶ **Table 4.1** (*Continued*)

Firefighting Operations	• Access
	• Rescue operations
	• Venting
	• Extinguishment
	—Equipment
	—Spatial design features
	• Protection from structural collapse
Structural Integrity endurance)	• Building structural system (fire
	• Compartmentation
	• Stability
Site Design	• Exposure protection
	• Firefighting operations
	• Personnel safety
	• Miscellaneous (water supply, traffic, access, etc.)
Fire Emergency Considerations	
Life Safety	• Toxic gases
	• Smoke
	• Surface flame spread
Structural	• Fire propagation
	• Structural stability
Continuity of Operations	• Structural integrity

Source: Adapted from *Principles of Fire Protection*, NFPA, 1988, Table 6.1.

ment. They also need to be able to support the weight of the apparatus; this is not always easily accomplished. Access roads that are needed for emergency use are often designated as fire lanes and vehicle parking is prohibited. If fire apparatus access cannot be provided or achieved, the building should be protected with automatic sprinklers.

Firefighting Water Supplies

The most effective method of fire suppression that we have is still water. Water needs to be supplied or available for the fire department to fight a fire in a building. In many municipalities and jurisdictions, a governmental entity (such as the city or county) or a quasi-governmental entity (such as a water purveyor or water utility) provides a water supply through a system of underground water mains and fire hydrants. If the municipality or jurisdiction provides the water, the designer or developer probably has little to provide in terms of additional water supplies.

If there is no permanent water supply at the site, one will need to be provided. Simply because no municipal water supply exists does not mean that there will be no fire; fire behaves the same whether there are hydrants or not. The lack of a water supply severely hampers firefighting operations. Fire insurance companies have recognized this fact for years. A building with an inadequate water supply may be difficult or expensive to insure.

▶ OBJECTIVES OF FIRE-SAFE BUILDING DESIGN

Designing a building is a balance of providing a facility that meets the needs of the owner and addresses the unique risks or operations that will occur in the building. The easiest way to accomplish these objectives is to limit the use of fire and ignition sources within a building.

Unfortunately, eliminating all sources of ignition is not realistic. Buildings typically use fire for heating and cooking processes, they have electrical systems and appliances that can cause ignition or overheating, they have processes that use fire (such as welding or open flames), they are exposed to accidental sources of ignition (such as lightning), and they are subject to unsafe human acts (such as careless smoking or arson). Fire safety design and construction of the building should consider the following areas of concern:

- Life safety
- Property protection
- Continuity of operations
- Environmental protection

Life Safety

Life safety should always be the primary consideration in the design and construction of any building. Many construction and building features can greatly add to the ability of occupants to safely evacuate a building in emergency conditions. Life safety is often achieved by providing systems for early warning of a fire, extinguishment of a fire, and proper egress for prompt exiting from the building.

Factors that should be considered include the number of occupants, the familiarity from the occupants with the building, the ability of the occupants to recognize fire hazards and take appropriate actions, and the length of time that the building will be occupied.

Property Protection

Property protection is the factor that many people generally think of in terms of fire protection: keeping the building and its contents from burning. Property protection is a very important fire safety objective; annual fire loss from building fires in the United States exceeds $8 billion [1]. Property protection can be achieved by installing fire-extinguishing systems, by providing compartmentation features to confine or limit fire spread within a building, and by constructing the building of materials that resist fire development.

Continuity of Operations

Continuity of operations considers the specific and unique functions of the building. The amount of downtime following a fire should be considered during the design process. Sometimes this downtime is referred to as business interruption. Some structures are so important to the community or the economy that adequate protection must be incorporated to protect the building's vital operation. Continuity of operations is best accomplished through the installation of automatic fire-extinguishing systems.

Environmental Protection

Environmental protection is another factor that needs to be considered in some buildings. Fires at some facilities can cause environmental damage through air pollution due to smoke or hazardous materials runoff from contaminated firefighting water. Although relatively rare, there are certain types of buildings where the firefighting strategy may be to "let it burn" as opposed to dealing with the pollution from firefighting water runoff. Environmental protection can be enhanced with the installation of automatic fire-extinguishing systems that limit the fire size and minimize the subsequent firefighting water runoff.

▶ KEY CONSTRUCTION TERMS

Certain terms and words are used to describe and classify construction types and materials. Many of these terms relate to how the materials behave when exposed to fire. Terms used by fire and building inspectors, architects, engineers, and insurance industry representatives when dealing with construction-related issues include the following NFPA definitions:

Combustible (Material). A material that, in the form in which it is used and under the conditions anticipated, will not ignite and burn; a material that does not meet the definition of noncombustible, or limited combustible. (NFPA *101*-2003: 3.3.135.1) [2]

Noncombustible (Material). Refers to a material that, in the form in which it is used and under the conditions anticipated, does not ignite, burn, support combustion, or release flammable vapors, when subjected to fire or heat. Materials that are reported as passing ASTM E 136, *Standard Test Method for Behavior of Materials in a Vertical Tube Furnace at 750 Degrees C,* are considered noncombustible materials. (NFPA *101*-2003: 3.3.135.3) [2]

Limited combustible (Material). Refers to a building construction material not complying with the definition of *noncombustible* that, in the form in which it is used, has a potential heat value not exceeding 8140 kJ/kg (3500 Btu/lb), where tested in accordance with NFPA 259, *Standard Test Method for Potential Heat of Building Material,* and includes (1) materials having a structural base of noncombustible material, with a surfacing not exceeding a thickness of 3.2 mm (⅛ in.) that has a flame spread index not greater than 50; and (2) materials, in the form and thickness used, other than described in (1), having neither a flame spread index greater than 25 nor evidence of continued progressive combustion, and of such composition that surfaces that would be exposed by cutting through the material on any plane would have neither a flame spread index greater than 25 nor evidence of continued progressive combustion. (NFPA *101*-2003: 3.3.135.2) [2]

Fire barrier wall. A wall, other than a fire wall, having a fire resistance rating. (NFPA 221-2000: 1.3.9.3) [3] Fire barrier walls are used to protect buildings from areas of increased hazard or risk or to provide protection for purposes of egress.

Fire resistance rating. The time, in minutes or hours, that materials or assemblies have withstood a fire exposure as established in accordance with the test procedures of NFPA *Standard Methods of Tests of Fire Endurance of Building Construction Materials.* (NFPA 220-1999: 2-1) [4]

Fire wall. A wall separating buildings or subdividing a building to prevent the spread of fire and having a fire resistance rating and structural stability. (NFPA 221-2000: 1.3.9.4) [4]

Flame spread rating. A relative measurement of the surface burning characteristics of building materials. (NFPA 801-2003: 3.3.16) [5]

Flame resistance. The property of a material whereby combustion is prevented, terminated, or inhibited following the application of a flaming or nonflaming source of ignition, with or without subsequent removal of the ignition source. Flame resistance can be an inherent property of a material, or it can be imparted by specific treatment. (NFPA 1977-1998: 1-3) [6]

Flame retardant. So constructed or treated that it will not support flame. (NFPA 79-2002: 3.3.45) [7]

▶ TYPES OF BUILDING CONSTRUCTION

Fire personnel must have a basic understanding of building construction to understand the impact of a fire on a structure and to safely operate on the fireground. As illustrated in Figure 4.1, building construction is impacted by the following three primary forces:

1. Compression: weight applied to a material; has the effect of pushing objects or shortening. When a person is standing, they are applying compressive loads through their body and onto the floor.

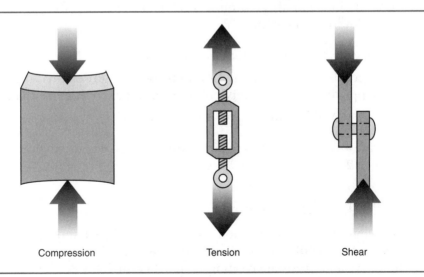

Compression   Tension   Shear

▶ **Figure 4.1**   Three Basic Forces That Act on Building Materials. *(Source: Building Construction for the Fire Service, NFPA, 1982.)*

2. Tension: essentially the opposite of compression; stretching, pulling apart, extending, or elongating. When a person hangs on something, they are applying tension forces on the object they are hanging from.

3. Shear: an opposite but parallel sliding motion caused by two objects in contact.

There are two basic types of construction: noncombustible and combustible. The primary differentiation between the two types of construction is whether the structural frame elements are noncombustible or combustible. Structural frame elements refer to the columns; the beams, girders, trusses, and spandrels having direct connection to the columns; and any other structural members essential to the stability of the building as a whole.

If the main structural frame element holding the building up is wood, the building should be classified as combustible. If the main structural frame elements are concrete, masonry, and/or steel, the building should probably be classified as noncombustible.

Within each of the two types there are several subclassifications. NFPA 220 [4] is used for classifying the type and classification of construction. The construction classifications are represented by a Roman numeral designation from Type I to Type V. The lower the Roman numeral, the more fire resistive the building is. Buildings with higher levels of fire resistivity are typically allowed by the model building code to be taller and larger in size.

NFPA 220 [4] also uses a series of three numbers following each type of construction. The three numbers refer to the hourly fire-resistive rating of the following three items:

1. First Arabic number: Exterior bearing walls
2. Second Arabic number: Columns, beams, girders, trusses and arches, supporting bearing walls, columns, or loads from more than one floor
3. Third Arabic number: Floor construction

The fire resistance ratings for the various types of construction are shown in Table 4.2.

Noncombustible Construction

Noncombustible construction has two types: Type I and Type II. The structural elements of Type I construction are required to be noncombustible and have a fire-resistance rating as specified in Table 4.2. There are two classifications of Type I construction: 443 and 332.

▶ **Table 4.2** Fire Resistance Ratings (in hours) for Type I through Type V Construction

	Type I		Type II			Type III		Type IV	Type V	
	443	332	222	111	000	211	200	2HH	111	000
Exterior Bearing Walls										
Supporting more than one floor, columns, or other bearing walls	4	3	2	1	0¹	2	2	2	1	0¹
Supporting one floor only	4	3	2	1	0¹	2	2	2	1	0¹
Supporting a roof only	4	3	1	1	0¹	2	2	2	1	0¹
Interior Bearing Walls										
Supporting more than one floor, columns, or other bearing walls	4	3	2	1	0	1	0	2	1	0
Supporting one floor only	3	2	2	1	0	1	0	1	1	0
Supporting roofs only	3	2	1	1	0	1	0	1	1	0
Columns										
Supporting more than one floor, columns, or other bearing walls	4	3	2	1	0	1	0	H²	1	0
Supporting one floor only	3	2	2	1	0	1	0	H²	1	0
Supporting roofs only	3	2	1	1	0	1	0	H²	1	0
Beams, Girders, Trusses, and Arches										
Supporting more than one floor, columns, or other bearing walls	4	3	2	1	0	1	0	H²	1	0
Supporting one floor only	3	2	2	1	0	1	0	H²	1	0
Supporting roofs only	3	2	1	1	0	1	0	H²	1	0
Floor Construction	3	2	2	1	0	1	0	H²	1	0
Roof Construction	2	1½	1	1	0	1	0	H²	1	0
Exterior Nonbearing Walls ³	0¹	0¹	0¹	0¹	0¹	0¹	0¹	0¹	0¹	0¹

▓ Those members that shall be permitted to be of approved combustible material.

¹ See A-3-1 (table).

² "H" indicates heavy timber members; see text for requirements.

³ Exterior nonbearing walls meeting the conditions of acceptance of NFPA 285, *Standard Method of Test for the Evaluation of Flammability Characteristics of Exterior Non-Load-Bearing Wall Assemblies Containing Combustible Components Using the Intermediate-Scale, Multistory Test Apparatus,* 1998, shall be permitted to be used.

Source: NFPA 220, *Standard on Types of Building Construction,* 1999, Table 3-1.

Type I. Type I construction used to be called "fireproof" construction; a term which fell into disfavor in the mid-1900s following a number of significant fires in these types of buildings. Since all buildings have combustible contents, the term "fireproof" is a misnomer. Type I has also been referred to as "fire-resistive" (or F.R.) construction. Type I-443 construction has noncombustible structural elements, 4-hour fire resistance rated exterior walls, 4-hour fire resistance rated structural framing elements, and 3-hour fire resistance rated floors.

Type II. In Type II construction the structural elements can be materials that are noncombustible or limited combustible. There are three classifications of Type II construction: 222, 111, and 000. The first two types require some level of fire resistance to protect the structure (2 hour or 1 hour), sometimes referred to as "protected noncombustible" construction. The third type (000) requires noncombustible or limited combustible structural elements

▶ **Figure 4.2**   Example of Type II-000 Construction.

but there is no fire resistance rating required. This type is known as "unprotected noncombustible" construction. Figure 4.2 shows an example of Type II-000 construction. The building has noncombustible masonry walls and exposed (unprotected) steel bar joists supporting the metal roof decking.

Type II-000 is probably the most common commercial type of construction used. A building with block, brick, or concrete exterior walls and which has bare steel columns and steel bar joists supporting the roof is Type II-000 construction. Many retail, grocery, and mercantile stores are of this type of construction. In many areas of the country, Type II-000 construction is cheaper than wood frame buildings (based on the cost per square foot to construct).

The concern from a fire safety standpoint is that the structural elements have no level of fire resistance. They are exposed to direct heat and flame contact in a fire. Most building codes severely limit the size and height of these "unprotected" buildings or require the installation of automatic fire sprinklers to increase the height.

Combustible Construction

There are three types of combustible construction: Type III, Type IV, and Type V. All of these types of construction use wood as part of the structural framing elements. Type III is actually kind of a hybrid between noncombustible and combustible construction. It was a fairly common type of construction in the late 1800s and early 1900s but is rarely used for new construction today.

Type III.   In Type III construction, the exterior walls of the building have a 2-hour fire resistance rating and are noncombustible or limited combustible. Interior structural framing elements, such as those supporting other stories and the roof are combustible (i.e., wood). These buildings often appear to be noncombustible from the exterior; they most often have masonry (block or brick) exterior walls. A closer interior examination is necessary to ascertain that interior structural elements are combustible.

Once again, there are two classifications of Type III: Type III-211 (protected) and Type III-200 (unprotected). Type III is also known as "exterior protected combustible" or "ordinary" construction. In modern construction techniques, there is nothing ordinary about this; to the contrary, it is quite an uncommon construction method.

Type IV.   Type IV is a unique method of construction also known as heavy timber. In order to qualify as Type IV construction, the wooden members that form the structural framing elements need to be certain minimum di-

▶ **Figure 4.3** Elements of a Building of Type IV (Heavy Timber) Construction. *(Source: Fire Protection Handbook, NFPA, 2003, Figure 12.2.2.)*

▶ **Table 4.3** Recommended Nominal Dimensional Requirements for BCMC Type IV(2HH) Construction

	Supporting Floors	**Supporting Roofs**
Columns	8 in. × 8 in.	6 in. × 8 in.
Beams and girders	6 in. × 10 in.	4 in. × 6 in.
Arches	8 in. × 8 in.	6 in. × 8 in., 6 in. × 6 in., 4 in. × 6 in.
Trusses	8 in. × 8 in.	4 in. × 6 in.
Floors	3 in. T & G or 4 in. on edge w/1-in. flooring	
Roofs		2 in. T & G or 3 in. on edge w/1⅛-in. plywood

Note: T & G is tongue and groove. For SI units: 1 in. = 25.4 mm.

Source: Fire Protection Handbook, NFPA, 2003, Table 12.2.4.

▶ **Figure 4.4**   A Variation of Type IV (Heavy Timber) Construction with Haunched Arches of Laminated Wood (Glue-Laminated Construction) and Beams Anchored to Arches by Steel Hangers. *(Source: Fire Protection Handbook, NFPA, 2003, Figure 12.2.3.)*

mensions. Typically the columns, arches, and trusses are at least 8 inches by 8 inches if they support other stories. These columns, arches, and trusses are allowed to be a minimum of 6-inch dimension if they are only required to support the roof assembly. Some examples of Type IV construction can be seen in Figure 4.3 and Figure 4.4. Also see Table 4.3.

Type IV is also sometimes known as *mill construction*, based on a common practice of constructing textile mills in New England in the mid-1800s. Type IV construction is still used today, although the construction techniques and materials have changed. One of the modern heavy timber materials is glue-laminated timbers. These are thinner strips of wood that are glued together under pressure to form arches, beams, girders, and columns. One of the most commonly seen applications of Type IV construction is in churches and places of worship.

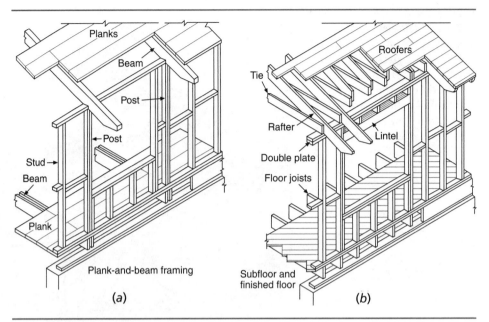

▶ **Figure 4.5** Two Variations on Basic Type V (Wood Frame) Construction: (a) Plank-and-Beam Framing in Which a Few Large Members Replace Many Small Members Used in Typical Wood Framing, (b) Conventional Wood Framing (Western or Platform Construction). Fire blocking is essential in concealed spaces. (*Source: Fire Protection Handbook, NFPA, 2003, Figure 12.2.4.*)

Type V. Type V is probably the most commonly used type of construction. It is commonly seen in houses, garages, low-rise apartments, low-rise hotels, smaller office and mercantile buildings, and restaurants. Type V construction uses 2 × 4-inch (or sometimes 2 × 6-inch) wooden studs to construct interior and exterior walls. There are some variations on the construction methods used in Type V construction, as illustrated in Figure 4.5.

Type V construction also has two classifications—Type V-111 (protected) and Type V-000 (unprotected). In modern construction techniques, the wooden studs are typically covered with gypsum wallboard. In some cases there may be a brick veneer covering also. If the building is wood frame but all of the wooden studs, joists, trusses, rafters, and beams are covered with gypsum wallboard or plaster, it is probably Type V-111. If wood studs, joists, trusses, rafters, or beams are visible (and exposed to flame contact), it is probably Type V-000 construction.

▶ **Figure 4.6**   Example of Mixed Types of Construction.

Classifying Types of Construction

Mixed types of construction are very common. An apartment building of Type V-111 construction may have an underground parking garage of concrete (Type I construction). Buildings are generally classified as to the lowest type of construction used; in the previous example, the apartment would be classified as Type V-111 construction. Figure 4.6 is an example of mixed types of construction. The building has an underground parking garage of Type I construction and apartments of Type V construction. Because of concern for structural collapse, most model building codes also prohibit construction types of lesser fire resistance to support construction types of greater fire resistance.

Assessing and classifying types of construction can sometimes be difficult and time-consuming. When in doubt as to which classification to use, the building should be classified to the lower type of construction. Also, some construction techniques give the appearance of fire-resistive elements that

▶ **Figure 4.7** Example of a Wood Frame Building with a Brick Veneer.

are only for aesthetics. Figure 4.7 shows a wood frame (Type V-111) building with a brick veneer. At first glance, this may appear to be Type II or Type III construction.

▶ BUILDING MATERIALS

Wood

One of the oldest and most commonly used construction material is wood. For the most part, wood continues to be in abundant supply. It is fairly economical and is relatively easy to work with; no specialized tools or training are typically required. Every building contains some amount of wood, whether it is for structural framing, trim, decorations, doors, roofing materials, flooring, or to provide support for other construction materials. Wood is often used in furniture and building contents also. Even in totally non-combustible buildings made of concrete, wood is often used to create the forms and to support the concrete until is has cured. The wood is then removed from the building.

The primary disadvantage of wood as a construction material is that it is combustible—it burns. Because it is a cellulose-based material, wood will burn in temperatures ranging from 500°F to 750°F (260°C to 399°C) [8]. A number of factors determine the ignition temperature of wood, including its moisture content and the species of wood. There are materials and treatments that can retard the ignition of wood and slow the rate of burning. Such wood is known as fire-retardant-treated lumber. Contrary to some claims, however, there is no way to make wood noncombustible.

Within buildings, the structural wood is often covered with other materials, such as stucco, plaster, or gypsum. These coverings protect the wood from high temperatures and can substantially reduce fire involvement of the wooden members.

When the wood is exposed to heat, a char forms on its surface. This char can actually be a good insulator when dealing with larger pieces of wood, which explains why heavy timber construction has a reasonably good fire safety record. This charring occurs on the surface of wood at a rate of about $\frac{1}{40}$ of an inch per minute [9]. Flashover in a building fire occurs at 1000°F–1200°F (538°C–649°C). At these temperatures, virtually all wood materials will have exceeded their ignition temperatures. See Chapter 3 for the definition of flashover.

Wood comes in many formats. It is used as a structural material (wood studs, roof joists, trusses, etc.), as siding and sheathing materials, as flooring and roofing underlayment (plywood sheets), as interior finish (paneling and wainscoting), as shelving and furniture (solid wood and particle board).

Steel

Steel is another common building material. It is used to form the skeleton or frame for many buildings. It can also be used for the exterior walls. Steel is often used in beams, girders, columns, and bar joists to support floors and roofs. Steel is more difficult to work with than wood. It requires specialized tools and fastening methods. Steel is often bolted, riveted, or welded to other pieces of steel to form a very sturdy structure.

Steel is noncombustible; it does not contribute fuel to a building fire. It does, however, expand when exposed to heat. As a general rule of thumb, steel will expand 0.06 to 0.07 percent in length for each 100°F (38°C) rise in temperature. Heated to 1,000 °F (538°C), a steel member 100 feet in length will expand about 9½ inches [10].

Another problem with steel is that it conducts heat. There have been a number of fires where steel allowed heat to ignite combustible materials that were close to it. A third problem with steel is that it loses its strength when exposed to temperatures that are fairly common in interior structure fires. Some research shows steel will lose 40–50 percent of its strength when exposed to temperatures in the 900–1100°F (482°C–593°C) [11]. See Figure 4.8. Failure of the structural steel can result in partial or total collapse of the building.

Most building codes will not allow a larger building having a relatively high combustible fuel load or large numbers of people to be constructed using unprotected steel. Most codes require that the steel be protected with fire-resistive materials or that the building be protected with sprinklers. This requirement comes about because of a fairly significant fire loss history in buildings with unprotected steel.

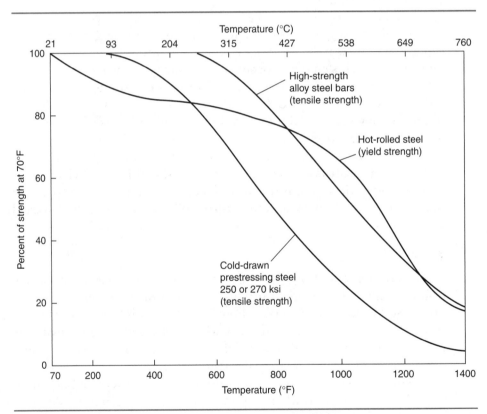

▶ **Figure 4.8** Strength–Temperature Relationships for Hot-Rolled, Cold-Drawn, and High-Strength Alloy Steels. *(Source: The SFPE Handbook of Fire Protection Engineering, 3rd ed., Figure 4-10.1. © Society of Fire Protection Engineers. Used with permission.)*

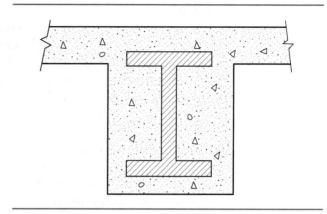

There are a number of methods of providing fire resistance to steel. For many years, asbestos was used to "fireproof" steel. Over the years in response to health concerns from asbestos, other materials have been used. Steel can be encased in concrete, covered with gypsum or other commercially available fire-resistive coatings. The thicker the application of coverings or coating, the greater degree of fire resistance is afforded. Figures 4.9 and 4.10 illustrate some examples of fire-resistive coatings of steel beams.

Concrete and Masonry

Concrete is a mixture of cement, water, and aggregate (rocks or stones). The strength and durability of concrete depends on the proper proportions of cement and water, size, type and amount of aggregates, and thorough mixing. Concrete has very good compressive strength but tends to be weak in tensile strength unless reinforcing materials are added to the concrete. Concrete is very fire resistant as it has the ability to absorb incredible amounts of heat because of the water and aggregate used.

▶ **Figure 4.10**   Spray-On Fireproofing. *(Source: Fire Protection Handbook, NFPA, 2003, Figure 12.4.17.)*

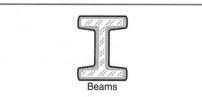

Beams

Like wood and steel, concrete is used in many forms in the built environment. There are three general types of concrete construction: precast, cast-in-place, and concrete masonry units. In order to add tensile strength to concrete, it is often reinforced by adding steel to the concrete. The reinforcing can be from steel bars or rods (often called *re-bar*) or a steel wire mesh. A relatively new method of reinforcing concrete is the use of synthetic or polymer fibers within the concrete mix.

Precast Concrete.   Precast concrete is typically made at a plant or other site, cured, and set into place at the construction site. Precast concrete is typically used for walls, floors, and roof assemblies. To allow concrete members to span larger areas without intermediate support columns, precast concrete is often produced with steel cables or steel rods imbedded in it. These steel members are put under pressure and then concrete is poured in and around it. This process is known as pretensioning; it is one of the two forms of prestressed concrete.

Cast-in-Place Concrete.   Cast-in-place concrete is often mixed in concrete plants and transported to construction sites. For larger projects, concrete may be mixed on site. The concrete is poured into forms or molds at the construction site. It can be plain concrete, reinforced concrete, or posttensioned concrete. Posttensioning is the other form of prestressed concrete. In posttensioning, conduits or pipes filled with steel cables are placed in the concrete molds. After the concrete has set and cured, tension can be applied to the cables to add additional tensile strength and load-bearing capabilities.

Concrete Masonry Units.   Concrete masonry units (CMU) are also sometimes called *cinder blocks*, which are hollow concrete blocks used for foundations and walls. They are very fire resistive and tend to maintain their integrity under fire conditions. They are often used to build fire walls required by the codes. CMUs can be reinforced with steel bars to add strength. They are joined to other CMUs with mortar made of concrete.

Glass

Glass is most commonly used as glazing for windows and doors. Glass is not very fire resistive but does resist the passage of smoke. The fire resistivity of glass is improved by adding wire mesh into the glass. In recent years, some newer acrylic and composite materials have been added to glass to increase strength and fire resistance. Glass blocks are also used to form interior partitions for certain applications.

Plastics and Composites

Plastic materials tend to behave fairly poorly in fire conditions; they often soften and distort when exposed to heat. Like other materials, many plastic products give off toxic fire gases during thermal decomposition. Plastics are not typically used for structural support in buildings. However, they are often used for insulation, trim, interior furnishings, and plumbing and electrical systems.

Fiberglass-reinforced plastic is one of the most common forms of plastic used in a building. It is used in shingles, bathroom and plumbing fixtures, siding materials, and is a good insulator. Fiberglass itself is noncombustible but the resin used in making fiberglass is combustible.

Gypsum and Plaster

Plaster was made from gypsum rock; it was crushed and ground into powder. Water was added to the powder to create a pastelike substance. This was placed onto wooden strips (also called *wood lath*), certain types of wallboard, or wire mesh (metal lathe) to create walls and ceilings. Plaster is very fire resistive but is difficult to work with as specialized tools and experience are needed to produce smooth and even walls.

A few decades ago, gypsum wallboard was developed. It is commonly called *Sheetrock*™ (a brand name of gypsum wallboard) or *drywall*, as opposed to the wet process formerly used in plastering operations.

Gypsum and plaster are used to construct fire walls and fire barrier walls and to provide fire-resistive protection for structural steel and wood. Since water is a major component of the gypsum production process, these materials are very fire resistant. One of the problems with gypsum and plaster fire walls is that they are fairly easy to penetrate with holes and openings.

Other Common Construction Materials

Brick and tile are two other common construction materials. Brick is often used to construct walls. Tile is commonly used as an interior wall or floor finish material. In older buildings, clay tile was used to form walls. Aluminum is sometimes used for a lightweight skin on a building.

▶ BUILDING FAILURES AND CONCERNS

Modern construction techniques attempt to limit the cost of construction by minimizing the materials used in a building. One of these techniques is the use of lightweight construction materials. Lightweight construction materials are used in both combustible and noncombustible types of construction.

Lightweight Wood Construction

Decades ago it was common to use full dimension, milled lumber for construction of a building. Full dimension lumber was milled to the exact sizes that were specified; a 2 inch by 12 inch wooden board measured 2 inches by 12 inches in size. In modern construction practices, "nominal" measurements are often used (the nominal measurement of a 2 × 4 wooden board is 1½ inch by 3½ inches). This represents a fairly significant reduction—almost 40 percent—in the total mass of the lumber.

These buildings were supported on 2-inch by 12-inch floor joists or roof rafters. It is now common practice to construct houses, hotels, apartments, and smaller commercial buildings with pre-engineered roof and floor trusses. This often means that the largest piece of lumber in the building is a 2 × 4 (actual dimension: 1½ inch by 3½ inch).

When arranged in a triangular pattern, these trusses have tremendous load-bearing capabilities and can span relatively long distances. Since they are made of standard-sized lumber, they are very economical for building.

There certainly is a downside, however. Wood chars at a rate of about ¹⁄₄₀ of an inch per minute. If the 2 × 4 truss is exposed to fire or intense heat, the charring may be occurring on three sides of the board at the same time. In 10 minutes time, the depth of char would be about ¼ inch per side of the board exposed, which essentially means that almost 40 percent of the cross-sectional dimension of the 2 × 4 has been damaged. There simply is not enough load-bearing capability remaining in the truss.

In addition trusses are often joined with gusset plates, which have small protrusions or fingers that are used to nail boards together. Gusset plates replace conventional nails. The fingers or protrusions on a gusset plate only extend ¼ to ½ inch into the board, so if the board is damaged by flame exposure and charring, there is no solid mass of wood for the gusset plate to attach to. In recent years there has also been an increased use of wooden I beams; also called composite wood joist construction. These are constructed by installing plywood boards within notches or grooves cut into 2 × 2 or 2 × 4 boards. Figure 4.11 shows an example of composite wood joist construction.

Lightweight Steel Construction

Just as smaller lumber is being used in combustible construction, smaller and lighter weight structural steel is being used in noncombustible construction. Rather than using standard-size steel beams and bar joists, it is now possible to calculate the size and weight of the structural steel; this practice

▶ **Figure 4.11** Typical Composite Wood Joist Construction. *(Source: NFPA 13, 2002, Figure A.3.7.1(b).)*

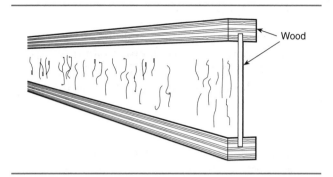

Wood

is called structural engineering. It also considers the other dynamics of a building: wind load, live loads, and so forth.

As the mass of steel being used is decreased, the impact of heat or flame impingement is greater. It takes far less heat to raise the temperature of ½-inch thick steel to its failure point than it does to raise ¼-inch thick steel. Less mass of material means less factors of safety in fire conditions.

Increased Fuel Loading

At the same time that construction has used lighter-weight construction materials, we are seeing an increase in the typical fuel loading or fire loading in a building caused by an increase in the use of synthetic materials, which tend to burn faster and have higher rates of heat release than cellulose-based materials [9]. Although plastics and similar synthetic materials tend to have higher ignition temperatures than wood, some are easily ignited and burn vigorously. Very high surface flame spread rates are possible—up to approximately 2 ft/s, or 10 times the rate of flame spread across most wood surfaces [9].

▶ IMPACT OF FIRE IN BUILDINGS

Once a fire occurs, there are four products of combustion. Because occupants in a building can be located within the same enclosure as the fire, they may be exposed to the following combustion products:

1. Flame/fire
2. Heat
3. Smoke
4. Toxic gas

Flame/Fire

It is common for larger buildings to have fire walls or fire barrier walls to provide separate compartments to limit fire spread and the occupant's exposure to flames. Relatively few fire fatalities are due to exposure to fire or flames; the other three products of combustion tend to be more dangerous and toxic.

Heat

Heat can be devastating to a structure, its contents, and its occupants. Figure 4.12 shows the effects of elevated temperatures on life safety and building materials. Heat rises in a building, so protecting from vertical heat and smoke travel is an important objective in fire-safe building design. Occupant tenability is severely limited in temperatures over a few hundred degrees. The structure itself can be severely impacted by temperatures exceeding 900 to 1100°F (482°C to 593°C). It is common for the temperature in a fully involved room fire to reach 1200°F to 1500°F (649°C to 815°C).

Smoke

Completely preventing exposure to smoke is extremely difficult to accomplish in a building. Some methods of smoke control involve providing ventilation into a space, removing air from a space, pressurizing adjacent areas, providing smoke-resistant construction, or a combination of these methods. Most of the successes that have been seen in the reduction of fire deaths have come from the installation of smoke alarms. When properly installed, smoke alarms give early warning of fire conditions before they become lethal and allow occupants sufficient time to egress from the building.

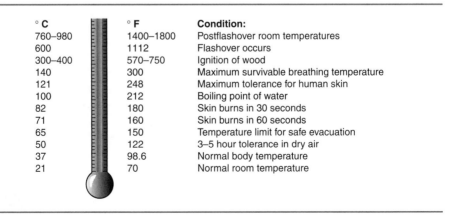

°C		°F	Condition:
760–980		1400–1800	Postflashover room temperatures
600		1112	Flashover occurs
300–400		570–750	Ignition of wood
140		300	Maximum survivable breathing temperature
121		248	Maximum tolerance for human skin
100		212	Boiling point of water
82		180	Skin burns in 30 seconds
71		160	Skin burns in 60 seconds
65		150	Temperature limit for safe evacuation
50		122	3–5 hour tolerance in dry air
37		98.6	Normal body temperature
21		70	Normal room temperature

▶ **Figure 4.12** Effects of Heat on Humans and Building Materials.

Toxic Gas

Toxic fire gases are often present in the smoke and heat generated in a structure fire. These gases can be lethal in relatively small quantities (see Chapter 3 for a more detailed analysis of these combustion products). There are many toxic fire gases: carbon monoxide, hydrogen sulfide, nitrous oxide, hydrogen cyanide, hydrogen chloride, carbon dioxide, acrolein, and formaldehyde are some of the more common gases. Some of these materials are asphyxiants, others are irritants, while others have anesthetic effects. In recent years, some technology has been developed that allows detection of certain fire gases; carbon monoxide being the most common of the detectable gases. Some communities and jurisdictions are requiring the installation of carbon monoxide detectors. Carbon monoxide detectors should not be considered a substitute for smoke alarms.

▶ RATINGS, TESTING, AND RESEARCH

Tests and Listings

In the early 1900s some testing criteria were developed to ensure standardized methods for evaluating the fire endurance of certain construction features in the building. This method utilizes a standard time-temperature curve as defined in ASTM E-119, *Standard Test Methods for Fire Tests of Building Construction and Materials* [12] and NFPA 251, *Standard Methods of Tests of Fire Endurance of Building Construction and Materials* [13]. See Figure 4.13. This test method is widely accepted and used by the majority of the testing and listing organizations (such as Underwriters Laboratories and FM Global). It is also the basis of the fire-resistance requirements in most building codes.

The ASTM E-119 test involves placing the wall, floor, roof, or door assembly into a furnace and subjecting it to the heat rates specified in the time-temperature curve for the duration desired. At the end of the test, a hose stream is sometimes applied to the assembly tested. In order to pass the test, the material or assembly must stay intact following exposure to the fire and hose stream. This procedure provides a relative test for the fire resistance of building materials but is not a precise measurement of fire resistance. Some critics argue that the standard time-temperature curve is archaic and not representative of the types of fire growth or heat seen in present-day structure fires.

Fire-Resistance-Rating Materials

Some of the earliest successes with fire-resistance-rated construction came from applying requirements to exterior building walls. This helped greatly

▶ **Figure 4.13** Time-Temperature Curve. (Source: NFPA 251, 1999, Figure 2-1.1.)

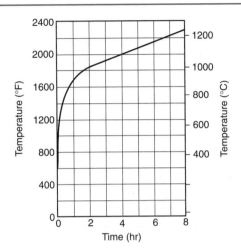

Note: The following are the points that determine the curve.

1000°F (538°C)............................at 5 minutes
1300°F (704°C).........................at 10 minutes
1550°F (843°C).........................at 30 minutes
1700°F (927°C)...............................at 1 hour
1850°F (1010°C)..............................at 2 hours
2000°F (1093°C)..............................at 4 hours
2300°F (1260°C)..............................at 8 hours
                                       or over

by reducing the risk of fire spreading from one building to another. This fire safety issue was common 100 or so years ago and continues to be a concern in communities where buildings with combustible construction are located close together.

In time, many progressive fire officials, building officials, and insurance interests determined that because fire-resistive construction worked to keep fires confined to the building, perhaps it would work to confine a fire to a portion of the building. With this in mind, it became common to build interior walls, floors, and ceiling from fire-resistive materials in an effort to confine the fire to a room, compartment, or area within the building. Once again, over time successes were realized. Rather than losing entire buildings to a fire, the loss could be contained to an area within the building.

Many building codes, including *NFPA 5000™*, *Building Construction and Safety Code™* [14] and NFPA *101, Life Safety Code* [2], have requirements for interior construction to limit fire spread within a building or to provide protection for occupants as they egress from the building. These interior walls must meet certain fire resistance requirements (typically 1- or 2-hour fire resistance rating).

In many cases, the installation of automatic sprinkler systems allows the authority having jurisdiction to waive or reduce the fire resistive requirements. Fire-resistive construction does a good job at confining fire spread within that compartment assuming that the protections are all in place. It is very common to see fire doors propped or wedged open and holes cut into fire-rated walls. These types of violations negate the effectiveness of fire-resistive construction.

Another consideration is that fire-resistive construction and compartmentation (fire walls and fire barrier walls) are based on the assumption that the compartment, but not the entire building, will be destroyed by fire. This risk may be unacceptably high. Automatic sprinkler protection can keep fire damage to a few hundred square feet as opposed to several thousand square feet.

Doors, windows, and other openings within fire-rated walls must be protected against fire spread just as the wall itself is. Other openings include penetrations for heating, ventilating, air conditioning, electrical systems, elevators, and fire protection systems. Fire doors are the most common means of protecting vertical and horizontal openings in fire-rated floors and walls.

To protect openings in walls, most model building and fire codes reference NFPA 80, *Standard for Fire Doors and Fire Windows* [15]. NFPA 80 uses the following designations for opening protection:

Class A—Openings in fire walls and in walls that divide a single building into fire areas

Class B—Openings in enclosures of vertical communications through buildings and in 2-hour rated partitions providing horizontal fire separations

Class C—Openings in walls or partitions between rooms and corridors having a fire resistance rating of 1 hour or less

Class D—Openings in exterior walls subject to severe fire exposure from outside of the building

Class E—Openings in exterior walls subject to moderate or light fire exposure from outside of the building

Building, fire, and life safety codes typically specify the fire resistance rating of the wall and opening. Table 4.4 is derived from the requirements found in *NFPA 5000* for opening protection in fire-rated assemblies.

Because egress is such a critical objective in achieving life safety in a building, the building, fire, and life safety codes often require a level of fire resistance to certain components of the egress system to afford occupants protection against the products of combustion. Exit corridors, exit stairs, horizontal exits, and smoke barriers are required to have a certain level of fire resistance and to have fire-rated opening protection (see Table 4.4 for examples).

Flame Spread Ratings

A great deal of effort is put into building codes to regulate the materials used in constructing the building. In many cases these materials are required to be noncombustible or protected behind a fire-resistive covering or coating. Unfortunately, fire loss history has shown us that the materials most often ignited or involved in early fire growth are not the building's structural elements. Typically, the materials first involved in fire ignition or spread are the contents, furnishings, and interior finishes.

The building, fire, and life safety codes specify that materials used as interior finish on walls and ceilings must meet certain standards to resist fire spread. These specifications are called the _flame spread rating_ or _flame spread index_. The flame spread rating of a material is determined by testing a sample of the material in accordance with NFPA 255, _Standard Method of Test of Surface Burning Characteristics of Building Materials_ [16], and ASTM E-84, _Standard Test Method for Surface Burning Characteristics of Building Materials_ [17]. The tests will give a flame spread rating and smoke development classification in accordance with the information in Table 4.5.

Flame spread ratings provide a relative measure of how materials will allow fire to spread across their surfaces. As references, cement asbestos board has a flame spread rating of 0 (Class A) whereas red oak flooring has a flame spread rating of 100 (Class C). The codes typically require a higher level of flame spread rating on materials used in assembly occupancies or where there are large numbers of persons. The flame spread requirements are more restrictive in protected egress components, such as exit stairways.

Fire and Fuel Loading in Buildings

In past decades, a great deal of research and testing has occurred on calculating the fire loading in a building. Some reasonable assumptions can be made based on the data presently available. The fire loading is expressed as the weight of combustible materials per square unit of floor area. It should include combustible structural elements, combustible contents, and interior finishes (walls, ceilings, and floors). The fire loading takes into account the type and quantity of materials in the various types of occupancies. Table 4.6 and Figure 4.14 show the fire severity for various types of occupancies. In Figure 4.14, the straight lines indicate the length of fire endurance based on amounts of com-

▶ **Table 4.4**   Fire Resistive Assemblies and Opening Protection

Component	Fire Resistance Rating—Walls and Partitions (in hours)	Fire Protection Rating—Fire Doors (in hours)	Fire Protection Rating—Fire Windows (in hours)
Elevator hoistways	2	1½	Not permitted
	1	1	Not permitted
Vertical shafts (including stairways, exit, and refuse chutes)	2	1½	Not permitted
	1	1	Not permitted
Fire walls	4	Two 3 hr	Not permitted
	3	doors	Not permitted
	2	3	Not permitted
		1½	
Fire barrier walls	4	3	Not permitted
	3	3	Not permitted
	2	1½	Not permitted
	1	¾ (45 minute)	¾ (45 minute)
Horizontal exit	2	1½	Not permitted
Corridors, exit access components	1	⅓ (20 minute)	¾ (45 minute)
	½ (30 minute)	⅓ (20 minute)	⅓ (20 minute)
Smoke barrier	1	⅓ (20 minute)	¾ (45 minute)
Smoke partition	½ (30 minute)	⅓ (20 minute)	⅓ (20 minute)

Source: Adapted from *NFPA 5000*™, *Building Construction and Safety Code*™, 2003 edition, Table 8.7.2.

▶ **Table 4.5**   Classification Table—Surface Burning Characteristics of Building Materials

Classification	Flame Spread Range	Smoke Developed
A	0–25	0–450
B	26–75	0–450
C	76–200	0–450

bustibles involved. The curved lines indicate the severity expected for the various occupancies, as shown in Table 4.6. There is no direct relationship between the straight and curved lines, but, for example, 10 lb of combustible per sq ft will produce a 90-minute fire in a C occupancy, and a fire severity following the time-temperature curve C might be expected.

Smoke Spread

Full-scale fire testing and research has given us the tools to predict or model smoke and fire spread in a building. Unfortunately, this technology is used in relatively few building designs.

Control of smoke that develops in a fire is important from both a life safety and property protection standpoint. In most residential buildings and smaller commercial occupancies, smoke is removed by breaking out windows and cutting holes in the roof. In larger factories and warehouses, there are often roof vents that can be opened automatically or manually by fire fighters to aid in smoke removal. In more complex buildings, such as prisons, arenas, atriums, and shopping malls, the building codes often require the installation of a smoke management system to keep the smoke layer high in a building to allow adequate time for egress of the occupants. Automatic sprinklers are considered one of the most effective means of smoke management as they limit fire growth; smaller fires mean less smoke development.

Smoke management is usually used to accomplish one of the following objectives:

- Maintain tenability in the means of egress system
- Confine smoke spread to the area or compartment of fire origin
- Assist firefighting operations by maintaining conditions outside the fire area that allow for firefighting and rescue functions
- Reduce property loss and minimize business interruption

Smoke management is accomplished through two primary methods. The first and more effective method provides an air pressure differential across physical barriers. The second method is to remove smoke by large volume air flows between physical barriers. A third method, often called *smoke dilution*, is used primarily for postfire smoke removal. Smoke dilution is typically not effective for life safety considerations. Smoke dilution is a process of replacing contaminated air with "fresh air." The process usually takes several minutes or hours and several "air changes," which is typically accomplished with the building's heating, ventilating and air-conditioning (HVAC)

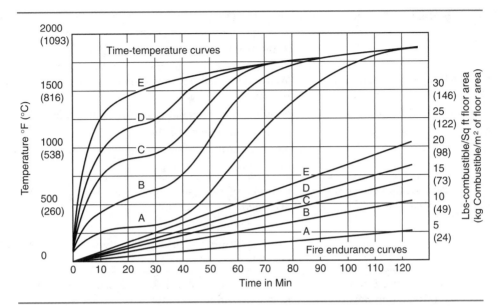

▶ **Figure 4.14**   Possible Classification of Building Contents for Fire Severity and Duration. *(Source: Principles of Fire Protection, NFPA, 1988, Figure 6.3.)*

equipment or with specialized, dedicated smoke removal fans. It is also called smoke exhaust, smoke purging, smoke removal, or smoke extraction.

▶ BUILDING SYSTEMS AND SERVICES

A number of building services and systems can add to the spread of fire or the products of combustion. Elevators can provide a vertical opening the height of the building. They can allow the spread of smoke or heat throughout a building. Elevator shafts also tend to have high air movement caused by the rapid travel of the elevator car; this can exacerbate smoke spread.

Escalators are like an unprotected vertical opening or chimney. It is common to see draft curtains around escalator openings to prevent smoke from migrating vertically. These can be effective when combined with automatic sprinklers. Figure 4.15 shows an example of protection for escalator openings.

HVAC systems can allow smoke spread throughout areas of a building. Where HVAC air ducts penetrate fire walls or fire barrier walls, automatic dampers are often required. Dampers close down all air movement at that point. Dampers can be activated by heat (called fire dampers), by smoke

▶ **Table 4.6** Fire Severity Expected by Occupancy*

Fire Severity	Type of Occupancy
Temperature Curve A (Slight)	Well-arranged office, metal furniture, noncombustible building Welding areas containing slight combustibles Noncombustible power house Noncombustible buildings, slight amount of combustible occupancy
Temperature Curve B (Moderate)	Cotton and waste paper storage (baled) and well-arranged, noncombustible building Paper-making processes, noncombustible building Noncombustible institutional buildings with combustible occupancy
Temperature Curve C (Moderately Severe)	Well-arranged combustible storage, e.g., wooden patterns, noncombustible buildings Machine shop having noncombustible floors
Temperature Curve D (Severe)	Manufacturing areas, combustible products, noncombustible building Congested combustible storage areas, noncombustible building Temperature
Curve E (Standard Fire Exposure-Severe)	Flammable liquids Woodworking areas Office, combustible furniture, and buildings Paper working, printing, etc. Furniture manufacturing and finishing Machine shop having combustible floors

*See Figure 4.14 for the temperature curves identified in this table.
Source: *Principles of Fire Protection*, NFPA, 1988, Table 6.5.

(called smoke dampers), or by both heat and smoke (called combination smoke/fire dampers).

Fire protection systems are often installed in larger buildings. The three most common types of fire protection systems are automatic sprinklers, standpipes, and fire alarm systems. Automatic sprinklers put water on a fire at the earliest stages and have been shown to be effective for life safety, property protection, continuity of operations, and minimizing environmental impact. Fire alarm systems are designed to rapidly notify people of a fire so that egress can begin promptly and are particularly effective in buildings that

▶ **Figure 4.15**   Sprinklers
Around Escalators.
*(Source: NFPA 13, 2002,
Figure A.8.14.4.)*

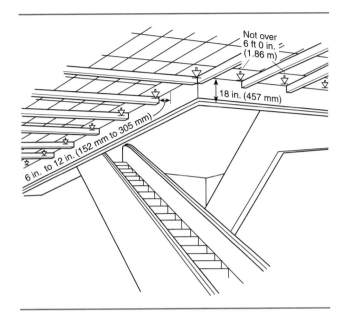

Not over
6 ft 0 in.
(1.86 m)

18 in. (457 mm)

6 in. to 12 in. (152 mm to 305 mm)

have a high life safety risk (Assembly and Educational) and buildings where
people sleep (Institutional and Residential). Standpipe systems are primarily intended for fire department operations in larger and taller buildings.
They allow the fire department to attach hose to a built-in water supply as opposed to having long lengths of hose throughout the building. Figure 4.16
shows a standpipe connection for fire department use. See Chapter 8.

▶ SPECIAL STRUCTURES

High-Rise Buildings

A few types of unique structures bear mentioning as they can present unusual fire suppression challenges. The first type is high-rise buildings. High-
rises are typically defined as 75 feet above the lowest level of fire department
vehicle access to the floor of highest occupiable story. The height of 75 feet
was selected as that is the maximum effective working height of 100-foot fire
department aerial apparatus.

Fires in buildings taller than that cannot be fought from the exterior and
must be attacked from the interior of the building. Fighting high-rise fires is
very resource-intense and dangerous. Since the late 1970s or early 1980s, most
of the building codes have required that high-rise buildings be protected with

▶ **Figure 4.16** Example of a Class I Standpipe Connection for Firefighting Water Supplies Within a Building.

automatic sprinkler systems. However a large number of existing high-rise buildings are not sprinklered as they were built before that requirement went into effect. Some jurisdictions have adopted retroactive sprinkler requirements for some or all buildings built prior to the current requirements.

Over the years the construction method for high-rise buildings has changed. The earliest high-rise buildings had very thick exterior walls to support the weight of the building itself. These buildings tended to be wider at the bottom than they were on the upper stories.

In more recent decades, high-rises were constructed using an interior steel framework or skeleton to support the building from the inside rather than supporting the building from the outside walls. In most modern high-rise buildings, the exterior walls are known as *curtain walls* as they hang or are attached to the building but do not support any of the building load, see Figure 4.17.

One of the concerns of these curtain wall high-rise buildings is that there have been a number of fires that have extended vertically from story to story on the outside of glass curtain wall high-rise buildings.

Large Warehouses

Some recent fire safety concerns have also come to light in buildings that contain high-piled combustible storage or merchandise. These can be large warehouses or what is often called *big box* retail stores that strive for a warehouse appearance. Typically 12-foot is the miniumum storage height considered as high-piled storage. Especially in the retail setting, these buildings contain a mixture of combustible merchandise along with flammable and combustible liquids and hazardous materials. These buildings also have wooden and plastic pallets for the storage and movement of merchandise. These pallets burn very rapidly. Over the years, specialized types of sprinkler protection have been developed for these high-challenge fires.

▶ **Figure 4.17**   Example of an Older High-Rise Building and a Newer Curtain Wall High-Rise Building.

▶ SUMMARY

Fire safety issues should be considered during the design of all buildings, structures, and sites. Considerations should be made for fire department responses and firefighting practices. Automatic fire suppression systems—especially automatic sprinkler systems—provide an effective means of limiting fire spread in a building.

A variety of different materials and methods are used in the construction of buildings. Some of these materials or methods can be substantially damaged in fires. There are five types of building construction classifications ranging from a fire-resistive Type I building of all steel and concrete construction to an unprotected wood frame (Type V) building.

The materials and furnishings within a building can also have an impact on fire growth and development. Certain construction features and automatic sprinkler systems can aid in confining fires or limiting fire growth in a building.

Group Activity

Draw a figure (stick figures are acceptable) illustrating the story of building construction using two chapter objectives.

Review Questions

1. What have been some of the significant improvements in the fire safety aspects of building design over the past several decades?
2. When should decisions be made in the design or planning process for a new building and why?
3. What should always be the primary consideration in the design and construction of any building?
4. What three primary forces impact building construction?
5. On what does the strength and durability of concrete depend?
6. List an advantage and a disadvantage of glass with regard to fire protection.
7. What is a major concern of using plastic products as a building material?
8. What are some building services and systems that can add to the spread of fire or the products of combustion?

Suggested Readings

Brannigan, F. L. *Building Construction for the Fire Service*, 3rd ed. Quincy, MA: National Fire Protection Association, 1992.

Cote, A. E., ed. *Fire Protection Handbook*, 19th ed. Quincy, MA: National Fire Protection Association, 2003.

The SFPE Handbook of Fire Protection Engineering, 3rd ed. Quincy, MA: National Fire Protection Association, 2002.

NFPA 5000™, *Building Construction Code.*™ Quincy, MA: National Fire Protection Association, 2003.

5 Fire Department Structure and Management

Richard Bennett
University of Akron

Learning Objectives

After completing this chapter, the reader should be able to do the following:

- Discuss and describe the scope, purpose, and organizational structure of fire and emergency service
- Describe the organizational structure of a typical fire/emergency service organization
- Define a mission statement and explain the role of the mission statement in setting the tone and direction of the organization
- List primary objectives for a mission statement for a fire/emergency services organization
- Describe the traditional functions of management and how they apply to today's fire service
- Describe six principles of effective management command systems
- Describe and list components of the emergency management cycle
- Describe the concept of customer service in the fire service
- Identify at least two future trends for fire/emergency services management

Any effective organization must have a solid organizational structure that drives the organization toward accomplishing its stated goals. The more than 30,000 fire departments in the United States rely on these organizational structures in protecting citizens and property. Much of this structure is derived from a history rich in tradition, as well as tried and proven practices of the past. According to NFPA statistics, in 2001, every 18 seconds a fire department responded to a fire somewhere in the United States. There are just over one million fire fighters in our country, three-fourths of whom are volunteer fire fighters at 48,500 fire stations. It costs more than $10 billion dollars each year for the operation of paid local fire departments. Add to that the cost of daily operations that devoted volunteer departments contribute through hard work and sacrifice, and we can begin to grasp the importance of public fire protection. When vehicles and people respond to local or regional emergencies, the structure supporting their efforts is not readily visible. This chapter explores the effective management and organizational structure of fire departments.

▶ OVERVIEW

Organizational Structure

Fire departments are organized around accomplishing the mission of saving lives and protecting property. The basic paramilitary organizational structure, ranks, and uniforms were borrowed from the military structure of our armed forces dating back to the Civil War. Figure 5.1 depicts the basic structure of a small fire department. Note that while small, medium, and large departmental structures differ in size and scope, they all perform the same basic functions, as shown in Figures 5.2 and 5.3.

Fire departments may provide fire suppression, fire prevention education, code enforcement, and emergency medical and rescue response services for the citizens. Fire department divisions include both line and staff functions. Line functions are often identified as those fire fighters involved in emergency response duties, such as responding to fires and other emergencies, including medical rescue and hazardous materials incidents.

▶ **Figure 5.1** Organizational Chart for a Small Fire Department. *(Source: Fire Protection Handbook, NFPA, 2003, Figure 7.1.2.)*

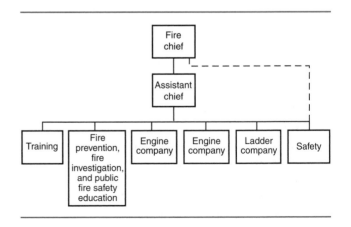

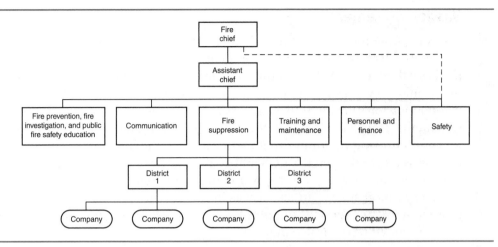

▶ **Figure 5.2** Organizational Chart for a Medium-Sized Fire Department. *(Source: Fire Protection Handbook, NFPA, 2003, Figure 7.1.3.)*

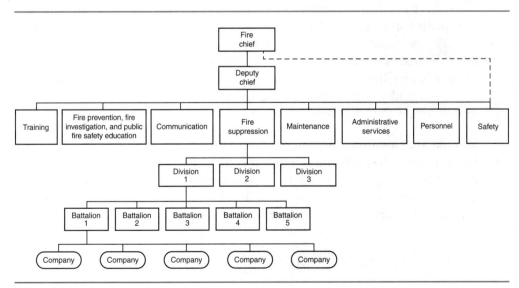

▶ **Figure 5.3** Organizational Chart for a Large Fire Department. *(Source: Fire Protection Handbook, NFPA, 2003, Figure 7.1.4.)*

Mission Statement

A mission statement is a written declaration of what the particular fire department will do in accomplishing its goals of protecting the residents in the territory they cover. Although the specific mission statements of fire departments may vary, they all strive to save lives and to protect property. It is also a good idea for individual fire fighters to create their own personal mission statements. Doing this helps define a fire fighter's values and principles, as well as identifying the things that matter most to the individual. Optimal efficiency is reached when the personal goals of individual members of the fire department are in line with the overall departmental mission.

The mission statement also assists in formalizing primary objectives for a fire/emergency services organization. These objectives may include, but are not limited to the following:

- Establishing organizational identity
- Determining the abiding purpose
- Setting a philosophy
- Determining core values
- Establishing ethical standards

Sample Statement 1.   One example of a mission statement comes from the German Township Volunteer Fire Department in Indiana: "The German Township Volunteer Fire Department is committed to providing the highest quality emergency services assistance to our citizens. Our mission is to reduce the loss of life and reduction of property loss through progressive community leadership, public education, fire suppression, hazardous materials response, and nontransport prehospital care within our legal jurisdiction and others as required by mutual aid."

Sample Statement 2.   Another example comes from the Plano, Texas, Fire Department: "To protect and enhance the quality of life in Plano through a comprehensive program of services directed toward public education, prevention, and control in the areas of fire, rescue, medical emergencies, hazardous materials incidents, and disasters."

▶ FIRE SUPPRESSION STRUCTURE

The mission of fire suppression is generally common among all fire departments. Suppression includes the major goal or mission of saving lives and protecting property. To accomplish this goal, fire departments are organized into smaller tactical units or companies. A company consists of fire fighters who operate one or more vehicles or apparatus under the direct supervision of a company officer. A fire station may house either one or more companies. The companies based out of a station are generally classified as either engine companies or ladder companies. There may also be rescue companies or other special apparatus such as emergency medical service (EMS) ambulances. Specifics of the types and functions of apparatus are discussed in Chapter 6.

Engine Company

The engine company is the most common type of company found in fire departments. Engine company apparatus carries a fire pump, hose, and other appliances like nozzles specifically designed to apply water to control or extinguish a fire. Additional equipment may be carried such as medical supplies and other things to assist fire fighters in providing basic life support duties. Other companies generally respond to assist the operations of the engine company.

Ladder Company

Ladder companies, sometimes referred to as truck companies, may respond to fires to perform supporting functions such as search and rescue. An aerial ladder or elevating platform distinguishes ladder company vehicles. Ground ladders are also necessary pieces of equipment carried on ladder trucks. Tools to be used in forcible entry, ventilation, salvage, and overhaul are also carried on these ladder trucks.

EMS Company

In many areas, the number of EMS runs greatly exceeds the number of fire calls companies respond to in any given year. EMS responds to medical emergencies acting as an extension of the local hospital system. Equipment is carried on the apparatus to assist the paramedics or emergency medical technicians (EMTs) in providing the advanced or basic life support functions necessary for the specific emergency they are responding to.

▶ STAFF FUNCTIONS STRUCTURE

While fire department line personnel aid in emergency response efforts, many individuals also provide additional services such as fire investigation, fire and life safety education, training, and other administrative staff.

Fire Investigation

The purpose of the fire investigation arm of local departments is determining the cause and origin of fires, in accordance with the state requirements. These fire fighters also investigate cases of arson (incendiary fires), and work with the state investigation agency and local law enforcement officials to successfully solve these cases. These fire fighters are sometimes trained as police officers, and therefore have the power to arrest suspects if necessary. Chapter 9 explores the subject of fire investigation in greater depth.

Fire and Life Safety Education

Many members of the fire department provide fire prevention education to the public, to help reduce the risk of fire. These programs reach out to business, industry, institutions, and schools, and target many age groups. Chapter 7 covers this area in greater depth.

Training

The training office coordinates and provides training and certification to fire department personnel. Most tasks performed by fire personnel are very technical and require a lot of hands-on skill. The training department works in concert with company officers to ensure that personnel are capable of performing the necessary job functions. Training offices also keep detailed records of personnel training that has been completed along with accounts of when recertification training is needed. This training may be done directly by the department or fire personnel may be sent to be trained or recertified by a local community college or other training entity. Often EMTs and paramedics are sent to a local hospital for training and recertifications.

Administration

Department staff coordinate and oversee a wide variety of staff functions and services provided by the other department components. These functions may include repairing apparatus or purchasing supplies and uniforms. Generally, administrative staff work a 40-hour week. They usually handle the nonemergency functions of local fire departments. Administration and the functions they perform are not visible on the fireground or at the scene of an emergency, but if these support functions are not done, those operating in emergency situations will not have the necessary support to do their jobs. Suppose for example, that fire fighters at a house fire suddenly realize that there are not any spare self-contained breathing apparatus (SCBA) air bottles at the scene. If the administrative function of refilling air bottles has been neglected, then those at the emergency scene will not have the necessary support. Conversely, when air bottles are readily available, the necessary repair and maintenance of this system is often forgotten by fire fighters.

▶ MANAGEMENT FUNCTIONS

Management refers to the tasks that must be performed to oversee the continued operations of an organization. These functions should be aligned with that organization's mission statement. Leadership is the function of providing direction and guidance to people as the organization goes about the business of accomplishing the organization's mission. Leadership is recognized and given legitimacy by those who are being led. A person can have management responsibility and not be a very effective leader. A good leader may not be promoted into management ranks, but he or she may be recognized by their peers and looked to when group decisions need to be made. In short, managers are not necessarily good leaders, and leaders are not always promoted into management. The obviously ideal situation is that a person is not only a good manager, but also a recognized and effective leader. See the section on leadership effectiveness.

The job of managers and leaders in any effective organization is to make sure that the resources of the organization are used to accomplish the mission and objectives of that organization. The most valuable of these resources is the people employed by the organization.

In the fire and emergency services, management represents those who oversee the day-to-day business done in accomplishing the organization's mission. Managers should help workers be more effective in their daily job duties. Effective management involves performing the following duties: planning, organizing, commanding, coordinating, and controlling.

Planning

Good management begins with good planning. There is wisdom to the old adage "those who fail to plan, plan to fail." Planning is generally divided into short-, medium-, and long-range goals. Short-range goals consist of things to be accomplished in up to one year. Medium-range goals set the target at 1 to 5 years. Long-range goals may range from 5 to 20 years. Can you imagine a football game without goal lines? How could you tell if a team scored? The lives of individuals and organizations mirror gaining first downs on the way to scoring touchdowns. The establishment of clear, measurable goals is a necessary part of good planning. Planning is a very valuable tool for the fire service. If a city is growing in the number of homes being built or has plans for annexing in other areas, it will be necessary to plan for additional resources that may be needed in the future. Many fire departments may see that a number of fire fighters are eligible to retire in the next 3 to 5 years. Plans should be made for future recruitment drives and any other activities that may be necessary if there are mass retirements.

Organizing

Once a plan has been established, the resources and activities of the fire department must be organized. Organizing breaks down the tasks into more manageable activities. For example, we have already identified the mission of the fire service as saving lives and protecting property. The planning phase began with identifying the short-, medium-, and long-range goals. Now we need to organize the fire fighters into companies. These companies need other resources such as training and equipment to do their jobs effectively.

Commanding

Once the personnel and equipment have been organized, someone must be responsible for leading, motivating, and encouraging teamwork. This is accomplished in emergency situations by the use of the Incident Command System (ICS). The specifics of the Incident Command System are covered in Chapter 10.

Coordinating

Available resources must be coordinated to ensure their effective and efficient use. Money, equipment, time, and personnel are the resources that must be managed. Budgetary restraints of recent years have forced managers to do more with less. An example of this is the need to coordinate training activities with public education demands. If a company is due to practice extrication from automobiles and a levy is on the ballot for emergency equipment or other departmental needs, it might be a good idea to coordinate some publicity about the training exercise with public service announcements supporting the levy.

Controlling

Controlling involves the monitoring of goals in order to determine if progress is being made, measuring the effectiveness of strategies and initiatives, and allowing for improvements based on best practices. All of these functions are closely related to the budgetary process.

Fiscal Management/Budgeting

The mission of any local fire department cannot be accomplished without money. Budgeting is a necessary tool in all of the phases of the management cycle. Two types of budgets are generally used by fire departments: capital budgets and operating budgets.

Capital Budgets.   Capital budgets detail large expenditures, which generally are expected to last more than one year. Equipment such as apparatus, fire stations, and fire hose are examples of items usually included in capital budgets.

Operating Budgets.   A fire department's operating budget represents the day-to-day expenditures necessary to accomplish the mission of saving lives and protecting property. The largest item in this budget in paid fire departments pays for the salary and benefits of personnel.

Measuring Productivity

Most communities spell fire safety, "big red truck." Because of this paradigm, the basic measure of productivity has been response times and the number of runs responded to annually. While the number of fire calls responded to and the amount of time that lapses between a call for help and engines are on the scene and have extinguished the fire is a good measure, the nature of fire departments is changing. In many communities, the number of fire calls has decreased. In fact, the number of EMS runs may constitute approximately 80 percent of all emergency runs. With the increasing need for hazardous materials teams and other prevailing issues such as homeland security, change is inevitable. The American fire service is becoming less of a fire response organization and more of an emergency response organization.

▶ BASIC PRINCIPLES OF COMMAND SYSTEMS

There are common guiding principles for fire department organizations. These principles may not be used in the same way in every fire department, but they should be seen in some form, regardless of jurisdiction. These principles are the following:

- Chain of Command
- Unity of Command
- Span of Control
- Division of Labor

- Delegation of Authority
- Discipline

Chain of Command

There must be a link between the levels of a fire department's organizational structure. It is critical for efficiency and effectiveness that the levels do not operate independently. All decisions made at the top of the department must be communicated and carried out uniformly. Rookie or probationary fire fighters generally report directly to their company officer. As the front-line supervisor, the company officer's responsibility is to address any concerns of fire fighters who report directly to him or her. If it is necessary, the company officer should be the one to take issues to higher levels. An exception to this principle may be a fire fighter who is on special assignment and may report directly to a department head. In cases where the officer is the object of the fire fighter's complaint, departmental policy should be followed regarding who is next in the chain.

Unity of Command

This principle says that each person in the fire organization has only one supervisor. In order for this principle to work, everyone in the department must have a clear understanding of both who their supervisor is and who they supervise. If this principle is not followed, fire fighters will receive conflicting orders. Unfortunately there is a very real possibility that this principle is violated on the scene of an emergency. A fire company may arrive at the scene of a multiple vehicle accident. The fire fighter's company officer sends her back to the truck for another tool. On the way back to the truck, another company officer stops the fire fighter and orders her to assist the incoming EMS vehicle with patient care. The confused fire fighter is spotted by the chief, standing motionless and ordered to retrieve a radio from the command vehicle. What is this fire fighter supposed to do? Similar situations have actually occurred on firegrounds across this country. Following the principle of unity of command will eliminate these types of situations.

Span of Control

There is a limit to the number of personnel any supervisor can effectively supervise. The ideal number to manage in an emergency response organization is three to seven employees. Figure 5.4 shows an effective span of control. Note that each company reporting to the battalion chief is also

▶ **Figure 5.4**   Effective Span of Control.

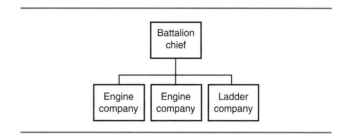

made up of three to seven fire fighters. Figure 5.5 is an example of an ineffective span of control. Qualified and experienced staff require much less supervision. Well-motivated staff require much less supervision.

Division of Labor

The principle of Division of Labor means that work must be divided into specific areas to prevent duplication of efforts and to ensure that adequate resources are allocated to complete assigned tasks. The organization of fire departments into fire companies is done to satisfy this principle of division of labor.

Delegation of Authority

Managers or supervisors are seldom involved in routine work assignments. The higher fire officers go in their department, the less hands-on work may be done. For this reason, it is often necessary to delegate authority to subordinates. Delegation of authority is guided by several key concepts: exception principle, scalar chain of command, decentralization, and parity principle.

Exception Principle.   Someone must be in charge. A person higher in the organization handles exceptions to the usual. The most exceptional, rare, or

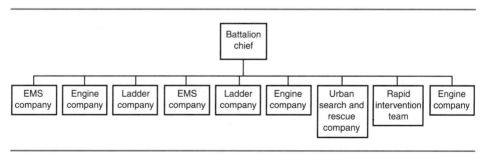

▶ **Figure 5.5**   Ineffective Span of Control.

unusual decisions end up at the top management level because no one lower in the organization has the authority to handle them.

Scalar Chain of Command.   The exception principle functions in concert with the concept of scalar chain of command—that is, the formal distribution of organizational authority is organized. The higher one is in an organization, the more authority one has.

Decentralization.   Decisions are to be pushed down to the lowest feasible level in the organization. The organizational structure goal is to have working fire fighters rather than managed fire fighters. An example of this is a company's officers making strategic decisions at the scene of an emergency.

Parity Principle.   Delegated authority must equal responsibility. With responsibility for a job must go the authority to accomplish the job.

Discipline

The concept of discipline is often viewed from a negative perspective. Discipline should be used to bring behaviors or actions that are counterproductive to the department's mission in line. When this is done effectively, the result is that individuals become more effective in accomplishing their duties. Effective organizations are made up of effective individuals working together as a team. There are several steps in the cycle of discipline. These steps are often referred to as the *progressive system* of discipline, as shown in Figure 5.6.

Verbal Warning.   The initial step in the progressive system of discipline is usually a verbal warning. A clear and concise explanation of what the fire fighter is doing wrong along with an explanation of the desired behavior

▶ **Figure 5.6**   Progressive System of Discipline.

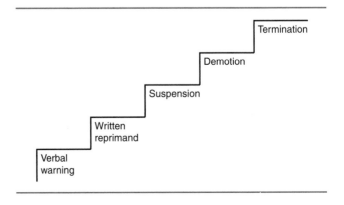

may be enough to get the individual on the right track. If the behavior that needs to be changed is a matter of the fire fighter not knowing how to perform a task, this issue may be training, not discipline.

Written Reprimand.   If the issue is not training and the behavior is not corrected, the next step is to give a written reprimand. Putting the situation in writing signifies that this matter is more than casual. It allows the supervisor to document the issue that needs to be addressed.

Suspension.   More serious infractions will require suspension of the fire fighter. Offenses for which a fire fighter can be suspended should be defined in rules and regulations of the department. Great care must be taken to administer suspensions fairly. Departments have caused themselves legal problems when fire fighters were suspended for violations for which other fire fighters were only given warnings.

Demotion.   Demotions are one of the more severe steps in the system of discipline. These are used when it is clearly documented that an officer cannot perform the duties necessary to be effective in that rank. Demotion should never happen for budgetary reasons. If the issue of performance can be solved by training, then training should be given to the officer. Unfortunately, many fire officers do not receive training prior to being promoted. The new professional development model addresses the need for training and education prior to promotion. That model is discussed later in this chapter.

Termination.   The last resort in the progressive system of discipline is termination. Violation of rules that require the employee to be fired (or the volunteer dismissed) should be spelled out clearly in departmental policy. There will be times when a fire fighter is not capable of effectively performing his job functions. To continue to allow the fire fighter or officer to operate in emergency situations may endanger citizens and other emergency response workers. In these situations, termination may be the only possible answer. Proper documentation must be done of the rationale for termination.

▶ MANAGEMENT AND MOTIVATION THEORIES

Scientific Management

Fredrick Taylor is known as the father of scientific management. He is given credit for creating principles that can assist any organization in operating efficiently. Taylor believed that the interests of management and workers were the same, therefore managers should not antagonize workers. The key was that management and the employees being supervised should work together, thereby enhancing the efficiency of work. Much of Taylor's work forms the basis for current theories of management.

Theories of Motivation

It is difficult to understand why one fire fighter needs intervention with the progressive system of discipline, while another fire fighter is highly motivated to work and study in preparation for promotional opportunities. Management theory is an attempt to understand the best methods and practices managers can use to assist employees in being effective members of their department.

Maslow's Hierarchy of Needs. Perhaps the best known pioneer in the area of motivation is Abraham Maslow. Maslow suggested that the primary motivator of people is their needs. He constructed a hierarchy of needs as illustrated in Figure 5.7. The levels of needs begin with the basic need for food, shelter, and clothing. Level two is the need for safety and security. The next level is the need for acceptance by others, followed closely by the need for esteem. The top level is self-actualization. It is at this level that fire fighters are concerned about their personal or professional development. An individual must satisfy needs of the lower level before addressing higher level needs. A fire officer must recognize the level of an individual's operation in order to determine the type of incentive or motivation needed.

McGregor's Theory X or Theory Y Managers. Douglas McGregor suggests that the way that fire officers manage fire fighters is based on whether they

▶ **Figure 5.7** Maslow's Hierarchy of Needs.

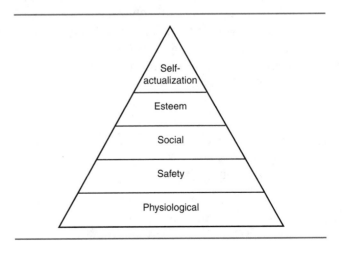

are Theory X or Theory Y managers. Theory X managers believe that fire fighters dislike their work and will therefore avoid work if possible. This type of manager looks for opportunities to control the behavior of workers by coercing or threatening them. Theory Y managers believe that work is as natural as rest or play. Because of this belief, efforts are made to create and maintain a supportive atmosphere. This work setting encourages fire fighters to seek out responsibilities and activities where they can be effective employees.

▶ HUMAN RESOURCE MANAGEMENT

Personnel Management

Recruiting fire fighters is generally not the sole responsibility of municipal fire departments. Notifying the department of existing vacancies and requesting the number of fire fighters needed to fill those vacancies is how fire departments begin the recruitment process. To assist with recruitment, the local fire department should do the following:

- Recommend recruitment standards based on the knowledge, skills, and abilities appropriate for entry level personnel
- Provide the training necessary for personnel to perform their assigned duties
- Certify that personnel are trained to the level of technical competence to protect and serve citizens

The training process does not stop once personnel have been hired. Training and professional development should continue throughout a fire fighter's career. Properly administered performance appraisals can be an effective tool in assisting personnel in personal and professional development. This process should be formal and systematic. The goal is to accurately assess how well employees perform their jobs in accordance with established standards. These standards should be equally applied to all personnel to avoid legal challenges. There has been a recent move to establish national standards for the fire service.

Promotional Practices

Fire fighters may seek promotional opportunities for a number of reasons. The possibility of having a greater impact on the department, the need for more challenging work, or an increase in pay may be some of the reasons fire fighters want to be promoted. This competitive process is sometimes decided by a fraction of a point, leading to the careful scrutiny of this process by interested parties. The biggest challenge of the fire department is to make sure that all of the components of the promotional process are valid, fair, and job related. Generally, written exams are a part of this process, to determine if the candidates for promotion have the necessary knowledge, skills, and abilities to perform in the position for which they are seeking promotion. This should not be a test of the general knowledge necessary to be a fire fighter. Those skills should be tested during the probationary period and evaluated in performance appraisals. Testing for promotion should include evaluation of the individual's skills at being a supervisor or a manager.

Assessment Centers

Assessment centers are sometimes used to determine how an individual might respond in a management position. Assessment centers generally include the following four components:

1. An emergency scene assessment
2. An in-basket assessment (ranking assignment given)
3. A training presentation
4. A performance counseling session

Some fire departments require college degrees for promotion to officer positions. The underlying assumption is that the academic curriculum prepares fire fighters to become officers. Degree requirements work better with practical experience that can only be gained from real-world situations. For fire fighters who are not near a community college or university, distance learning programs are increasingly available.

▶ CUSTOMER SERVICE

There is an interdependence between the members of the local fire department and its community. Without the residents of that community, there would not be a need for local fire departments. It is vital for fire departments to understand the concept of customer service. The residents of that community should be treated as valued customers.

Customer service is a common concept in the private sector. There businesses compete to get prospective customers to choose and buy their product. In the public sector, fire departments do not have competition. That is not a good reason to take residents of the communities we protect for granted. Local citizens pay fire fighters' salaries and therefore should be treated as valued customers. Every contact members of the fire department make with a resident is an opportunity to show how much that customer is appreciated. One area customer service can be practiced is in loss control. Attempts to reduce property loss and damage resulting from fires and other emergencies represent the best customer service local fire departments can give. Customer service is as simple as having someone looking for a business in your district follow you to the business or drawing them a map and directions on how to get there. Embracing the concept of customer service in the fire service was advocated by Alan Brunacini, chief of the Phoenix, Arizona, Fire Department. See *Essentials of Fire Department Customer Service* for more information [1]. He suggests the following basic concepts for good customer service[1]:

1. *Always be nice.*   Treat people with respect, and show patience and consideration. What people will remember and cherish is your performance and attitude. Did you solve their problem? Did you treat them with respect?

2. *Treat everyone as a customer.*   Many times the problem is simple and the solution is straightforward for us, but for the patient and the family, things may be very complicated. Factor their emotional needs into the customer-service process.

3. *Remember that our customers are everywhere.*   They watch us work with interest. They wander into fire stations looking for assistance and directions. They may be part of a group on a visit. Show them respect too.

4. *Do not forget our own.*   While we are showing all this kindness and consideration to strangers, keep in mind those we work with too. Show mechanics and payroll clerks the same kind of respect as well. They may return the favor sometime when you need help.

5. *Keep working on improvement.*   We have to be constantly working to improve just to be able to respond in this not-so-perfect but constantly changing world.

Identifying the Customer[2]

The notions of "customer" and "customer service" are used in this chapter to examine the relationship between fire department officials and the folks that they work with, regardless of who those folks are or who they represent.

---

[1]This text has been provided by Clinton Smoke, the author of Chapter 1.
[2]The remainder of the material in this customer service section was provided by Charles Kime, the author of Chapter 7.

It is extremely important for fire department officials to define who the customer is in each relationship, as this often sets the tone for what needs to be done and how it should be accomplished. The following discussion is intended to sort out some of these rather sticky issues. The list of customers is almost inexhaustible yet some of the main players can be identified. They include citizens at-large; architects, developers, builders, and contractors; business owners and managers; tenants and landlords; elected officials and the city manager's office, fire fighters, the department and the union; and other city, county, state, and federal agencies.

This list is not meant to be inclusive, but it represents a good number of the different types of people with whom fire department officials must work. The list also provides a visual picture of the heterogeneity of different points of view that these customers most likely will represent, all of which can make for a very interesting career. The Koffron Customer Service Tetrahedron shows the tangential relationships among these critical players. Barbara Koffron, Fire Marshal for the Phoenix, Arizona, fire department, argues that, "Like the tetrahedron of fire, the customer service tetrahedron, shows how each group is dependent on the other to achieve good customer relationships" and reminds fire department members of the need to work with all of the players.

Identifying Customer Expectations

Fire department officials should expect that each customer might have a different set of expectations on any given project and it is the job of the fire department officials to sort this out. For example, when it comes to fire prevention activities, fire department officials should be aware that the developer typically wants to fast track a project and eliminate as much time and cost in the review process as possible. The fire fighter wants to make sure that firefighting access is provided and built-in fire protection devices are accessible. And the elected official wants the project to add to the community's tax base and provide jobs, while the manager's office just wants things to go smoothly; interpretation—"Keep everybody happy with no complaints." As projects progress through the system, different problems arise and solutions to these problems must address all of these conflicting and competing expectations, and do it at the same time.

Answering complaints raises even bigger red flags for an astute fire marshal. Is the complainant the customer? Is the business owner the customer? Is the contractor the customer? Is the community the customer? This often requires a bit of detective work before the best solution is selected. The following examples should illuminate this problem and expand on how different some of the expectations are among those folks encountered during fire department activities.

Basic Customer Service Principles

Understanding who the customer is and what he or she wants does not have to be an Alfred Hitchcock thriller. Brunacini [1] identifies eight essential elements of good customer service that any fire department can apply. However, at least three basic customer service principles should be applied to all fire department activities. These principles revolve around respect, time, and choice and are imbedded in Brunacini's [1] essential elements. Fire department customers expect service, and the service they expect is not different from the service they receive at their favorite restaurant, their hair stylist, or their car dealer.

Respect. Customers want to be treated with respect; respect as people, respect as professionals, and respect as equals. Respect is being nice to the customer, listening to the customer, and asking probing questions that can help gain better insights into the customer's problem. Respect is best displayed by behavior more than words, although what is said and the inflections used are important. Respect is not talking down to the customer or ignoring someone because of some outward appearance, whether it is the way they dress, the accent that they speak with, or their gender. Using the Golden Rule often sums up respect; "Do unto others as you would have them do unto you."

Time. They want a reasonable amount of time to get things in order and if possible they like to be given a chance to demonstrate that they care. Customers prefer to be informed (translation: educated) and then be given time to fix things before someone does a formal inspection and writes a summons for each violation. It is interesting to note that when a fire department official asks the customer how long he or she needs to do the work required to make a correction, the customer, more often than not, will ask for much less time than the fire department official was going to set. There are always those that want a millennium but they are typically outliers.

Choice. Customers also like to choose for themselves. Choice is as American as apple pie and often the fire department official that gets the customer involved in choosing the solution is far more successful than the official who simply tells the customer what he or she has to do.

Outcomes of Good Customer Service

Customers who comply with fire and life safety codes and try their best to be fire safe with a happy heart are more likely to keep a watchful eye out for unsafe situations than those who are brow beaten into compliance. Often, those that have to be forced to comply only do so as long as they are being watched, and no fire department in America has the resources to be everywhere all of the time. Only with the cooperation of others can acceptable levels of fire safety be achieved and maintained over time. Good customer service creates an environment for a better understanding through education and can establish a symbiotic relationship where all parties work together toward finding creative solutions to achieve the desired outcomes. Finding creative solutions or alternatives is the engineering phase and relies on good customer relations when it is needed.

▶ TRENDS

It can no longer be business as usual for the fire service. Fire service managers are not the only members of their organizations who need to understand the dynamics of human behavior. Every member of the organization can benefit from the interpersonal skill areas mentioned later in this section. The mastery of these areas is just as important to fire fighters as the technical skills that are relied on so heavily each day. The result will be better trained individuals who see themselves as valued parts of effective organizations.

Fire and emergency service leaders must confront the need for change as they move toward the future. The tragic events of September 11, 2001, and the resulting emphasis on homeland security is evidence of the fact that change is necessary. These changes are likely to involve a greater emphasis on emergency management and professional development issues. A deeper understanding of topics like emotional intelligence and interpersonal skills will be important as they will be used more in the selection of fire fighters and promotion of fire officers.

Emergency Management Cycle

Recent evaluations of fire officers' duties indicate the need for fire chiefs to be all-risk managers. Regardless of whether a community is facing a terrorist incident or a natural disaster such as a flood or a severe storm, the fire chief will be expected to provide assistance to whatever unified management team is used to deal with these situations. To that end, fire officers need to be aware of the emergency management cycle.

Mitigation.   The emergency management process begins and ends with mitigation. Mitigation is an ongoing attempt to limit or eliminate the effects of disasters. Activities that prevent an emergency or lessen the impact of unavoidable emergencies will become more of a focus for future fire and emergency service leaders. The fire service participates in mitigation when providing fire and life safety education. If people adhere to the safety measures suggested by fire personnel, emergencies can be prevented.

Preparedness.   The next integral part of the emergency management cycle is preparedness. This phase coincides with the planning cycle discussed earlier under management functions. Fire chiefs have been involved in planning for response to disasters. Terrorist threats and natural disasters make planning for new situations now necessary. Hazardous materials teams that go through mock disasters are performing part of the preparedness function of emergency management.

Response.   Fire and emergency services have always been a part of response. Efforts to minimize the hazards created by an emergency are not new to the fire service. What may be new in some localities is an emphasis on interagency cooperation. As stated earlier, the mission or goal of the first responder is to save lives, minimize property damage, and enhance the beginning of recovery from the incident.

Recovery.   The last phase is recovery. Efforts traditionally done in salvage and overhauling fit into this category. Recovery returns the infrastructure and other systems to minimum operating standards and guides long-term efforts designed to return life to normal or improved levels after a disaster. The fire company that uses tarps to cover and protect undamaged contents from water damage is assisting in the recovery phase.

Emotional Intelligence

Research continues to be done in the area of employee motivation. Some of the more fascinating research is being done on emotional intelligence. Initially it was believed that what separated those who were successful and or promoted in their jobs was their intelligence quotient (IQ). It is now believed that IQ does not play as significant a role as we originally thought with regard to success at work as compared to the worker's emotional intelligence. Emotional intelligence speaks to fire fighters' ability to get along with others and to be able to know themselves well enough to control their emotional status while achieving goals. The ability to delay gratification is a key component of emotional intelligence, which is closely tied to interpersonal skills.

Interpersonal Skills

Interpersonal skills are the basic core of abilities that aid individuals in getting along with others in their environment whether at home or at work. These skills include conflict resolution, giving and receiving feedback, appreciation of cultural diversity, and personal leadership effectiveness.

Conflict Resolution

Conflict is unavoidable in groups that work together. Effective teams are made up of individuals who are able to move past conflict toward collaboration. Fire and safety organizations expect employees at all levels to work together, sometimes across functions. In the future, these groups will be expected more often to make decisions that were formerly the responsibility of management. This new paradigm will make evident the need for fire fighters at every level to have the skills to deal successfully with conflict.

Feedback

Situations change so frequently on the fireground that feedback is necessary to correct problems or change strategies. But when there is a nonemergency situation, there is also a need for the skills of giving and receiving feedback. In emergency situations, the commands given sometimes are communicated harshly. But if this same tone is carried over to the everyday engine house communication, a problem will likely result. All of the members of the organization should learn guidelines for both giving and receiving feedback in positive and constructive ways. Doing so will aid in getting information to the right person at the right time and identifying problems before they get out of hand. The end result will be the building of stronger work relationships. These relationships will, in turn, encourage mutual respect throughout the organization.

Cultural Diversity

Census data indicates America is becoming more multicultural. According to the U.S. Department of Labor, the percentage of women, older workers, Asians, and Hispanics will continue to increase through the year 2005. It will be absolutely imperative that those who serve this diverse population be trained to acknowledge, appreciate, and affirm America's historically underrepresented groups. Members of these cultures will be tied to the fire service in the future, whether it is in providing protective services or as members of fire departments.

Leadership Training

Effective organizations are made up of effective people working together. The extent to which individual members of groups feel empowered and recognize their importance to the success of their company has everything to do with the individuals' view of themselves as effective leaders. Every member of an organization should be trained in leadership effectiveness. It is then up to the individual whether they choose to use these skills to pursue promotion or to help their current work team function better.

Professional Development

The fire service is beginning to create a professional development model similar to other professions. The purpose for this model is to establish a career track from entry level fire fighter to fire chief, which includes educational requirements beyond the technical skills normally associated with this profession. These efforts have culminated in model course curriculum and professional development models. See Chapter 12 for a discussion on the Fire and Emergency Services Professional Development Model.

SUMMARY

In 2002, public fire departments responded to more than 1,708,000 fires. The vast majority of these fire departments are small volunteer departments. The organization and management of fire and emergency services are similar, regardless of their size. These structures support the emergency responses that protect and serve our citizens daily. Future trends and changes will test the organizational structures of emergency response organizations. The men and women of America's fire service are well able to meet these challenges while providing quality customer service.

Group Activity

Individually decide which management principle you believe is the most critical for fire departments (i.e., chain of command, span of control, etc.). Discuss your choice with the members of your work or study group. Which principle did most people agree on? Why? Which principle did the least number of people agree on? Why?

Review Questions

1. Suppression includes the major goal or mission of saving lives and protecting property. To accomplish this goal, fire departments are organized into smaller _____ units or _____.

2. What kinds of emergency calls may exceed fire calls?

3. What does good management begin with?

4. What is the difference between capital budgets and operating budgets?

5. What are the common guiding principles for fire department organizations?

6. There are several steps in the cycle of discipline. These steps are often referred to as the _____ _____ of discipline.

7. What did Maslow suggest was the primary motivator of people?

8. What are sometimes used to determine how an individual may respond in a management position?

9. What are the four components of the emergency management cycle?

10. What are interpersonal skills?

Suggested Readings

Edwards, Steven T. *Fire Service Personnel Management.* Englewood Cliffs, NJ: Prentice Hall, 2000.

Grant, Nancy, and David Hoover. *Fire Service Administration.* Quincy, MA: National Fire Protection Association, 1993.

Maslow, Abraham Harold, Robert Frager, and James Fadiman. *Motivation and Personality.* New York: Addison-Wesley, 1987.

McGregor, Douglas, and Warren G. Bennis. *The Human Side of Enterprise: 25th Anniversary Printing.* New York: McGraw-Hill/Irwin, 1985.

Taylor, Frederick W. *Scientific Management.* Mineola, NY: Dover Publications Inc., 1911.

6 Fire Department Facilities and Equipment

Robert Tutterow
Charlotte, North Carolina Fire Department

Learning Objectives

After completing this chapter, the reader should be able to do the following:

- Describe the common types of fire and emergency facilities, equipment, and apparatus
- List the basic fire service facilities
- Describe the functions of each basic type of fire service facility
- Describe the personal protective equipment worn by fire fighters
- Describe the basic hand and power tools and equipment used by fire fighters
- Describe the types and functions of apparatus used by the fire service
- Describe the role of communications technology in fire and emergency services operations

It is practically impossible to think about a fire department without visual images of its equipment or facilities. This has been true through the ages. The uniqueness of fire equipment and facilities creates an attraction to both children and adults. A fire station is a beacon of safety in neighborhoods throughout the country. And a child's adoration of a fire truck creates positive images of a fire department that last a lifetime. This chapter provides an overview of the equipment used by fire departments to provide service to their communities. Included in this overview are facilities, personal protective equipment, hand and power tools, apparatus, and communications.

FACILITIES

Fire stations and other fire service facilities can be very simple or very complex depending on the functions of the facility. Facilities can range from being a simple apparatus bay in a rural area to a campuslike complex in a suburban or urban area. There are no minimum standards or requirements, aside from the appropriate building codes, for the design and construction of fire department facilities.

Fire Stations

Fire stations are in almost every community in the world and larger communities have several fire stations. They are a major investment in the life safety of a community. The European communities and other parts of the world tend to have fewer stations per square mile than North American communities. Their fire stations tend to be larger with more apparatus and fire fighters assigned to a station than in North America. See Figure 6.1.

Fire stations are located based on two criteria. The first criterion is time and distance, that is, what are the response times and what is the distance? Many insurance companies base their rates on the distance a structure is from a fire station and water supply. And some insurance companies base their rates on past losses so time becomes more critical. A common response

▶ **Figure 6.1**   Charlotte, North Carolina, Fire Station No. 33. *(Courtesy of Meg Sorber.)*

benchmark is for a department to respond to 90 percent of emergency calls within 8 minutes. NFPA 1710, *Standard for the Organization and Deployment of Fire Suppression Operations, Emergency Medical Operations, and Special Operations to the Public by Career Fire Departments* [1], and NFPA 1720, *Standard for the Organization and Deployment of Fire Suppression Operations, Emergency Medical Operations, and Special Operations to the Public by Volunteer Fire Departments* [2], provide requirements for response staffing and time.

The second criterion is call volume, or how many responses one facility can handle at a time. In areas where call volumes are very high, it may become feasible to build another fire station.

Fire stations are unique in that they can be several types of facilities. For example, in some form they can be offices, garages, restaurants, homes, workshops, dormitories, classrooms, civic halls, urgent care facilities, and fuel stations. The areas of a fire station often are referred to by more than one name and an area can have multiple uses. A review of the different areas of a fire station follows:

Apparatus Bay.   The apparatus bay shown in Figure 6.2 is the area where vehicles are stored in the readiness state. The size of the bay depends on the number of vehicles to be housed. The size of a fire station is often related to its number of doors such as a two-bay, three-bay, four-bay station. It is very common for stations to have a rear door to enter the station with the front

▶ **Figure 6.2**   Large Apparatus Bay. *(Source: Fire Protection Handbook, NFPA, 2003, Figure 7.16.2.)*

door for exiting the station. Bay floors are typically slip resistant to prevent falls. There should be a means to capture vehicle engine exhaust and disperse it to the outside installed in the apparatus bay.

Dormitory. The dormitory is where personnel sleep. It can be referred to as the bedroom or bunkroom. Fire departments differ on how the dormitory should be configured. Some dormitories are one big open room; others have partitions to separate the beds; and some have individual rooms for each bed, as shown in Figure 6.3. The decision is usually based on gender issues. It is traditional for the officers to have separate sleeping quarters from the fire fighters. The dormitory can also be a place to study. Some departments provide storage space for personal items near the bed.

Kitchen/Dining. Fire station kitchens such as the one in Figure 6.4 are usually designed for heavier use than household kitchens. Maintaining excellent sanitary conditions is of primary importance so it is common to see appliances and sometimes countertops made of stainless steel. If the station has 24-hour staffing, there will typically be three or four shifts with each shift having its own food storage lockers and refrigerators. *Kill switches* are usually

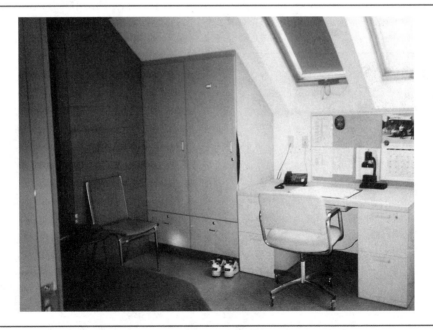

▶ **Figure 6.3** Sleeping Room with Desk and Space for Clothing. *(Source: Fire Protection Handbook, NFPA, 2003, Figure 7.16.3.)*

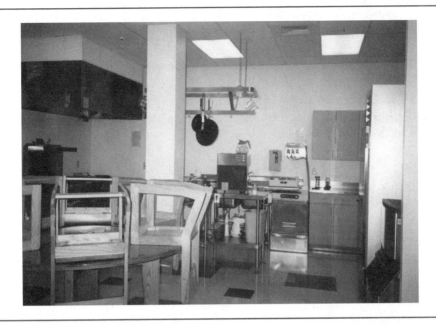

▶ **Figure 6.4**   Kitchen Area in a Multiple-Company Station. *(Source: Fire Protection Handbook, NFPA, 2003, Figure 7.16.4.)*

conspicuously located to quickly cut the power to cooking units if fire fighters must respond to a call while cooking. The dining area size depends on the number of people who normally eat at the station. Some stations also use the dining area as a classroom for individual company training. Traditionally the dining table is a place for informal gathering and fellowship.

Day Room.   The day room can also be called the living room, ready room, or lounge. It closely resembles the family room or den of a house. Recliners, chairs, and sofas are usually facing the TV/VCR/DVD system. The room is usually complemented by carpet, lamps, end tables, and coffee tables to provide a homelike atmosphere. Commercial or institutional grade furnishings are typical to withstand the heavy use. The TV/VCR/DVD provides both entertainment and training. Training through videos, subscription fire service TV networks, and local dedicated cable or satellite channels is commonplace.

Study.   Some stations include a room for study. It may be referred to as the library or work room. The study is usually the quiet room in a fire station. One or more computers with Internet and intranet connections are standard along with study materials. There is usually a desk or study cubicles.

The study provides a place for fire fighters to study for their continuing education and promotional requirements.

Training Room. Depending on the needs of the fire department, there may be a training room in the station. The room will have multiple desks or tables and chairs along with the instructional aids found in most classrooms including multimedia equipment as shown in Figure 6.5. The training room may also serve as a community room.

Office(s). Office space is usually available for the officers of the station. If the station is a headquarters station, it could have multiple offices. The office space is usually typical of any office with desk, chairs, phone, filing cabinets, computer, printer, and fax machine, or other equipment. The office provides an area to do administrative work, store records, and have supervisor and employee meetings. If there is a chief's office, it is usually larger than the other offices with the appearance of an office of a CEO in that the furnishings are nicer than other offices. Usually there is space for meetings for two to five people.

▶ **Figure 6.5** Training Room with Computers. *(Source: Fire Protection Handbook, NFPA, 2003, Figure 7.16.6.)*

Exercise Room.   Exercise or fitness rooms became popular in fire stations during the 1990s with increased emphasis on fire fighter wellness. A typical exercise room has at least an aerobic-type exercise machine such as a stair-climber, treadmill, or stationary bike and a universal weight machine. Free weights are also common in exercise rooms.

Decontamination Room.   To prevent the spread of infectious diseases, most fire stations have a room dedicated to the cleaning of equipment that has been exposed to body fluids. This room is not used for anything else and decontamination should not be performed in any room but this one. This room is often located adjacent to the apparatus bay and totally isolated from any other rooms. It usually includes a double sink with hands-free valve control, a nonporous cleaning surface, and industrial washer/extractor.

Laundry Room.   Most stations include a laundry room for cleaning of station work uniforms and bed linens. The laundry room includes a front-loading washer and dryer. Protective clothing is *not* cleaned in the laundry room to prevent the possibility of cross-contamination.

Restrooms and Showers.   Restrooms and showers in fire stations are typically configured to address gender issues. If the area is designed for more than one person to use at a time, there are usually separate rooms for males and females. If the area is designed for only one person to use at a time, the area may be gender neutral. Restrooms and showers are made of institutional type materials for ease of cleaning.

Locker Space.   There are two types of lockers common to fire stations: protective clothing and personal. Protective clothing lockers are for storing a fire fighter's head-to-toe personal protective equipment. They are ventilated and can often be found on the apparatus bay floor, as shown in Figure 6.6. Because ultraviolet rays cause deterioration of protective clothing fabric, these lockers are often isolated in a separate room where lighting is easier to control. Personal lockers are for fire fighters to store their personal clothing, station work uniforms, bed linens, toiletry items, and other personal belongings. These lockers are often located in a room adjacent to the shower area referred to as the locker room. Some fire stations have additional personal lockers at bedsides.

Watch Room.   The watch room is an office-type room (usually small) adjacent to the apparatus floor. It usually contains a desk, chair, and radio. If the fire department prints out response reports for "tear and run" as alarms are received, the watch room is usually where the printer is located. It is also used to log all the daily activities of the station.

▶ **Figure 6.6** Lockers with Ventilation to Dry Personal Protective Equipment. *(Source: Fire Protection Handbook, NFPA, 2003, Figure 7.16.5.)*

Storage Rooms. Storage rooms are varied depending on a station's activities. They can include rooms for storing such items as office supplies, janitorial supplies, fire hose and equipment, educational literature, and emergency medical supplies. Separate outside storage buildings are used to store yard maintenance equipment and flammable or combustible liquids.

Community Room.   A community room is typically a large room that is made available to the community for various activities such as club meetings, parties, and receptions. These usually have multiple tables and chairs with their own kitchen and restroom facilities. Community rooms are a way in which the fire station becomes more accessible to its customers.

Other Rooms.   There are many other possible rooms or areas found in fire stations. A station may have a shop area for maintenance and repair of fire equipment. Some stations have a room containing air compressors for recharging breathing air cylinders. A headquarters station will often have a conference or boardroom for business meetings.

Training Facilities

Training facilities are typically located in areas remote from residential areas because of noise and smoke. They range in size usually in accordance with the size of the fire department(s) it serves.

Classroom buildings are common to training facilities and usually have multimedia equipment. The classroom building often has many of the characteristics of any school classroom building.

A training tower, as shown in Figure 6.7, is often the defining structure of a training facility. It typically ranges in height from three to seven stories and can be constructed of wood, steel, or concrete. The tower provides opportunities for hands-on training for many training evolutions. In addition to high angle training, each floor can be configured for training scenarios that reflect incidents likely to occur within the department's area.

A live-fire training building is another structure common to training facilities. These buildings are specialized structures built to withstand fire temperatures of 2000+°F (1093°C) and further withstand rapid cooling when a fire is extinguished. They are constructed with a combination of reinforced concrete, precast-prestressed concrete, and masonry. In addition, panels are added in an offset manner to the interior walls to absorb the heat and protect the structural walls from the temperature extremes. The panels are typically made of calcium aluminate concrete with a lightweight high-carbon content, refractory blocks, or steel liner. Newer burn buildings now use propane or natural gas as the fuel source. The gas and resulting flame and heat can either be manually or computer controlled.

Most training facilities are located on expansive grounds to provide props for nonstructural firefighting training including driving courses, pump operator drafting, flammable liquids training pits, air/crash rescue, propane or natural gas leaks, vehicle extrication (including roadway and rail), collapse rescue, and trench rescue.

▶ **Figure 6.7**   Training Tower. *(Photo courtesy of Meg Sorber.)*

Other Facilities

Large metro fire departments often have additional support facilities. These can include administrative building, vehicle maintenance shop, communications center, fire prevention center or bureau, fire investigation facility, logistics (warehouse), reserve apparatus storage, museums, wellness centers, and building maintenance. Often one or more of these facilities are jointly shared.

▶ ## PERSONAL PROTECTIVE EQUIPMENT

The fire service has always used some form of personal protective equipment (PPE) as illustrated in Figure 6.8. In the beginning, the purpose of PPE was mainly to keep fire fighters dry (and warm during cold weather). Today, this

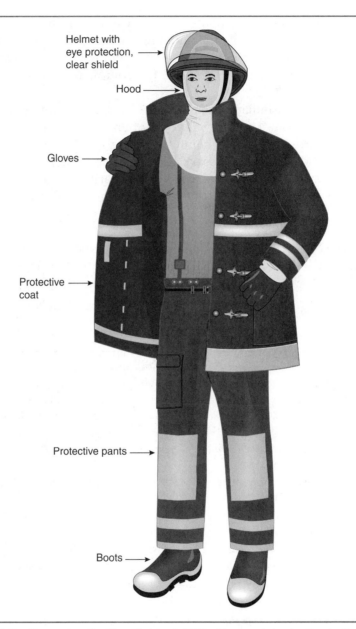

Helmet with eye protection, clear shield

Hood

Gloves

Protective coat

Protective pants

Boots

▶ **Figure 6.8** Example of a Fire Fighter's PPE. *(Source: Introduction to Employee Fire and Life Safety, NFPA, 2001, Figure 8.9(a).)*

equipment protects fire fighters from weather extremes, supplies them with clean breathing air while in contaminated atmospheres, protects them from the hazards of falling debris, protects them from hazardous materials, and protects against infectious diseases. In addition, it sounds an alert when a fire fighter is in trouble.

Most PPE is similar throughout the world. One notable exception is with head protection. The fire helmet is often the symbol of a fire fighter and most countries use a style helmet unique to that country. The U.S. fire service uses both a contemporary and a traditional helmet. Ironically, the traditional style helmet is gaining in popularity. Some other countries use helmets similar to crash helmets with additional side impact protection. The crash helmet design lends itself to better integration with the rest of the fire fighter's ensemble.

The National Fire Protection Association (NFPA) has several standards covering the design and performance of PPE (see Table 6.1). All elements of the PPE must be third-party certified that they meet the requirements of the ap-

▶ **Table 6.1** NFPA's PPE Standards

NFPA 1971—*Standard on Protective Ensemble for Structural Fire Fighting*, 2000 (This standard includes coats and trousers, helmets, hoods, footwear, and gloves.)

NFPA 1851—*Standard on Selection, Care, and Maintenance of Structural Fire Fighting Protective Ensembles*, 2001 (This standard covers the care and maintenance of the items identified in NFPA 1971.)

NFPA 1981—*Standard on Open-Circuit Self-Contained Breathing Apparatus for Fire and Emergency Services*, 2002

NFPA 1852—*Standard on Selection, Care, and Maintenance of Open-Circuit Self-Contained Breathing Apparatus (SCBA)*, 2002

NFPA 1989—*Standard on Breathing Air Quality for Fire and Emergency Services Respiratory Protection*, 2003

NFPA 1982—*Standard on Personal Alert Safety Systems (PASS)*, 1998

NFPA 1983—*Standard on Fire Service Life Safety Rope and System Components*, 2001

NFPA 1977—*Standard on Protective Clothing and Equipment for Wildland Fire Fighting*, 1998

NFPA 1976—*Standard on Protective Ensemble for Proximity Fire Fighting*, 2000

NFPA 1991—*Standard on Vapor-Protective Ensembles for Hazardous Materials Emergencies*, 2000

NFPA 1992—*Standard on Liquid Splash-Protective Ensembles and Clothing for Hazardous Materials Emergencies*, 2000

NFPA 1994—*Standard on Protective Ensembles for Chemical/Biological Terrorism Incidents*, 2001

NFPA 1951—*Standard on Protective Ensemble for USAR Operations*, 2001

NFPA 1999—*Standard on Protective Clothing for Emergency Medical Operations*, 2003

---

plicable standard. As a result of the NFPA standards and third-party certification, it is very rare that a fire fighter is injured or killed because of faulty PPE—assuming the PPE has been properly maintained and used as it is intended.

All repairs to PPE must be done in accordance with the manufacturer's recommendations. For safety and liability reasons, fire fighters must *never* modify, remove, or add anything to PPE without the *written* permission of the manufacturer.

Coats and Trousers

Coats and trousers (turnouts or turnout gear) are constructed of three layers: outer shell, moisture barrier, and thermal barrier. The outer shell resists flame, heat, and abrasion. The moisture barrier provides protection from water and other fluids. The thermal barrier provides additional insulation and protection from heat. There are several performance requirements of turnout gear. Two of the most critical test requirements that must be understood are the thermal protective performance (TPP) and the total heat loss (THL).

TPP Test.   The TPP test is designed to somewhat replicate flashover fire condition heat and flame exposure. The test uses radiant heat and direct flame contact to a swatch of all three layers of the garment. The layers of the garment must provide a minimum TPP rating of 35. In layperson terms, this means a fire fighter has approximately 17.5 seconds of exposure to this environment before suffering from second degree burns. The real life exposure time is determined by dividing the TPP rating in half.

THL Test.   The THL test is designed to measure the breathability of the turnout gear layers. At the time of this writing, turnout gear must have a breathability of not less than 130 watts per centimeter squared as measured on a guarded sweating hot plate. (This minimum number is expected to rise with future revisions of NFPA 1971, *Standard on Protective Ensemble for Structural Fire Fighting* [3].) Typically, higher TPP values will cause lower THL values and vice versa. Fire departments must weigh the benefits of each requirement during their decision-making process.

Other Tests.   Several other tests are required to ensure that fire fighters are receiving an established level of protection. These tests are the following:

- Flame resistance
- Heat/thermal resistance
- TPP
- Tear resistance

- Cleaning/shrinkage
- Water penetration resistance
- Viral penetration resistance
- Total heat loss
- Retroreflectivity and fluorescence
- Breaking strength
- Thread melting
- Seam strength
- Water absorption resistance
- Liquid penetration resistance
- Corrosion resistance
- Label durability
- Overall liquid penetration
- Conductive and compressive heat resistance

Maintenance. Care and maintenance of coats and trousers have recently emerged as important to protecting the health and safety of the fire fighter as well as protecting the fire department's product investment. Coats and trousers should *never* be left in direct sunlight for extended periods of time. Exposure to the ultraviolet rays of the sun can severely weaken the fabric. Turnout gear must be inspected on a regular basis and at least once per year.

Coats and trousers are to be cleaned when needed and at least every 6 months. To prevent cross-contamination, turnout gear is washed separately from other items, preferably in machines dedicated to turnout gear cleaning. They are *never* cleaned using chlorine bleach because chlorine weakens the fabrics. Front-loading washer/extractors perform a better job of cleaning and cause less wear and tear on the fabrics than top-loading machines. It is preferable to allow the garments to air dry rather than placing them in a tumble dryer.

Helmets

Fire helmets are used to protect the fire fighter from head impact with objects, falling objects, and water. The components of a helmet are shell, energy absorbing system, retention system, fluorescent and reflective trim, ear covers, and either a face shield, goggles, or both. The shells of helmets are usually made of fiberglass, Kevlar®, thermoplastic, composites, or leather. Helmets are subjected to a variety of tests to assure their performance. The key tests are penetration, electrical insulation, heat, and flame resistance. Helmets can be cleaned with soap and warm water. Solvents are not be used on faceshields or goggles.

Hoods

Fire fighters' hoods are designed to provide limited protection to the head, face, and neck areas that not protected by the helmet or self-contained breathing apparatus (SCBA) facepiece. Hoods must have a TPP of 20 as compared to 35 required by coats and trousers. Hoods do not have a moisture barrier. The care and maintenance of hoods is the same as that for turnout gear.

Gloves

Firefighting gloves provide protection against heat, flame, cuts, punctures, abrasion, water, blood-borne pathogens, and other liquids. There are tests that ensure that gloves provide against these elements. The minimum TPP for gloves is 35, just like turnout gear. Because of the cut and puncture protective properties of firefighting gloves, hand and finger dexterity are somewhat compromised. The outer shell is typically made of leather (cow, pig, elk, etc.), however, a few synthetic materials are used. They must have a moisture barrier. Gloves are available with wristlets or with gauntlets. Choosing between wristlets or gauntlets is best determined by examining the interface area between coat sleeve at the wrist area and the glove. If the coat has an extended wristlet, then a gauntlet type glove is appropriate. If the coat does not have an extended wristlet, then the wristlet style glove must be used.

Because of the cost and limited dexterity of firefighting gloves compared to other gloves, fire fighters and fire departments might be tempted to purchase or use gloves that have not been tested and certified by an independent third party agency as being compliant with NFPA 1971. Fire fighters and fire departments should *never* assume this safety and liability risk. Gloves should be washed in a utility sink rather than by machine. Air drying is better than machine drying.

Footwear

Firefighting footwear (boots) is primarily designed to provide protection from physical and thermal elements. Boots are subjected to a series of tests to ensure their protective properties. Key tests include puncture resistance, cut resistance, slip resistance, flame and heat resistance, and impact tests. Boots consist of sole with heel, upper with lining, an insole with puncture resistance, and a crush-resistant toecap as illustrated in Figure 6.9. They are

▶ **Figure 6.9** Identification of Footwear Terms. *(Source: NFPA 1971, 2000, Figure A-1-3.42.)*

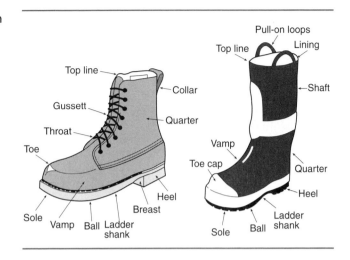

available in rubber, leather, and combination rubber/fabric materials. In recent years, leather and leather fabric combinations have gained in popularity because they are lighter and typically provide a better fit and feel. Boots should be cleaned in a utility sink and should be air dried.

Self-Contained Breathing Apparatus

Perhaps the single most important piece of PPE available to the fire fighter is the SCBA, as shown in Figure 6.10. Fire departments with enforced mandatory SCBA-use regulations, combined with effective search and operational tactics, often never experience respiratory injuries. The hazards of breathing smoke are well known and SCBA units are designed to prevent this needless exposure. SCBA are usually rated for 30-, 45-, or 60-minute durations and are positive pressure in design. Positive pressure means the pressure inside the facepiece, in relation to the pressure surrounding the outside of the facepiece, is positive during both inhalation and exhalation. Rated duration is almost always longer than actual times. For example, the average real-use time of 30-minute unit is around 14 minutes. SCBA are designed to operate at 2216 psi, 3000 psi, or 4500 psi. A low-air warning is given when the air supply drops below 25 percent.

SCBA typically consist of a backframe, waist/shoulder harness, first-stage regulator, second-stage regulator, and face mask and air cylinder. On modern SCBA the second-stage regulator attaches directly to the face mask. Fire fighters should be fit-tested annually to ensure that there is a good seal around their face masks. Air cylinders must be hydrostatically tested in accordance with the U.S. Department of Transportation every 3 or 5 years, depending on the material used in constructing the cylinder.

▶ **Figure 6.10** Fire Fighter
Wearing SCBA. *(Photo
courtesy of Meg Sorber.)*

In 2002, three important changes were incorporated into NFPA 1981,
*Standard on Open-Circuit Self-Contained Breathing Apparatus for Fire and Emer-
gency Services* [4].

1. They must have two independent End-of-Service-Time Indicator
   (EOSTI). Each EOSTI must have a sensing mechanism and a signaling
   device and at least two human senses must be stimulated.
2. They must have a Heads-up-Display (HUD) to alert wearers of their remaining air supply.
3. They must be equipped with a Rapid Intervention Crew/Universal Air
   Connection (RIC/UAC) System, which consists of a male and female fitting (coupling) to allow refilling of the air cylinder in the event a fire
   fighter goes down. The coupling must be compatible among all SCBA
   manufacturers.

In addition to NFPA requirements, all SCBA used by fire departments
must be certified by the National Institute of Occupational Safety and Health
(NIOSH).

The air quality of the breathing air compressor and air storage units (cascade systems) must be tested on a quarterly basis. OSHA requires that fire departments have a written respiratory program that must include training and maintenance.

Personal Alert Safety Systems Devices

Personal alert safety systems (PASS) devices are designed to sound an alert if a fire fighter is in trouble. There are three modes for all PASS devices: off, sensing, and alarm. They are activated in one of two ways. First, they automatically sound a prealert if they sense no motion for 20 seconds. This is followed by an automatic full alert sound if the fire fighter is motionless for 30 seconds. Second, they can be manually activated by the fire fighter if he or she is in trouble. PASS devices can be stand-alone units or they can be integrated into the SCBA. If integrated with the SCBA, they are automatically turned on when the SCBA is in use. If stand-alone, fire fighters must be vigilant in remembering to turn them on. Several fire fighters have died while wearing PASS devices that were turned off.

Life Safety Components

Life safety components include life safety rope, life safety harnesses (classes I, II, and III), escape rope, belts, water rescue lines, and auxiliary equipment. Life safety rope is rope solely dedicated for the purpose of supporting people. Life safety harnesses are a system component secured about the body to support a person. Escape rope is a single-purpose, one-time use emergency escape rope. Water rescue lines or throwlines are floating one-person ropes for retrieving or securing personnel in water rescue situations. Auxiliary equipment includes an assortment of hardware used to interface the life safety components. NFPA requires minimum load capacities, classifications, fiber sources, and construction type for life safety rope. Utility rope and other hardware commonly found in the consumer marketplace are never to be used for life safety situations.

Wildland PPE

NFPA provides minimum requirements for the following elements of wildland PPE: protective clothing, helmets, gloves, footwear, and fire shelters. Basically, wildland PPE is lighter weight than structural firefighting PPE. For example, protective clothing does not require a thermal or moisture barrier. The care and maintenance is basically the same as for structural firefighting elements.

Proximity PPE

Proximity PPE is very similar to structural PPE except that it is aluminized to reflect radiant heat commonly found in flammable liquid and gas fires. Personnel who fight aircraft fires also use proximity PPE.

Hazardous Materials and Chemical/Biological PPE

NFPA provides minimum requirements for PPE used in liquid-splash as well as requirements for vapor protection. In addition, there is a standard for PPE for responses to chemical/biological (terrorism) incidents. This standard provides for three levels of protection based on the nature of the event.

Urban Search and Rescue PPE

Urban search and rescue (USAR) includes incidents such as structural collapse, vehicle extrication, confined space entry, trench/cave-in rescue, and rope rescue. USAR PPE items include protective clothing, helmets, gloves, footwear, and eye/face protection. They are very similar to wildland PPE except that USAR protective clothing must have a moisture barrier.

EMS PPE

Many fire departments are directly involved in some level of emergency medical service (EMS) care, even if it is only first aid. The hazards posed by infectious disease have been well documented. NFPA has minimum requirements for the following elements of EMS PPE: garments, gloves, and face protection devices.

▶ HAND AND POWER TOOLS AND EQUIPMENT

The equipment used by fire fighters to accomplish their job is highly varied. It ranges from simple hand tools to high technology items such as thermal imaging cameras. As most equipment is unique to the fire service, most of it cannot be found at a hardware store, which makes prices higher than the public is aware. A description of the hand and power tools and equipment follows and some of those tools are shown in Figure 6.11.

Hose

Fire hose is used to transfer water by performing two functions, supplying water to the pump on a fire apparatus and applying water to the fire. NFPA 1961, *Standard for Fire Hose,* provides the minimum requirements for fire hose [5].

▶ **Figure 6.11**   Tools and Equipment on a Fire Apparatus. *(Photo courtesy of Meg Sorber.)*

Attack Hose.   Attack hose typically comes in 50 to 100 ft sections with threaded couplings. The diameter of the hose can vary from ¾ in. to 2½ in. The amount of water a hose can transport is directly proportional to its diameter.

Hose used for primary attack is usually 1½ in., 1¾ in., 2 in. or 2½ in. in diameter. The most popular choice for interior attack on single-family dwellings and other similar size structures is 1¾-in. hose. The most popular choice when there is heavy fire involvement in larger structures and for defensive operations is 2½-in. hose. Attack hose is collapsible and there are two common types of construction. One type consists of one or two woven fabric jackets with a rubber liner. The other type consists of single weave fabric with rubber impregnated through the weave locking the fabric into a single construction with a rubber inside and a rubber outside.

Booster hose is used when small volumes (less than 40 gal. per minute) of water are required, such as to suppress a grass fire. It is not designed as a primary attack hose for structural firefighting. Most booster hose is rubber covered, rubber lined, fabric reinforced, and not collapsible. However, light-weight synthetic booster hose is available. The diameter of booster hose is $\frac{3}{4}$ in. or 1 in. and it typically comes in 100-ft sections.

Supply Hose.   Hose used to supply a fire pump can range from $2\frac{1}{2}$ in. to 5 in. The most popular size is 5 in. because of its efficiency in providing larger volumes of water with minimum friction loss. Hose that is $3\frac{1}{2}$ in. or larger in diameter is referred to as large diameter hose or LDH. Typically supply hose is used in 100 ft sections with sexless nonthreaded couplings. This type of hose is often used to connect a pump on a fire apparatus to a fire hydrant. The most common type of construction for supply hose is the single-weave fabric with rubber impregnation.

Another type of supply hose is hard suction hose. Hard suction hose is noncollapsible and is used to draft water. Drafting is when the pump, by creating a vacuum, can draw water from a source such as a lake or portable tank. Suction hose typically comes in 10 ft long sections with threaded couplings and ranges in diameter from $2\frac{1}{2}$ in. to 6 in.

Nozzles and Appliances

Nozzles and appliances are used to integrate the fire hose into a system where it can be used by the fire fighter. Nozzles are attached to the end of attack hose and various appliances or adaptors are used to develop a water distribution system. Three types of fire streams are used for fire suppression: solid, fog, and broken stream. The nozzle determines which stream is produced.

A solid stream is produced from a fixed orifice, smooth bore nozzle. The solid stream nozzle is designed to produce a stream as compact as possible with minimal spray. The solid stream is used to penetrate into a fire and does not produce as much steam as a spray pattern.

A spray stream is a fire stream of water droplets that can range from a very fine fog to a coarse spray. Spray nozzles have an adjustment to change the discharge pattern from a straight stream to at least 100 degrees. The spray stream provides more water surface to absorb heat. Manually adjustable spray nozzles have a selection ring and automatic nozzles use the shutoff valve to regulate the flow.

A broken stream is a fire stream that is coarsely broken as it exits the nozzle. The broken stream is part spray stream and part solid stream.

Various appliances are used to help control the water delivery system. Among the most common are ball, gate, butterfly, or clapper valves. Other appliances include wyes, siameses, water thiefs, hydrant valves, reducers, double males, double females, and LDH appliances.

Ground Ladders

Ground ladders are often necessary to gain access to areas on the fire-ground. They are found on both pumper and aerial fire apparatus and occasionally on specialty vehicles. Fire department ground ladders are designed to support much heavier weight than ladders found at a local hardware or home improvement store. NFPA 1931, *Standard on Design of and Design Verification Tests for Fire Department Ground Ladders,* provides minimum requirements for the manufacture of ground ladders [6]. NFPA 1932, *Standard on Use, Maintenance, and Service Testing of Fire Department Ground Ladders,* provides the minimum requirements of fire departments for the care and use of ground ladders [7]. The following five types of ladders are commonly used by fire departments:

Single Ladders. A single ladder is nonadjustable in length and consists of only one section. It is sometimes called a wall ladder. The most common lengths of straight ladders are from 12 ft to 20 ft.

Roof Ladders. A roof ladder is a single ladder that has folding hooks attached to one end. When deployed, the hooks are placed over the roof ridge to prevent the ladder from sliding off a roof. Roof ladders range in length from 12 ft to 24 ft with 14 ft and 16 ft being the most common.

Folding Ladders. Folding ladders are single ladders with hinged rungs that allow the two beams of the ladder to rest against each other. The hinged rungs allow the ladder to collapse so it can be carried through narrow passageways. A typical application is to gain access into an attic area of a structure, and hence, it is often referred to as an *attic ladder.* They range in length from 8 ft to 16 ft with 10 ft being the most popular.

Extension Ladders. An extension ladder consists of more than one ladder section and is adjustable in length, that is, they have a base or bed section and one or more fly sections. Extension ladders are used to gain access to upper stories of a building and roofs and are common on both pumper and aerial ladder fire apparatus. They range in length from 12 ft to 50 ft with 24 ft and 35 ft lengths being the most popular. All extension ladders that are 40 feet or longer must have staypoles attached to each beam of the bed section to assist with raising and stabilizing the ladder. These ladders are referred to as *pole ladders* or *Bangor ladders.*

Combination Ladders. Combination ladders are designed so they can be self-supporting and used like a stepladder, or they can be used as a single ladder or an extension ladder. One common use of a combination ladder is to provide access across security fences. They range in length from 8 ft to 14 ft.

Forcible Entry Tools

Forcible entry tools are used to gain access to a structure that is locked or blocked. In some cases, they can be used to create an opening. There are four basic types of forcible entry tools: cutting, prying, pushing/pulling, and striking.

Fire Extinguishers

Fire extinguishers are carried on all types of fire apparatus and are found in most commercial facilities. NFPA 10, *Standard for Portable Fire Extinguishers*, provides the minimum requirements for fire extinguishers [8]. The three common types of extinguishers used by fire fighters are: dry chemical, carbon dioxide, and water. Dry chemical extinguishers are used for Class B-C or Class A-B-C fires. Dry chemical extinguishers, shown in Figure 6.12, are the most popular type because of their versatility. Carbon dioxide extinguishers, shown in Figure 6.13, are used for Class B-C and offer the advantage of no chemical or water residue. Water extinguishers are used for Class A fires. Refer to the *NFPA Guide to Portable Fire Extinguishers* [9] and Chapter 8 for further information.

▶ **Figure 6.12**   Three Sizes of Handheld Ordinary Dry Chemical Extinguishers. *(Source: NFPA Guide to Portable Fire Extinguishers, NFPA, 2003, Figure 1-23.)*

▶ **Figure 6.13** Three Sizes of Carbon Dioxide Fire Extinguishers. *(Source: NFPA Guide to Portable Fire Extinguishers, NFPA, 2003, Figure 1-25.)*

Lighting

Lighting equipment can be powered by either the 12/24 low-voltage apparatus system, the apparatus line voltage system, or batteries. Lighting equipment is either mounted or portable. Any well-designed fire apparatus has mounted scene lighting designed to illuminate an emergency scene. These lights can telescope upward as well as swivel and tilt. A good lighting system has portable lights to place on the scene to provide 360-degree illumination. Small portable lights are placed inside a structure during the overhaul phase to provide a safer working environment.

Handlights are required for every fire fighter doing an interior attack regardless of day or night ambient light conditions. Handlights are battery operated and many are often rechargeable.

Electrical Systems

Electrical systems are common for all pumpers and aerials to provide electricity to the fireground. An electrical system consists of the power source (usually a generator), electrical cord (usually on reels), junction boxes, and a variety of electrical male and female plug adapters. NFPA 70E, *Standard for Electrical Safety in the Workplace*, provides requirements for an electrical system [10]. Examples of tools and equipment powered by the electrical system include lights, smoke ejectors, positive-pressure ventilation fans, and hand-operated power tools.

Ventilation Tools

Two basic tool groups are used for ventilation. The first group consists of tools needed to create openings for heated air, smoke, and gases to escape. In the past, the axe was the tool of choice for creating openings. However, today the most common tool used for family dwellings is a ventilation saw. A ventilation saw is a chainsaw modified to cut through most roofing materials. For metal roofs and other situations not suited for a ventilation saw, a gas-powered circular saw can be used.

The second group consists of tools used to assist in removing contaminated air from the inside of the structure. At one time, placing a fan (referred to as a *smoke ejector*) in a doorway or window and sucking the contaminated air from the structure was the common practice. Today, a gasoline or electrically powered blower is the tool of choice. These blowers are placed in a doorway and force the air through the ventilation opening. The practice is referred to as positive-pressure ventilation (PPV). Gas engines, electric motors, and water pumps can drive PPV fans or blowers.

Ropes and Harnesses

Two types of rope are used in the fire service: life safety rope and utility rope. NFPA 1983, *Standard on Fire Service Life Safety Rope and System Components*, provides the minimum requirements for life safety rope and harnesses [11]. Life safety rope is used to support fire fighters and victims at emergency incidents and training. Most utility rope is constructed of synthetic materials and is never used for life safety.

There are three classifications of life safety harnesses. Class I harnesses are designed to wear around the waist and thighs or under buttocks with a design load of 300 lbs Class II harnesses are of the same design, but with a design load of 600 lbs Class III harnesses fasten around the waist, thighs or buttocks, and over the shoulders with a design load of 600 lbs.

Air-Monitoring Equipment

Air-monitoring equipment is becoming common equipment on every fire call and is found on most fire department vehicles. The emergence of carbon monoxide (CO) detectors in households has increased the demand for this equipment. As CO detector activations increase, fire departments respond to validate the activation and take appropriate action. In addition, fire departments are now monitoring the air of structure fires during the over-haul stage to determine the proper time to doff self-contained breathing apparatus. The most common air-monitoring equipment is for measuring CO. Four-gas monitors that measure CO, oxygen, hydrogen sulfide, and lower flammable limits (LFL) are also frequently used. Most air-monitoring equipment is designed to alert the user of a go, no-go situation based on pre-determined levels. Hazardous materials teams carry more sophisticated air-monitoring instruments that provide numerical readings.

Rescue Tools

Depending on the type of rescue, many tools can be used. In the fire service, rescue has two related meanings. One is removing victims from burning buildings, which is the underlying mission of structural firefighting. The other is extracting victims from car crashes, structural collapses, confined spaces, water/ice, industrial machinery, high angles, and other treacherous situations. Most, but not all, fire departments are involved in victim extrication. This overview of rescue tools concerns extrication tools.

Hydraulic Tools. Hydraulic tools are the tool of choice for removing people entrapped or pinned in vehicle accidents. The basic tools are spreaders, shears, and extension rams. A small gas-powered motor usually powers the hydraulic pump for these tools, although electric motors are available. These tools can create forces ranging from 10,000 lbs to 40,000 lbs. Spreaders are used to create or enlarge openings. The tips of most spreaders will open to various lengths up to 32 in. Shears are used to cut and remove material to gain access to a person. They are capable of cutting almost any metal object and have an opening of up to 7 in. Extension rams, ranging from 12 in. to more than 60 in. are typically used to enlarge openings. Hand-powered tools can also perform many of the tasks performed by hydraulic tools, but they are not as effective and their operation is usually much slower.

Air Bags. Air bags are another key tool for removing trapped people. They are available in high-pressure, medium-pressure and low-pressure bags. Air bags are available in various sizes ranging from 6 in. by 6 in. up to 36 in. by 36 in. When deflated, high-pressure bags can be as little as 1 in. thick; when inflated they can be more than 20 in. high. Medium- and low-pressure bags can lift objects more than 6 ft from their original position. Many other tools are used in extrication, including hand jacks, block and tackles, come-alongs, pneumatic chisels, reciprocating saws, cribbing, and shoring.

Thermal-Imaging Cameras

Fire departments started using thermal-imaging cameras in the 1990s, and they are now common on many emergency scenes. Thermal-imaging cameras focus on infrared light emitted by all objects in view. In effect, thermal-imaging cameras allow fire fighters to see through smoke. They are usually hand-held, or they can be mounted on the helmet. Understanding how to interpret the image requires training and familiarization because a thermal image is not the same type image as produced by a video camera.

Salvage Tools and Equipment

The tools and equipment for salvage can be extensive, ranging from small hand tools to water vacuums. Common to all salvage equipment is the salvage cover. These are tarpaulins (approximately 12 ft by 18 ft) used to cover furnishings to prevent water damage. Similar in nature are floor runners (approximately 3 ft by 20 ft) used to prevent damage to floor coverings. Carryalls are used to carry debris from buildings. In previous years, these items were made of coated canvas or synthetic materials. Today, synthetic materials are the most common because they are inexpensive, lightweight, and easier to maintain. Welding blankets make great carryalls for carrying hot embers.

Other major tools used in salvage include water vacuums for water removal, sprinkler shutoffs or wedges, mops, squeegees, and an assortment of packaging materials for storing smaller important items. The inventory of salvage equipment is directly related to the fire department's approach to loss control and customer service.

EMS Equipment

Basic and advanced first-aid items have become a part of the fire service because many fire fighters are emergency medical technicians (EMTs) or paramedics. The EMS equipment carried on most apparatus is designed to administer first-responder patient care. This equipment can range from bandaging materials to defibrillators.

▶ APPARATUS

Fire apparatus are an icon to a community. Almost every child is fascinated with a fire truck. Fire apparatus are usually diesel driven and are used to transport equipment and personnel to an emergency scene. The eight fire apparatus designations are: pumper, initial attack, aerial, quint, mobile water, mobile foam, specialty vehicle, and wildland vehicle. NFPA 1901, *Standard for Automotive Fire Apparatus*, provides the minimum requirements for the design of these type vehicles, except for wildland [12]. The minimum requirements for wildland vehicles are found in NFPA 1906, *Standard for Wildland Fire Apparatus* [13].

Pumpers

The pumper is the most common type of apparatus and can be found in almost every fire station. A pumper consists of the cab and drivetrain, a fire pump, water tank, hose bed (except for the European designs), and storage compartments for firefighting equipment and tools shown in Figure 6.14. The chassis for a fire pumper can be either a commercial truck chassis or a custom made chassis. The more expensive custom chassis provide more flexibility in design and are more durable.

Pump.   The pump must have a minimum capacity of 750 gpm (gallons per minute). Pumps are rated in 250 gpm increments, with 1250 gpm and 1500 gpm being most common. However, there are pumps that are rated at 1750

▶ **Figure 6.14** Pumper Midship. *(Photo courtesy of Kevin Roche.)*

gpm and 2000 gpm. Pumpers with higher capacities are usually designed for oil refineries and industrial applications. The pump must be able to deliver the following percentage of its rated capacity at the net pump pressure indicated at draft: 100 percent at 150 psi (pounds per square inch), 70 percent at 200 psi, 50 percent at 250 psi.

Fire pumps are either single-stage or two-stage, with single stage being the most common. A single-stage fire pump consists of intakes that lead to an impeller that sends the water to a manifold for discharge. A two-stage pump consists of two impellers that allow the pump to be used in either a volume or pressure mode. When operated in the volume mode, equal amounts of water enter each impeller and are discharged at the same pressure. When operated in the pressure mode, water is pumped from the discharge of the first impeller (first stage) into the intake of the second impeller (second stage). Two-stage pumps are best suited for high-rise buildings without internal building pumps and for long relay pumping.

Regardless of size or type, the pump must have discharges. All pumpers must have a minimum of two $2\frac{1}{2}$-in. discharges rated at 250 gpm each. The balance of the pump capacity is achieved with any combination of the following discharge sizes: 3-in. 375 gpm, $3\frac{1}{2}$-in. 500 gpm, 4-in. 625 gpm, $4\frac{1}{2}$-in. 750 gpm, and 5-in. 1000 gpm.

All pumps must have a pressure relief valve or pressure governor unit. These devices limit the pressure rise to no more than 30 psi between the ranges of 90 psi and 300 psi. In addition, all pumps must have an intake relief valve that relieves excess pressure that might develop in the suction side of the pump.

Water Tank.   The water tank, sometimes referred to as the booster tank, must carry at least 300 gal. Most municipal fire apparatus have a 500-gal tank while apparatus serving rural and suburban areas often have 700 to 1000 gal tanks. The tank must have a tank-to-pump valve and piping capable of flowing at least 500 gpm. Most tanks are made of a noncorrosive material such as plastic or fiberglass with a lifetime warranty against leakage. All tanks must have baffles or swash partitions to stabilize the water load while the apparatus is in motion. The tank must also have a vent and overflow. The vent is used for displacing the air in a tank when it is being filled, and the overflow is designed to discharge the excess water behind the rear wheels of the apparatus. See Figure 6.15.

Hose Bed.   The hose bed must have a minimum of 55 cu. ft. Most hose beds have a capacity to carry at least 1000 ft of larger diameter (4-in. or 5-in.) hose

▶ **Figure 6.15** Rear View of a Typical Pumper. *(Photo courtesy of Alan Saulsbury.)*

as well as ample other space for attack or supply hose. Most hose beds have adjustable dividers to allow the fire department flexibility in determining its hose load needs.

Storage. Pumpers must have at least 40 cu. ft of weatherized equipment storage. Because of the array of tools and equipment needed to perform the mission of a fire department, most pumpers far exceed that, with many well in excess of 300 cu. ft. A well-designed pumper will take advantage of any available space for storage.

Foam-Making Systems

Foam-making systems are now prevalent on many pumpers and wildland vehicles. They are occasionally found on other types of apparatus and have been a mainstay of airport crash/fire/rescue vehicles.

Classes of Foam. Foams used for flammable liquid fires (Class B) typically form a film over the fire, thus removing the oxygen supply. Increasingly, fire departments are using foam for Class A fires such as wildland and structure fires. Class A foam reduces the surface tension of water causing it to penetrate burning material and accelerate the extinguishing process. Class A foam can be combined with compressed air to further enhance the extinguishing capabilities. The combination of compressed air, water, and class A foam are referred to as CAFS (compressed air foam systems).

Foam Delivery.   There are several methods of delivering foam. Eductors can be installed at the pump discharge in the hose line. The eductor siphons foam through a pickup tube that is inserted into a foam concentrate container. The foam can be premixed in a tank before it is applied. An around-the-pump proportioner where the eductor is mounted between the pump's discharge and intake can be used. Balanced pressure systems are an alternative. Most popular today are direct injection methods where foam concentrate is injected directly into the pump discharge. These type systems have a separate foam tank or a dedicated foam cell within the water tank.

Initial Attack Vehicles

Initial attack fire apparatus, such as the one shown in Figure 6.16, are smaller than pumpers but similar in nature. They are designed to make the initial suppression attack, as the name implies. An initial attack vehicle must have

▶ **Figure 6.16**   Initial Attack Fire Apparatus. *(Photo courtesy of Bob Barraclough.)*

a pump capable of flowing a minimum of 250 gpm and have a minimum water tank capacity of 200 gal. The minimum hose bed area is only 17 cu. ft and the minimum equipment storage compartment area is only 22 cu. ft. Initial attack vehicles are not prevalent but offer a fire department a choice when a larger pumper is not required.

Aerial Fire Apparatus

Aerial fire apparatus come in various configurations. They can be an aerial ladder, an elevating platform, or a water tower. All of them perform at least all or part of the following functions: positioning personnel, positioning equipment, discharging water, and providing a means of egress. All aerial fire apparatus consist of a turntable at the base to provide for 360-degree operation. All aerial fire apparatus also have outriggers to provide stability when the aerial device is extended. Aerial fire apparatus carry a larger assortment of ground ladders than other apparatus and often carry specialized equipment for rescue, ventilation, and salvage operations.

Aerial Ladders.   An aerial ladder consists of two or more sections. They typically are available in extended lengths of 75 ft, 85 ft, 100 ft, 110 ft, and 135 ft. They must have a rated weight capacity in the maximum extended position of at least 250 lbs. Aerial ladders that are rated higher than that are rated in 250-lb increments.

Elevating Platform.   An elevating platform has an area at the end of the aerial device for fire fighters to stand and operate. The platform itself must have a guardrail around it. The minimum weight capacity without the waterway charged is 750 lbs. There are three types of elevating platforms. In the first design, the platform is extended telescopically in much the same manner as an aerial ladder. In the second design, the platform is mounted on an articulating boom. In the third design, the platform is mounted on a boom that is both telescopic and articulating as in Figure 6.17.

Water Tower.   Water towers are elevated streams that must flow at least 1000 gpm. The water tower can be either articulating or telescoping and typically has a maximum extension range of 50 ft to 75 ft. Note that most water towers are installed on pumpers. However, if the water tower has a ladder for egress, it must meet the requirements of an aerial ladder.

▶ **Figure 6.17**   Aerial Platform. *(Photo courtesy of Meg Sorber.)*

Quint

A quint, as shown in Figure 6.18, is a combination pumper and aerial fire apparatus and can perform either role. It is defined as an apparatus with a pump, water tank, hose storage area, and an aerial or elevating platform with a waterway. Quints have become more popular in recent years because of the versatility they provide.

Mobile Water Supply Apparatus (Tanker)

Mobile water supply vehicles (often referred to as tankers) are used to transport water to the scene of fires or other emergencies. They must have a minimum water tank capacity of at least 1000 gallons. The tank must be able to off-load at a rate of 1000 gpm for 90 percent of its capacity. The water must be able to be discharged to either side as well as the rear of the vehicle. The tank must also be designed so it can be loaded at a rate of 1000 gpm.

Mobile water supply vehicles are found in almost every fire station in rural areas and also strategically placed in urban and suburban areas to supplement fire hydrants. Often a fire department will purchase a mobile water supply vehicle that also meets the requirements of a pumper.

▶ **Figure 6.18**   Typical 75-Foot Quint. *(Photo courtesy of Alan Saulsbury.)*

Specialty Fire Apparatus

There is a wide diversity of fire apparatus designed for fireground support and special operations. These operations can include, but are not limited to, rescue, hazardous materials, lighting, electric power generation, and air supply for SCBA, command, rehabilitation, and urban search and rescue. Common to most of these vehicles is the large amount of compartment space they provide for equipment storage. Specialty vehicles are typically self-motorized. However, trailers and platforms-on-demand (PODs) are also used. PODs are common in the European fire service and are occasionally found in North America.

Wildland Fire Apparatus

Wildland fire apparatus such as the one shown in Figure 6.19 are designed for fighting vegetation fires. They are equipped with a pump and water tank and have the capability of pumping while moving (pump-and-roll). Generally, these are smaller vehicles equipped with all-wheel drive, a small water tank, forestry hose, a small-volume, high-pressure pump, a number of portable water extinguishers, and various hand tools for wildland firefighting.

▶ **Figure 6.19**   Wildland Fire Apparatus. *(Photo courtesy of Helen Cuccaro.)*

▶ COMMUNICATIONS TECHNOLOGY

Communications technology has always and will likely always be a challenge to the fire service. Communications consist of receiving requests for service, transmitting aid to the requests, and emergency scene coordination.

Communications Centers

At the core of communications technology is the communications center where requests for service are received, fire personnel and equipment are notified of the request, and emergency scene activities are supported.

Communications centers can handle a single governmental agency (city or county) or several governmental entities. Usually the single-source providers are in larger municipalities or countywide fire departments whereas the multiple-source providers are in smaller or mid-size communities. Methods of consolidating communications centers vary, but the two most popular are forming a joint powers authority (JPA) to provide this service as an independent agency or one or more jurisdictions contracting with another jurisdiction to provide this service. In many cases, the communications centers handle service requests for multiple emergency response agencies such as fire, police, EMS and others. Larger communications centers with bigger staffs tend to have more backup capability during peak demand periods than smaller centers. In addition, combined communications centers provide an increased level of service at a reduced cost.

Communications centers are designed to provide continuous service 24 hours a day, 7 days a week, 365 days a year. They are designed to function during natural disasters such as floods, earthquakes, blizzards, hurricanes, and tornadoes. They must be protected with backup emergency power generators with an adequate amount of fuel. Particular emphasis is placed on fire protection and security. With the emergence of the homeland security, a communications center must be protected against unwanted entry.

The interior of a communications center houses highly sophisticated electronic equipment shown in Figure 6.20. In years gone by, personnel who worked in communications centers were called dispatchers. Now they are called telecommunicators because of the wide range of expertise required to operate the equipment and understand emergency scene support requirements. Typically, each telecommunicator sits at a console housing all the equipment (computers, etc.) necessary to receive and transmit a request for service. Other equipment often found in a communications center includes:

▶ **Figure 6.20** Typical Communications Center. *(Photo courtesy of Meg Sorber.)*

audio recording equipment for recording calls and radio transmissions, weather monitoring equipment, and status boards.

Receiving Calls.   In the United States, most calls for service are received from citizens through the nationwide 911 emergency phone number system. The National Emergency Number Association estimates that 150 million calls were made to 911 in 2000, and 45 million of those calls were made with cell phones. Today, most areas have E911, which automatically provides the phone number and address from which the call is being made.

Before 911, most cities in the United States used alarm boxes as a means of notifying the fire department. These boxes were located along the streets where the public could easily access them to request service. When the box was pulled, the communications center was alerted. In some cases, the appropriate fire stations were directly alerted. Each box was assigned a unique number with the appropriate fire companies preassigned to that box. Some of the boxes were coded telegraph-type boxes and others contained telephones. In a few cases, the box contained both. For the most part, these boxes have become obsolete.

Alarm notification systems are very common in businesses as well as private dwellings. *NFPA 72*®, *National Fire Alarm Code*®, covers the application, installation, location, performance, and maintenance of fire alarm systems and their components [14]. Fire alarm systems are classified as follows:

- Household
- Protected premises
- Supervising station
  - Central station
  - Remote supervising station
  - Proprietary supervising station
- Public fire reporting systems
  - Auxiliary fire alarm systems—local energy type
  - Auxiliary fire alarm systems—shunt type

There are multiple ways fire can be detected to activate an alarm notification system. The oldest is the heat detector, which began with the emergence of sprinkler systems in the 1960s. A sprinkler can be considered a combined heat-activated fire detector and extinguishing device when the sprinkler system is provided with a waterflow detector connected to the alarm system. Heat detectors are usually located near the ceiling and thus are usually slow to activate. Fixed-temperature detectors are another way of activating an alarm system. They activate at prescribed temperature. Often this type of system is used with fusible links in sprinkler heads. Rate-of-rise detectors activate an alarm system if the temperature rises rapidly. Gas sensing detectors and flame detectors are used in some applications. There are even combination systems available.

For life safety as well as early detection, smoke detectors are commonplace. There are different types of smoke detectors including ionization, photoelectric, light obscuring, light scattering, and cloud chamber. The various types of heat and smoke detection device systems can perform any of the following functions:

- Close fire doors
- Control dampers in HVAC systems
- Open valves to release water into sprinkler systems
- Operate fixed extinguishing systems which can apply water and other extinguishing agents
- Open automatic drains
- Release drip tank covers
- Activate smoke removal systems
- Activate automatic audible emergency evacuation instructions

Processing Calls. Telecommunicators process calls and deploy the appropriate fire companies based on a running card file. Except for the smaller communications centers, the running card file is maintained electronically through CAD (computer-aided dispatch). A running card file contains a dispatch plan for every area of the jurisdiction served. The most common method of indexing is to assign numeric designations to all intersections or to grids superimposed on a map of the jurisdiction. There is a specific running card to each intersection or grid—similar to the unique number assigned to the alarm pull boxes. When a call is received, the telecommunicator alerts the predetermined appropriate companies based on the running card or CAD information. Requirements for CAD systems are covered in NFPA 1221, *Standard for the Installation, Maintenance, and Use of Emergency Services Communications Systems* [15].

Many larger communications centers use AVL (automatic vehicle locator) to dispatch the nearest fire company. An AVL is a device that makes use of the Global Positioning System (GPS) to track the location of vehicles. These devices combine GPS technology, cellular communications, street level mapping, and an intuitive interface with the ostensible goal of improving response times. AVL systems generally include a mobile radio receiver, a GPS receiver, a GPS modem, and a GPS antenna. This network connects with a base radio consisting of a PC computer station as well as a GPS receiver and interface. GPS uses interactive maps rather than static map images on the Web.

Radio Systems

Radio systems consist of base station transmitters, receivers, and repeaters. Several bands of radio frequencies are available. These bands are VHF low band, VHF high band, UHF 450 MHz, and UHF 800 and 900 MHz. Each band has its advantages and disadvantages. The selection of a band depends on factors such as frequency availability, area to be covered, type of terrain, number of radios, frequencies used by surrounding agencies, and mutual aid agreements. The number of frequencies a radio system needs is directly proportional to the size of the agency or agencies using them. The larger the agency, the more channels are needed.

Radio Paging Systems.   This system uses a paging encoder and individual pagers. Pagers can receive voice messages, alphanumeric text messages, or both. In this system the communications center transmits a signal to the pager. The pager alerts the wearer by either an audible or vibratory alert. This system is very effective for personnel who are not constantly monitoring radios. It should be noted that NFPA 1221 prohibits the use of commercial paging systems for emergency response notification because the authority having jurisdiction cannot control the commercial systems.

Mobile Radios.   Mobile radios are vehicle mounted and have high power output. Typically, mobile radios are used while the vehicle is moving.

Portable Radios.   Portable radios are carried by individuals and have low power output. Portable radios are battery powered. Typically, portable radios are used when persons are away from mobile radios when they are involved in on-scene activities.

Mobile Data Terminal (MDT).   MDTs are small computer terminals that are mounted in vehicles or handheld. These units generally consist of a microcomputer that converts digital information into a signal that a radio can transmit or receive. Typically, these units have preprogrammed messages that can be sent with the touch of a key. For example, the touch of a key can indicate the status of a unit. The screens of MDTs can be configured to provide resources to the user such as hydrant locations and hazardous materials storage.

Cellular Phones.   Cell phones are used to supplement service delivery. Many fire department staff members use cell phones to transmit administrative messages not directly tied to an incident. For example, a command officer can use a cell phone to talk directly to an agency or individual(s) who will provide assistance after the fire department has completed its service. This frees up radio airtime for more critical messages.

Personnel Monitoring Systems.   For several years, the fire service has used PASS devices discussed earlier in this chapter. Most recently, communications technology has emerged to a level that command staff can monitor several critical aspects of personnel operating on an emergency scene. The availability of GPS allows global positioning devices to monitor the exact location of fire fighters. The remaining air levels for SCBA can be remotely monitored. Vital body signs including pulse and body core temperature can also be remotely monitored.

▶ SUMMARY

Technological advancements coupled with ongoing revisions to minimum standards always keep the interest level in fire department equipment and facilities very high. It is imperative that the fire service stays abreast of industry development to maximize their service as well as their expenditures. Several user groups in the United States such as SAFER (Southern Area Fire Equipment Research) and F.I.E.R.O. (Fire Industry Equipment Research Organization) meet periodically to discuss the latest in equipment trends and developments. Other ways to stay abreast include attending trade shows, reading periodicals, studying manufacturer's literature, and researching through the internet.

Group Activity

Imagine that the group is a team planning for the building of a new fire station. Make a detailed plan for the station and list the vehicles and equipment to be purchased for an ideal, state-of-the-art facility.

Review Questions

1. On what two criteria is the location of fire stations based?
2. What are some basic fire service facilities?
3. What are the component layers of turnout gear?
4. What kind of tools are used to gain access to a structure that is locked or blocked?
5. What are two types of rope used in the fire service?
6. What is the most common type of apparatus that can be found in almost every fire station?
7. An _____ _____ has an area at the end of the aerial device for fire fighters to stand and operate.
8. What is at the core of communications technology?

7

Preventing Fire Loss

Charles H. Kime
Arizona State University

Learning Objectives

After completing this chapter, the reader should be able to do the following:

- Explain the significance of important conferences and public education programs to contemporary fire prevention
- Define the 3 E's—education, engineering, and enforcement—and how they impact fire prevention and public education programs
- Describe the role of fire prevention in the community
- Illuminate the importance and types of fire prevention inspections
- Explain the process by which codes and standards are enforced and the application of the 3 E's
- Describe some of the ways in which engineered fire protection systems are changing the way fire prevention is managed in the built environment and the impact they have on reduced property and life loss from fire
- Explain the role of fire and building codes in fire prevention
- Explain how codes, standards, regulations, and ordinances are developed, promulgated, and adopted
- Discuss the expanding role of public education in the delivery of effective fire prevention programs and how other organizations are involved in fire and life safety education
- Describe the role of fire investigations in fire prevention activities and how fire-problem experience is used to build more effective education and inspection programs, provide data used to make changes in fire and building codes, and benchmark fire prevention and public education achievements.

▶ BACKGROUND

A strong fire prevention program has long been thought to be one of the fundamental principles of fire protection [1]. Fire prevention in modern industrialized societies grew out of the concerns of the insurance industry to protect its insured assets. From these beginnings the modern approach to fire prevention has developed, expanding to encompass modern societies' concerns for life safety as well as property protection and then subsequently to more sophisticated approaches to public education and customer service. This chapter briefly discusses these subjects and presents an introductory exposure to some of the fundamental principles of contemporary fire prevention practices, especially within local government.

The notion of fire prevention is not a new concept, especially if one views it within the scope of Western civilization. Yet, the approach we take to fire prevention today, especially within the United States, is a relatively young discipline, especially true when one considers the scientific and engineering developments applied to the discipline and code making process, particularly within the last one or two decades. Three historical events have helped shape the direction fire prevention is taking in the United States. The first such event was the President's Conference on Fire Prevention [2], held in 1947. This conference convened an august body of national leaders to focus on the number of people injured or killed in the United States; the loss of property, goods, and natural resources; and the personal responsibility everyone must share to reduce the fire problem, including the consideration for strengthening existing laws related to negligence. The second event was the first of four Wingspread Conferences, held in 1966. Three additional Wingspread Conferences have been held in 1976, 1986, and 1996. National fire service leaders were convened at each conference in an effort to bring the best and brightest minds together who understood the fire problem and to produce reports that focused on the critical fire issues they believed the nation should address. Each report had several recommendations specifically focused on issues of preventing fire or providing fire education. The third event was the work done by the National Commission on Fire Prevention and Control, which wrote the report, *America Burning* [3]. This report has arguably made the most significant impact on the fire service of any single conference or report. These events have shaped, and continue to shape, the evolution of fire prevention and fire and life safety education in the United States.

President's Conference on Fire Prevention

President Harry S. Truman was the first president of the United States to call a conference that would focus on the fire problem in the United States. The President's Conference on Fire Prevention was held May 6 to 8, 1947, in Washington D.C. [2], where the president himself addressed the assembled body. His message was clear. First, the number of people injured or killed in the United States each year must be reduced. Second, the loss of property, goods, and natural resources cannot be ignored. Third, it is important to consider the personal responsibility everyone must share to reduce the fire problem, including the consideration for strengthening existing laws related to negligence.

Six committee reports were heard and discussed at the conference, followed by the presentation of a report called the Action Program, with recommended actions that should be taken and a guide for organizing a community to address fire safety. It is interesting to note that much of what is in these documents is just as pertinent in our current environment as it was almost six decades ago. The 3 E's model of education, engineering, and enforcement grew from this conference. Although the notion of education spans the spectrum from public education to the education of a single businessperson, it nevertheless holds a special place in the trilogy of effective fire safety concepts used by fire departments in their fire prevention programs. Much of what is discussed in this chapter either grew from this conference or was reinforced by the findings and actions of this conference [2].

In brief, the recommendations included better building designs, the use of technology (including automatic fire sprinklers and detection systems), and the revision and strengthening of fire and building codes. A major focus was placed on education at all levels, including the education of teachers in fire safety and the inclusion of fire safety design and engineering in the appropriate college curricula. Fire departments were encouraged to use fire fighters for inspection and fire prevention activities including thorough investigation of fire cause. Additional recommendations were made for the states to follow up on fire safety programs and the organization of fire safety committees. Needless to say, this was an historic conference that set the stage for future discussions and was a prelude to the 1973 National Commission on Fire Prevention and Control report, *America Burning*, 26 years later [3].

Wingspread Conferences

The first Wingspread Conference on fire in America was held in Racine, Wisconsin, at the Johnson Wax foundation's Wingspread Conference Center in February 1966 [4]. The fire service has held three additional Wingspread

Conferences: Wingspread II was held in 1976, Wingspread III was held in 1986, and Wingspread IV was held in 1996 [4]. A report was written for each conference that highlights the critical issues and the recommendations made by the conference participants. Each report contains recommendations related to fire prevention and public education. The 12 recommendations in the 1966 Wingspread I report include three recommendations for the public's fire-related complacency toward life and property loss, communication problems between the public and the fire service, and behavior patterns that influence the fire problem [4]. Ten years later, in 1976, the 11 recommendations in the Wingspread II report included a need to build better liaisons with building design professionals and the fire service and recommendations that arson should be considered a major issue, the fire service should lead the way in encouraging widespread use of smoke detectors, and the fire service should assume more responsibility and play a leadership role in managing fire loss [4].

By 1986, Wingspread III concluded that the American society is unwilling to use proven technology to reduce the fire problem and public fire education must be systematically organized based on human behavior to achieve its objectives [4]. Wingspread IV, 1996, emphasized a focus on customer service, a reiteration of the need to commit to fire prevention and public education, and the need to ". . . support adoption of codes and standards that mandate the use of detection, alarm, and automatic fire sprinklers, with a special focus on residential properties" [4].

Clearly 25 percent of the recommendations made by each of the four Wingspread conferences reflected the need for more focus and emphasis on fire prevention and public education programs, adding a new dimension in 1996, that of customer service. The participants at each of the conferences were fire service leaders who brought leading edge contemporary thought to each conference, building consensus on the national fire issues within the context of the period. These issues reported in the Wingspread Conferences, are examples of the enduring nature of the need to find more effective ways to design fire prevention and public education programs that can mitigate, if not eliminate, the nation's terrible record of devastating injuries, life loss, and property loss from fire.

America Burning

President Richard M. Nixon established a national commission to study America's fire problem in 1972. The National Commission on Fire Prevention and Control sent its report, *America Burning* [3] to the president on May 4, 1973. In the transmittal letter that accompanied the report to the president, Richard E. Bland, Chairman of the National Commission on Fire Prevention and Control wrote that the recommendations in the report emphasized ". . . fire prevention through implementation of local programs . . . and built-in fire safety—measures which can detect and extinguish fire before it grows large enough to cause a major disaster" [3] (transmittal letter to the President). Further, the letter to the President expressed the need to continue a federal focus on the fire problem and that the Commissioners hoped their report would provide ". . . guidelines for local, state, and national efforts to reduce the life and property loss by destructive fire in the United States."

America Burning [3] resulted in the passage of Public Law 93-498, the Federal Fire Prevention and Control Act of 1974, which was signed by President Nixon on October 29, 1974 (U.S. Fire Administration Web site, 5/22/03). This historic document set in motion a movement in the United States that resulted in the creation of the United States Fire Administration and the National Fire Academy, which placed America's fire problem in the most prominent national spotlight it had ever known. The U.S. Fire Academy estimates that it has trained almost one and one-half million students since it began delivering fire-training programs in 1975 (U.S. Fire Administration Web site, 5/22/03). These courses range from programs in-residency at the National Fire Academy in Emmitsburg, Maryland, to short courses delivered in the field. Additionally, the National Fire Academy has taken a leadership role in working with the fire service and fire service organizations to develop programs designed to reduce property and life loss from fire in the United States.

The impact *America Burning* [3] has had on fire prevention and public fire education efforts has been enormous. The number of civilian lives lost in America from fire exceeded 7000 and fire fighter deaths averaged 150 a year in the late 1970s. Although the property loss from fire and the death rates from fire in the United States continue to be among the highest of any industrialized democracy outside the former Soviet block, civilian deaths have dropped to about 4000 per year and the number of fire-related fire fighter deaths has dropped to approximately 100 per year. Even with the introduction of better engineering and fire prevention practices, the direct property loss continues to hover around $10 billion per year.

▶ TRADITIONAL LOSS MODEL—THE 3 E'S

For more than 50 years, fire prevention officials have used the 3 E's (education, engineering, and enforcement) model for fire prevention program design and implementation in the field. This model emphasizes education and considers enforcement a last resort, yet sometimes necessary to gain compliance with society's fire prevention regulations. The model suggests a linear approach to fire prevention efforts, especially when fire inspectors are working with businesses and the public. Today's environment, however, presents a sophisticated and complex set of circumstances and concerns that make it difficult for inspectors to focus on only one approach before moving to the next.

Education

Education has long been considered the first priority and the most desirable way to get businesses, government, and the public to practice good fire prevention. The underlying principle of education seems intuitively correct in today's social environment and consistent with the philosophy that the fire service is best served when working in collaboration with other sectors of the community. It is often desirable to form coalitions with other government agencies and private businesses, thus bringing additional resources to bear through cooperative efforts. Through these cooperative efforts, collaborations, and educational pursuits, businesses and citizens become more knowledgeable about the destructiveness of fire and the devastating injuries, fatalities, and property loss that results from unwanted fires.

The goal of any educational process is to enhance a person's knowledge about a problem that results in behavior modification, which in turn will reduce or mitigate the predictable effects of unwanted fires. Teaching a businessperson or citizen about fire prevention can translate into fewer fires, or at least reduce the magnitude of the fires that do occur and thereby reduce the number of injuries, deaths, and the amount of property loss. That sounds very idealistic, especially if you are the businessperson or citizen being asked to do the changing, yet it is imperative that fire prevention officers and fire departments continue to work toward these very reasonable goals of fire safety education when working with businesses and the public. Equipping them with better knowledge about ways that they can make their businesses and homes safer from fire directly benefits them and their communities.

This notion of education as a first priority is especially important when conducting a fire inspection. One of the most common questions asked of a fire inspector is, Why do I have to do this? This question is typically followed by a myriad of reasons why they do not believe they need to do what is being asked of them. Reasons range from it is too costly to it will disrupt the logistics of my business to we are very safety conscious, and we have been in business for umpteen years and have never had a fire. The burden of explanation is then up to the fire inspector, whose success often is a function of her or his persuasive powers. Yet, experience shows that the powers of persuasion are not always enough or the person is just not able to see how he or she can achieve the goals of good fire prevention practices, stay within budget, and not irreparably disrupt his or her business or life.

Equally important is educating the fire inspector. "Because the code says so" is not an acceptable response when asked why something should be done. Fire inspectors should be educated (and trained) about the provisions in the code, what they require, and the underlying intentions, that is, what the code is intended to accomplish. Often if the inspector understands why a provision was put into the code, he or she can better explain (persuade) why it should be followed. This in turn can be a positive way to segue into a discussion about alternatives, that is, engineering alternatives.

Engineering

Engineering is the second approach used to achieve fire safety and the goals of a prevention program. However, remember that our world is not neatly packaged and conducive to tidy one, two, or three approaches. The sophisticated and complex environment of a modern industrialized society with its highly technological capabilities and sophisticated informational systems presents many special circumstances and concerns that make it difficult for inspectors to focus on only one approach at a time. Almost every gadget, product, service, or system that makes our lives more convenient, comfortable, and enjoyable presents a new and ever-present fire and life safety challenge for the fire service in general and fire inspectors in particular.

Often, fire inspectors are asked, How can I do what you ask me to do and still stay in business? Most of the time the business managers or owners believe they are as safe as they need to be (or want to be) and ask the question rhetorically—that is they do not expect an answer and just want to be left alone. However, it is the obligation—yes, responsibility—of the fire inspector to engage them in a discussion of the possibilities about how they might modify the situation to meet everyone's objectives or at least reach a reasonable compromise. Engineering solutions run the gamut from adding automatic fire sprinklers to providing more exits or constructing separation walls. However, often the solution is as simple as the minor rearrangement of inventory or separation of particular products. Engineering solutions for major structures during the design phase are yet another level of sophistication and are discussed further under the heading of Performance-Based Designs.

Enforcement

Enforcement is generally considered the last resort in the 3 E's model. Typically fire fighters and fire inspectors would rather use the more positive means of education and engineering than resort to enforcement with all of its negative connotations. However, most experienced fire marshals recognize that it is often the threat of enforcement that drives a business or citizen to jump on the education or engineering bandwagon. Fire departments are generally given the authority to enforce the fire regulations through a state law or municipal ordinance. In some types of occupancies, compliance with the state and local fire codes is a condition of their business license. For example, many health care occupancies are mandated by law to meet local codes; liquor licenses in many states require that the business comply with all applicable codes and ordinances, including the fire code.

Noncompliance can carry a range of consequences from fines to closure. Yet, most of the time fire departments are able to work with the business and gain compliance over time, without total disruption of the business. Fire codes can be one of the most powerful tools in the government's toolbox and this fact is not lost on the business community and many politicians. Businesses often comply during the educational process or are willing to work with the fire department and find ways to re-engineer their business, knowing that if they do not comply, the fire department has the power to shut them down. Politicians, however, occasionally ask the fire marshal to enforce the fire code in an undesirable occupancy to achieve a political objective rather than a fire safety objective. The use of enforcement powers in these instances is at the very least unethical and at the most unlawful. Fire departments should avoid getting sucked into this type of political quagmire and only enforce the fire code for the purposes for which it was intended.

▶ CONTEMPORARY PROGRAMS

As fire prevention bureaus and programs were developed early in the 20th century, they focused primarily on issues related to fire. Large fires, especially fires that caused many deaths and injuries; small fires that created panic in overcrowded public assembly occupancies; and other catastrophic events that injured or killed people have been driving factors in fire prevention and fire and life safety education program development and fire inspection programs. Although fire departments responded to natural disasters, building collapses, and people trapped in buildings or machinery who needed heavy rescue efforts, most fire departments considered that their primary mission was to combat fire first, and prevent it where they could, mostly within the built environment. By the middle of the 20th century, many fire departments experienced their citizens and communities asking them to address other community problems. Red Cross first-aid training became more widely used, expanding to what we know today as emergency medical technician (EMT), which then led to the development and implementation of what we now know to be advanced life support (ALS) programs. It was this exposure to injuries and deaths through the delivery of emergency medical services (EMS) and not related to fire that principally caused the fire service to think of its services in a much broader sense. The result of this evolution has been the development and implementation of a wider range of fire and life safety programs that address a range of issues from gun safety and water safety and drowning to car seat safety and "buckle up" campaigns. Contemporary fire and life safety programs reach across the community and involve many different groups to address the pain and suffering that the average fire department sees on any routine day in America.

Beyond Fire Inspections

Fire departments in the earlier years of the 20th century focused mostly on fire inspections. This focus seemed natural because originally insurance companies typically conducted fire inspections as a way to ensure some measure of safety compliance and thereby protect the assets that they had insured. Major industrial plants later began to create fire brigades to protect their properties. These brigades were staffed by regular employees of the plant who would respond to fires inside the plant, because many of the plants were in areas that had small volunteer fire departments or no fire department at all. Some plants even had plant fire inspectors whose job it was to periodically inspect areas of the plant that were deemed most susceptible to accidental fire. Most of the inspectors in the earlier years were tradesmen like electricians, plumbers, steamfitters, and boilermakers who understood how some of the systems were built and were able to tell if they were in good working order.

During the first quarter of the 20th century local government began to get involved with fire inspections through local fire departments. Fire departments historically responded to fires and rescue situations, and their communities gradually began to view the fire department as the one agency in their community that they could count on to solve a myriad of problems, some only remotely related to an actual fire. When an overcrowding condition resulted in loss of life, it was the fire department that was called to help with the rescue and first aid at the incident and then later asked to find ways to keep such an awful thing from happening again.

Life Safety Programs

Life safety programs include a wide range of programs designed to prevent and educate the public about situations and activities that can cause injuries or death aside from the dangers from fire. These programs result from the lessons learned from the responses fire departments make to varied life-threatening situations and are designed to mitigate or eliminate the causes or, at the very least, educate the public about the problems. Life safety programs emerged more prominently in the 1970s as many jurisdictions became concerned with injuries and the loss of life from a variety of sources beyond the traditional fire injuries and deaths such as natural disasters, major accidents, and complicated rescue scenarios. But as the fire service began to assume the responsibilities of EMS, it became acutely aware of other life-threatening situations and began to look for ways to prevent them. Most of these efforts were organized within existing fire prevention bureaus either as a separate function or integrated into existing fire prevention programs.

Life safety programs became especially relevant to infants and youths because fire fighters were able to relate to their own children but these programs were also an outgrowth of the increasing responses by fire departments to routine auto accidents, injuries on construction sites, and even EMS responses to doctors' offices. Most doctors do not have the medical staff on duty that is properly trained or equipped to handle medical emergencies outside of their narrow specialty. The reality of these changes resulted in the expansion of the NFPA *Learn Not to Burn*® *(LNTB)* program into areas of concern that transcended basic fire safety. For example the Phoenix, Arizona, Fire Department expanded its efforts beyond the scope of fire in its Urban Survival program in an effort to reduce injury and life loss among the nation's most vulnerable population. The NFPA developed the *Risk Watch*® program to address these contemporary issues. Yet, looking back, these types of efforts had been recommended some three decades earlier. The *Learn Not to Burn* and *Risk Watch* programs are discussed in more detail in the section Fire and Life Safety Education.

Fire Education

Although education has long been a hallmark of fire prevention programs, it was often relegated a second-class position within fire prevention bureaus that seemed to focus more on inspection and code enforcement. During the 1980s, public fire education seemed to get more attention with a broader focus on the job of fire prevention bureaus. Where education was considered a first step in a progressive inspection program, it meant educating the business owner to understand why she should comply with the fire code. Education was coming into its own with a focus on the public and especially the youth in America. Later efforts became more directed to seniors and other at-risk groups.

Although some efforts had been advanced within certain occupancy groups, for example, health care (particularly hospitals), efforts in many of the other occupancy classes was not common. Another occupancy group that emerged with special problems that needed special attention was the entire category of high-rise buildings and the unique problems they presented to the average fire department. Hazardous materials occupancies was another group that emerged to receive special attention. Many fire departments today have created special sections within their bureaus to address the myriad problems presented by rapidly changing technological challenges that range from military weaponry to pharmaceuticals.

► FIRE PREVENTION ORGANIZATIONS

Fire prevention is typically a function of state and local fire departments (i.e., city, town, or county). As such, departments are responsible for providing services that are in concert with their community requirements and the larger government functions as a whole. Within a city it is expected that the fire prevention codes apply to everyone and are administered on an even-handed basis, even though some properties within the city may be outside the jurisdiction of the fire department's fire prevention codes. Often, these properties come under the jurisdiction of another level of government or a different department in the city. Some examples of these anomalies might be public schools, factory-built buildings, facilities owned or operated by another level of government (especially if they are federal, state, or county owned), or federally regulated buildings. Prisons are yet another example.

State Fire Marshal's Office

State fire marshals typically represent the state in activities that range from code enforcement to public fire education to fire investigations. The governor of each respective state usually appoints the state fire marshal; and as political appointees, many of them are in office for only a few years. The background of state fire marshals varies, although most of them have some type of fire service or law enforcement background. Typical experiences of fire marshals prior to appointment include fire fighter, fire protection engineer, retired career fire fighter (or chief officer), volunteer fire fighter, and police officer or public safety officer. The level of authority and range of responsibilities varies among the states but fire marshals are an integral and important part of the national fire service.

Many of the state fire marshals are responsible for training that includes basic firefighting, hazardous materials response, wildland firefighting, and the inspections of public utilities. Although much of this training is done by large fire departments for their own members, there is often a working relationship between the state and the larger departments to share expertise and resources. The office of the state fire marshal typically acts as the repository for fire data for the state, collects fire data, and conducts data analysis on the fire problem in the state. The state fire marshal's office is often the contact point for federal fire grant monies that come from the Federal Emergency Management Agency (FEMA), the U.S. Fire Administration (USFA), or other Homeland Security agencies. Grants come through the state fire marshal's office to local fire departments, whether career or volunteer, municipal or district. State fire marshals also advise governors, state legislators, and members of Congress on fire issues that affect the states and the fire service at large.

Local Fire Departments

Fire prevention is typically one of the major responsibilities of a fire department along with fire suppression, personnel, budgeting, apparatus and station maintenance, and other activities. Fire prevention responsibilities typically are a function of the fire marshal within the fire department, and although it is a major function, it is often relegated a relatively low priority within the department, but priority varies from department to department. Many fire chiefs and fire fighters consider the fire inspection and code enforcement functions of fire prevention as negative and therefore much prefer to keep it to a minimum, thus they are reluctant to allocate a lot of resources to it. The educational functions of fire prevention, however, are considered to be more positively received by the community and are often better funded and better staffed.

These feelings of positive versus negative perceptions in the community however, do not negate the responsibility the fire department has to provide fire prevention programs. It behooves the fire department to invest in this activity at a level that can achieve the desired outcomes of meaningful fire prevention inspections and achieve a level of code compliance that is acceptable to the community; to do less is asking for serious problems if a disaster happens. Fire departments should also remember that their actions are transparent to the community and how they manage their own facilities and processes is often viewed as the model for others to follow. Remember what mom said: "Don't ask anyone to do anything you would not do yourself." This advice means installing smoke alarms and automatic fire sprinkler systems in fire stations, using flammable liquid cabinets, and the proper storage of hazardous materials just to name a few items. The community and the rest of the city is also looking at the types of construction projects that a fire department manages and is keenly interested in any exemptions the department requests from building or fire codes. Again, it is important to set the standard at home that will be applied to others—model the way.

Citizens and Citizen Boards

Citizen or community groups are also popular as advocates and can help form coalitions with the business community, thus bringing together a broader range of community resources and commitments to achieve the objectives of fire and life safety. Advisory boards, for a variety of issues, are another approach often used. Fire code advisory boards are typically required, but vary in structure and authority from community to community. These boards are responsible for reviewing codes and code amendments or revisions and then making recommendations to the elected body. Most elected officials want input from the citizens and affected industries before they vote on the adoption of new or revised codes.

Communities

Fire prevention has a unique role within the community. On the one hand, the community wants the fire department to provide good (meaning high quality) services to the community, while on the other hand, many members of the community are somewhat conservative and hold the view that government regulatory actions should be limited. The fire department must maintain a keen balance in getting the job done within the community while keeping the respect and support of its individual businesses and citizens. Working with the community on nonenforcement fire prevention activities often builds community confidence that the fire department needs to manage an effective code enforcement program. Another positive outcome of this seemingly negative situation is it can provide the incentive for the department to assess its methodology of code enforcement, which can also set up some heated discussions inside the department about its fire prevention goals and objectives. This debate can be very positive and healthy and can lead to a better organizational understanding of the mission of fire prevention and its value to the community.

Businesses

The role that fire prevention plays within the business community is often precarious, requiring a great deal of diplomacy and finesse. The business community, generally, respects those who are professional and know their business. Since profit is the bottom line in the private sector, and time is money, businesses expect fire prevention officials to get to the bottom line in a courteous yet expeditious manner. Business managers and owners want to know what they have to do, how long they have to do it, and most importantly why they have to do it. The last point usually creates the most conflict between the business and the fire official.

It is important that fire prevention officers are ready to adequately discuss and thoroughly explain the provisions of the code that they are enforcing and why it is important; this is the education part of the model. Often, the next objection is related to costs. Sometimes it involves the logistics of the business (i.e., the flow of goods and services during the routine business they conduct with their customers) or simply the time requirements to bring the business into compliance. At this point the education phase meets the engineering phase, and business owners and managers expect the fire official either to answer their questions, in other words their demands (often it is an unreasonable expectation to meet their demands), or work with them to solve their problem (the most reasonable expectation).

It is important for the fire official to understand the intent of the codes and the level of authority they have to seek creative solutions. It is often a good idea to get the supervisor (sometimes as high as the fire marshal) involved. If specialists are available (e.g., hazardous materials, life safety, fire protection engineers), they should be called in to help with any solutions that are out of the ordinary. The authority to enforce the codes is always available but should be considered a last resort. This authority is usually discussed during the education phase but should only be discussed as part of the overall explanation of why the fire department is there and what they are responsible for as spelled out in the law and in the codes. More often than not, authority is not even brought up during the education phase, as most businesses already understand, at least generally, that the fire department has the authority to enforce the codes and standards in their business. It cannot be expressed strongly enough how vitally important it is for the fire department to maintain a professional relationship with the business community.

Community Partnerships and Coalitions

Community partnerships and coalitions have strengthened the fire service within the community, making it more effective in a variety of ways. Partnering with the complex and many factions in the juvenile justice system have been another interesting challenge. The players in this arena range from judges, prosecutors, and defense lawyers to counselors, probation officers, and detention officers and to educators, fire fighters, and investigators. It is interesting to note that one department reported that when most of the individuals were initially contacted to discuss the juvenile fire setter problem in the youth they worked with, most of them denied having youth who had a fire-setting problem. Then after learning the magnitude and pervasiveness of the problem they have become some of the fiercest supporters of youth fire setter programs and in some instances have continued their affiliation with the fire service even as they changed jobs and careers.

▶ ROUTINE FIRE PREVENTION ACTIVITY

Fire inspections are generally considered to be the main focus of fire prevention, however the normal routine of a contemporary fire prevention bureau goes beyond just pounding the pavement. Some of the other focal points include general inspections, permit or license inspections, target hazard inspections, and self-inspections.

General Inspections

General inspections have historically been the main focus of most fire prevention bureaus. From the 1920s through the 1950s, most fire inspectors were given a "beat," usually consisting of some geographical area like a block of streets where they were responsible for inspecting all of the buildings on each side of the street. The inspectors would start their day at the bureau and then proceed to their assigned area where they would go door-to-door, introduce themselves as an inspector from the local fire department, and explain that they were there to conduct a fire inspection. All of this was done unannounced. The fire protection community generally felt that it was important to conduct unannounced fire inspections because that was the best way to discover, or uncover, the normal fire safety practices at the business. Departments that sent on-duty engine companies out to conduct fire inspections typically used the same approach, however businesses along the block could not miss "Big Red," or could hear the unmuffled exhaust system of the old fire trucks, thus they had a few minutes to clean things up or try to remove any obvious hazard.

Most contemporary fire departments have discovered that it is far better to call businesses and make an appointment for the inspection. Making appointments accomplishes two important things. First, it gives the business an opportunity to build the inspection into its routine that day and is less disruptive to its overall operation. Remember time is money, and fire inspections take time; therefore inspections are a cost to the business, not part of its profit for that day. Second, it does give it an opportunity to straighten things up and put its best foot forward. After all, most of us like to straighten up the house before we have company and businesses are no different; they simply want to put their best foot forward. As professionals, we should give them the respect and courtesy of giving them the time to prepare their businesses for our visit.

Inspections conducted by appointment allow the inspector to put her or his best professional face on too. Files can be pulled and previous inspection histories reviewed before the inspection, enabling the inspector to become familiar with what they might find and to identify some critical areas that should be examined. If the problem still exists or has recurred, then this point needs attention. Examining the files also gives the inspector an opportunity to review any of the more technical code provisions that might apply and time to talk to any specialists in the office before conducting the inspections. All of these activities help the inspector get organized for the inspection, which makes the inspection process more efficient and conveys to the business owner a high level of professionalism. Figure 7.1 shows a sample inspection form.

Permit or License Inspections

Some businesses are required to have routine fire inspections conducted by the business license or special use permit they hold. Licenses and permits run the gamut from liquor licenses to health care facilities. Hazardous materials occupancies typically have a range of permits; the state's department of environment quality or a city ordinance or some other state law may apply. Needless to say, these permits or licenses are granted to businesses that meet the particular requirements of the permit or license and typically require that they must meet all other applicable codes and ordinances, including fire codes and fire safety ordinances.

In any case, these occupancies must meet all of the regular fire safety codes and ordinances in addition to any specific rules that might apply only to their business. These inspections tend to be more technical and typically

Inspection Checklist
Inspection Procedures

PREINSPECTION CHECKLIST
Equipment: _____

General
❑ Identification (photo ID)   ❑ Business work hours

Clothing
❑ Coveralls   ❑ Overshoes   ❑ Boots

Personal Protective Equipment (PPE)
❑ Hard hat   ❑ Safety shoes   ❑ Safety glasses
❑ Gloves   ❑ Ear protection   ❑ Respiratory protection

Tools
❑ Flashlight   ❑ Tape measure(s)
❑ Pad (graph paper) and pen or pencil   ❑ Magnifying glass

Test gauges
❑ Combustible gas detector   ❑ Pressure gauges   ❑ Pitot tube or flowmeter

Plans and Reports
❑ Previous reports   ❑ Violation notices   ❑ Previous surveys
❑ Applicable codes and standards

Notes: _____

SITE INSPECTION
Property Name: _____
Address: _____

Occupancy Classification

❑ Assembly   ❑ Educational   ❑ Day care
❑ Health care   ❑ Ambulatory health care   ❑ Detention and correctional
❑ One- and two-family dwelling   ❑ Lodging and rooming   ❑ Hotel/Motel/Dormitory
❑ Apartment   ❑ Residential board and care   ❑ Mercantile
❑ Business   ❑ Industrial   ❑ Storage
❑ Mixed

(Page 1 of 2)

▶ **Figure 7.1** Inspection Checklist. *(Source: Fire and Life Safety Inspection Manual, NFPA, 2002, Form A-1.)*

Hazard of Contents
- ❑ Light (low)
- ❑ Mixed
- ❑ Ordinary (moderate)
- ❑ Special hazards
- ❑ Extra (high)

Exterior Survey
- ❑ Housekeeping and maintenance

Building construction type
- ❑ Type I (fire resistive)
- ❑ Type IV (heavy timber)
- ❑ Type II (noncombustible)
- ❑ Type V (wood frame)
- ❑ Type III (ordinary)
- ❑ Mixed

Construction problems
Building height _____ feet _____ stories
- ❑ Potential exposures
- ❑ Outdoor storage
- ❑ Hydrants

Fire department connection
- ❑ Vehicle access
- ❑ Is it obstructed?
- ❑ Is it identified?
- ❑ Drainage (flammable liquid and contaminated runoff)
- ❑ Fire lanes marked

Building Facilities
- ❑ HVAC systems
- ❑ Gas distribution systems
- ❑ Conveyor systems
- ❑ Electrical systems
- ❑ Refuse handling systems
- ❑ Elevators

Fire Detection and Alarm Systems
See Form A-8.

Fire Suppression Systems
See Form A-10.

Closing Interview
- ❑ Imminent fire safety hazards
- ❑ Housekeeping issues
- ❑ Maintenance issues
- ❑ Overall evaluation

Items to be researched:
- ❑ _____
- ❑ _____
- ❑ _____

Report
- ❑ Draft
- ❑ Review
- ❑ Final

Notes: _____
_____
_____
_____

(Page 2 of 2)

▶ **Figure 7.1** *(Continued)*

require the inspector to have some special training in the specific area. Since these inspections are mandated by the licensing or permit issuing agency and since these agencies are often part of a higher level of government than the city or town, they may take precedence over other routine inspections. This becomes especially acute where budgets are slim and resources are scarce, yet they also offer the department an opportunity to collect some fees by making them a part of a special set of permit inspections. There are myriad rules for all of these licenses and permits that require fire prevention bureaus to dedicate some amount of time to understand them before jumping into a fee system that appears to be a cash cow, at least on the surface, to overcome budget shortfalls. It is good to keep in mind that the purpose of fire prevention in the community is to provide a reasonable level of safety from fire; not to produce revenue for the general fund.

Target Hazard Inspections

Target inspections have long been a concern of fire departments because they typically present special tactical challenges to the department. Again, from the 1920s through the 1950s or 1960s, target hazards included industrial and manufacturing plants, rail yards, lumber yards, paint stores, dry cleaning establishments, and gas stations, just to name a few. However, over time some of these occupancies have been removed from the list of target hazards while others have been added. For example, gas stations were once considered a target hazard that should be inspected quarterly. Today, they are a low priority for most departments after the initial inspections following construction or major remodeling and in some cases are put on the list of occupancies that are candidates for self-inspections.

Many of the target hazards are required to have a license or special use permit, which in turn requires regular fire inspections, while some do not. It was held during the 1960s through the 1980s that fire companies should inspect target hazards as part of their fire prevention responsibility. This idea became especially prevalent with the increase in hazardous materials occupancies in many cities and what seemed to be a corresponding reduction in fire prevention inspection staff. Although this seemed to be a good idea, at least on the surface, many fire departments found that suppression fire fighters did not have the education and training necessary to conduct such technical inspections and the inspection part of the job fell back on the fire prevention bureau. Yet the suppression units are called on to respond to emergencies in these occupancies, so special programs have been designed to train suppression forces to conduct tactical and strategic inspections along with fire prevention specialists who conduct the code inspections. This approach has proved to be very beneficial; by integrating the fire department resources, yet letting them focus on the areas they are best suited for; it presents a very effective and professional program to the business community.

Target Hazard Inspection Priorities

Priorities for targeting hazard inspections are typically set based on the fire experience in these occupancies. Although these priorities reflect the fire experience reported at the national level, they are often adjusted based on the local experience. It is common that major fires that cause large life loss are often the catalyst for a surge in target hazard inspections. The most recent example is the nightclub fire in West Warwick, Rhode Island, that killed 100 people. "The West Warwick nightclub fire at The Station, which erupted during a pyrotechnic display on February 20, ranks as one of the deadliest fires in nightclubs and other social assemblies in U.S. history. NFPA has been asked by the Rhode Island state fire marshal's office to assist with the analysis of this tragic incident" (NFPA Web site, www.nfpa.org, 2003).

Historically there have been a number of occupancies most fire departments identify as target hazards such as lumber yards, fuel depots, rail yards, high-rise buildings, hazardous materials occupancies, and heavy industrial and manufacturing businesses, all based on actual fire experiences. Some communities should consider shopping malls and large warehouses target hazards based on the suppression capability they have, whereas most large municipal fire departments handle these occupancies in a more routine fashion. Special target hazard programs might be in order for a jurisdiction that finds it has a number of cultured marble manufacturing businesses after responding to a fire in one; especially if it discovers that most of these businesses are operating in buildings that are not properly protected for hazardous operations. This one experience might be enough to cause the fire department to make cultured marble manufacturing a priority target inspection. Another example is a community that has a rash of small fires in paint spray booths that would go virtually unnoticed by fire prevention if it were not for the data provided by fire investigators. Here fire investigators can work closely with fire inspectors using the data collected from actual fire experience to develop the most appropriate target hazard inspection program.

Self-Inspections

Most fire departments do not have enough staff in the fire prevention bureau to do everything they are required to do by law, let alone do things that the department and the suppression units need them to do. One of the relief valves adopted by some departments to try to meet their obligations and at the same time relieve their staff is the adoption of a self-inspection program. The effectiveness of these programs is controversial to say the least, but they are an honest attempt to meet the responsibilities of the department with limited staff. Also, these types of inspections are normally directed at small businesses that do not have any special hazards.

Educational materials are generally developed, which can be mailed to the business, that give the business owner or manager some general fire safety information and a check sheet that can be used to check for hazards, as shown in Figure 7.2. Some programs require the business to mail the check sheets back to the fire department to verify the inspection has been completed; some programs include some random physical follow-up visits by the department to verify what was done and answer any questions the business might have. These inspections are not to be confused with industrial and manufacturing plants that have in-house staff responsible for conducting routine inspections in the plant.

It is important to keep in mind that self-inspection programs are useful, but they can generate the very work that they are supposed to mitigate. When forms that note violations are completed and returned to fire prevention, the department is responsible for doing due diligence and conducting a follow-up inspection. There needs to be some form of response to the self-inspection report sent to the department, and businesses need to be clearly informed that they are not exempt from complying with the provisions of the code just because they did a self-inspection.

▶ ENGINEERED SYSTEMS

Engineering has played a prominent role in fire prevention and fire protection for more than one hundred years. Automatic fire sprinklers were introduced in New England textile mills prior to the turn of the 20th century. Since then, engineers have developed automatic devices that detect fire and have found sophisticated ways to suppress fire without water. Construction engineers, civil engineers, mechanical engineers, electrical engineers, and fire protection engineers have worked on fire safety problems for decades, seeking ways to provide better fire protection systems. These engineered systems range from small devices to major mechanical systems used in buildings to the engineering of actual construction components that can take advantage of new technology, new materials, and creative designs. The fire service has moved from a time when construction was fairly standard, using traditional building and fire codes, to a place where sophisticated engineering solutions

This checklist is for the sole use of the business owner/occupant to enhance your awareness of fire and life safety in your place of business and is not intended to regulate any portion of the City of Phoenix Fire Code.

	Yes	No
• Is the address clearly visible and marked in large numbers to be seen from the street?	☐	☐
• Is the fire extinguisher in a visible and accessible location?	☐	☐
• Is the fire extinguisher classification 2A10BC or a greater classification?	☐	☐
• Is there at least one fire extinguisher within 75-feet travel distance from anywhere in the business?	☐	☐
• Has the fire extinguisher been serviced within the last year?	☐	☐
• Is there only one thumb turn or key lock on each exit door? (no slide-bolts)	☐	☐
• Does required exit door(s) have a sign that reads "This door to remain unlocked during business hours"?	☐	☐
• Are exits identified and/or lit, if illuminated?	☐	☐
• Is the rear exit door marked with an exit sign?	☐	☐
• Do the electrical panels have a 30-inch clearance in front for easy access?	☐	☐
• Are all stored materials stacked so they are at least 2 feet below the ceiling?	☐	☐
• If you have emergency lighting, does it work?	☐	☐
• Are areas outside and around the building free of dry weeds, debris, or trash?	☐	☐
• Are extension cords less than 6 feet long and used only for temporary wiring?	☐	☐
• Does electrical outlet(s) have two (2) or less appliances plugged into it?	☐	☐
• Do the telephones have 911 stickers on them?	☐	☐
• Are all large commercial dumpsters (garbage containers) at least 5 feet away from combustible walls, window or door openings, or combustible roofs?	☐	☐
• Is there less than 6 gallons of combustible or flammable liquids stored on site?	☐	☐

If the answer to any of these questions is *No,* you should correct the situation so that the answer is *Yes.* A *Yes* answer would indicate item is in compliance. It is the responsibility of the occupant to comply with the checklist by making all necessary corrections if needed.

Our goal is to increase your awareness of Fire and Life Safety, reduce your risks and liability, and to make your business a safer place for everyone in the community.

▶ **Figure 7.2** Example of Self-Inspection Checklist for Businesses. *(Source: Adapted from Urban Survival for Business Safety Checklist, Phoenix Fire Department, Division of Urban Services.)*

are used (and needed) to meet the challenges presented to them by a market driven by creative designers and end users. Many of the early engineering developments were only allowed by the codes on an exception basis, often requiring independent testing for an individual application. However, over time, many of these engineered systems were recognized by the codes as acceptable alternatives and in some cases were used to replace previous requirements. The engineering community, particularly the fire protection engineering community, has advanced the engineering discipline to a level that allows the use of performance-based codes. In collaboration with the fire service, the architectural community, and others who are stakeholders in developing the built environment, performance-based codes offer the opportunity to apply the best technology, materials, designs, and construction techniques available on the market from throughout the world.

Traditional Designs

Traditional building codes and architectural designs have always relied on some level of systems engineering. Many of the existing code provisions were specially designed engineered systems before they were adopted and made part of the standard code provisions. For example, although it no longer does so, the original automatic fire sprinkler standard, NFPA 13, *Standard for the Installation of Sprinkler Systems,* called for black steel pipe of particular size and design [5]. Then came hydraulically calculated systems, originally considered an engineered system for special situations, that became the accepted standard and a regular part of the fire code. Over the years the codes have more universally accepted engineered systems from the design community, which continues to press for the adoption of the latest technology and engineering design. It seems that no matter how hard code developers try, they can never keep up with the changing technologies and new materials available. Coupled with the competition by design companies to find better, improved, innovative, and more interesting ways to build buildings, the challenges presented to most fire departments have been almost overwhelming in the past two decades. Engineered detection and suppression systems are only a few of the challenges as building owners and architects create larger open spaces and where heating, ventilation, and air conditioning systems (HVAC) become more sophisticated, often used as the smoke removal system.

Unique Structures

While so-called traditional buildings become more complex and challenging, a greater number of buildings fall into the "unique structure" category. The most obvious of these buildings are sports stadia. Whereas some are just large open structures that present unique egress challenges and built-in detection and suppression challenges in and around the enclosed spaces, others present unique problems. Bank One Ballpark in Phoenix, Arizona, is a domed baseball stadium with a retractable roof. The Mall of America in Minneapolis, Minnesota, presented a very real set of design and protection challenges to architects, engineers, and regulators in trying to apply the traditional codes to a building that seemed to violate many of the very fundamental principles of fire safety design. Basketball arenas, airport terminals, high-rise buildings with oversized (as defined in traditional codes) atriums, are only a few of the types of structures any fire department in America could be faced with.

These structures are not confined to the so-called big cities, as more and more unique buildings are being built in rural and suburban areas. Large warehouses and super malls can also be challenging to work with, from a fire safety point of view, and it is not uncommon for developers to choose more rural locations in attempts to escape more stringent requirements in an adjacent larger city.

Performance-Based Designs

Performance-based designs set the general goals, objectives, and intended outcomes that have to be met in a building regarding fire and life safety features and allow the architects to work with a design team to address each issue rather than just using the prescriptive approach (a specific set of solutions) outlined in the traditional fire and building codes. The challenges to traditional building design and the greater number of unique structures in combination with the maturing of the fire protection engineering discipline have all been party to the advent of performance-based design. Architects have been challenging the code development community for decades to allow them to be more creative with the design and use of their projects.

Many architects have expressed frustration with the traditional code requirements that limited the use of certain materials or assemblies. They argue that they can build a safe building that provides for the functions their clients need and do it in unique and attractive ways, but they are too limited by existing regulations. Additionally they argue that the prescriptive codes that tell them exactly what to do, where and when to do it, and how to do it are not universally accepted or adopted, which further complicates what they have to do from project to project. All of the foregoing concerns, the developments over the years in design capability, and the movement toward deregulation and international trade have provided for the proliferation and acceptance of performance-based design throughout the world.

The Society of Fire Protection Engineers (SFPE) has responded to the challenge through its involvement in research and publications and by working to develop its discipline to the same level other engineering disciplines have achieved, for example, civil, mechanical, and electrical to name a few. A specific example of their work is the *SFPE Engineering Guide to Performance-Based Fire Protection Analysis and Design of Buildings* [6]. The society published this book for fire protection engineers who are engaged to help owners and designers take advantage of performance-based codes. Fire protection engineers designed this guide for fire protection engineers and it lays out the process that should be followed to successfully apply performance-based design criteria. Although this guide specifically states its purpose is to help fire protection engineers, it is also a very valuable reference for all stakeholders in a project and should be especially valuable to the fire service by helping them understand the process that should be used.

▶ CODE DEVELOPMENT, ADOPTION, AND IMPLEMENTATION

The United States is the only industrialized nation in the world that does not have a national (federal) authority responsible for the development, promulgation, adoption, and application of fire and building codes. Other countries find this situation very strange, yet in the United States, most people accept it as part of our heritage. It was common years ago for most of the major cities in the United States to write their own fire and building codes, but over the years, most, if not all, have adopted one of the model fire codes. Model building and fire codes were originally developed by three regional model code organizations.

Code Development

The Building Officials and Code Administrators (BOCA) wrote and published the National Building Code with a companion fire code that was used primarily in the eastern part of the nation. The Southern Building Code Congress International (SBCCI) wrote and published the Southern Building Code and companion fire code used primarily in the southern part of the United States. The International Conference of Building Officials Code (ICBO) wrote and published the Uniform Building Code and adopted a companion fire code that was adopted primarily in the western part of the United States. The original and primary intent of each of these groups was to write a building code. The fire codes came later with BOCA and SBCCI working with fire marshals to write a companion fire code that principally focused on maintenance issues.

The Uniform Fire Code (UFC) was originally written and published by the fire prevention officers in California and subsequently published by the ICBO through the efforts of the Western Fire Chiefs Association (WFCA). The WFCA is a division of the International Association of Fire Chiefs (IAFC). The importance of this background is to better understand who was writing and publishing the codes and who was able to vote on the adoption of a code by a code-making body. Although some minor deviations existed, the general approach was the same. Voting members were building officials and only later in the 1980s and 1990s were fire officials allowed to be voting members of the code development committees of the Uniform Building Code.

The NFPA also wrote and published a fire code, NFPA 1, *Fire Prevention Code*, but the process used to develop and adopt this code by the NFPA is considerably different than the process used by the three building code organizations. NFPA's *Fire Prevention Code* also meets the requirements set forth by the American National Standards Institute (ANSI), which qualifies it to be called a *consensus code*. ANSI is a private, nonprofit organization that administers and coordinates the U.S. voluntary standardization and conformity assessment system. This designation is significant in that ANSI requires that the process of development and adoption is open to any interested party and provides a selection process for participation. The selection process requires that various disciplines are represented on the committees that write the code, that any member of the NFPA can vote on its adoption, and NFPA membership is open to anyone who wants to be a member. This process is considered to be much more open than the process used by the model code groups because voting membership is limited to building officials and some fire officials. The codes are developed by a code committee and are voted on by the general membership at an open meeting. Even though architects, engineers, scientists, and other professionals can make presentations to the code committee and be member of work groups in the model code processes, they cannot vote on the final product like they can in the NFPA process.

The three regional model code groups no longer write or publish building and fire codes. Instead they have merged to create the International Code Council (ICC), which is tasked with writing a single building code, the International Building Code (IBC) and a single fire code, the International Fire Code (IFC). Much of this was in response to pressures put on the code development community to create a model building code and a model fire code that would be universally accepted throughout the United States. The new NFPA 1, *Uniform Fire Code*™ [7], is a joint venture between NFPA and the Western Fire Chiefs Association that incorporates provisions from both the NFPA's *Fire Prevention Code* and the WFCA's *Uniform Fire Code* to create one document.

Code Adoption

The development process in each of the model code organizations results in a model code that can be adopted by any jurisdiction in whole or in part. The next part is the tricky part of the American system. The code adoption phase is the decision point where each individual state or local jurisdiction is responsible for setting public regulatory policy that it will enforce in its state or community. Think of these codes as only another set of books on the shelf at a bookstore, ready for purchase. These codes are not law, statute, or ordinance until adopted by a political jurisdiction and made part of its regulatory process. However, it is much more complicated than that because each jurisdiction tends to believe that the characteristics and problems it must address in its community is unique to it and requires unique solutions (i.e., it needs to modify its codes) to deal with them.

Fire code and building code compatibility is a major concern of most fire marshals and building officials, that is, the requirements of the fire code must not conflict with the requirements of the building code and vice versa. The ICC fire and building codes are developed as companion codes (companion means the codes are compatible with each other) primarily based on a melding of three original model codes published by the BOCA, ICBO and SBCCI. The ICC family of codes is called the I Codes. The NFPA has developed and now publishes a building code, *NFPA 5000™*, *Building Construction and Safety Code™* [8], so it has a companion set of codes that are compatible with each other and compatible with NFPA *101®*, *Life Safety Code®* [9], and all of the NFPA standards. The NFPA family of codes is called the C3 codes. All of these efforts are driven by the need to have compatibility of codes in any given jurisdiction so the local jurisdiction can adopt a "family" of compatible codes, rather than a single code.

The response to these efforts are mixed at the local level by fire marshals and building officials alike. Each jurisdiction has a current code that they use, but in most cases, a new edition will not be published as it has been in the past because the model code organizations are no longer maintaining their own separate codes and even though 16 states had adopted NFPA 1 [7], NFPA did not have a building code. But the landscape has changed and the options in the future are different. Because local jurisdictions are not normally bound to adopt the latest edition of a code, many of them are waiting to see how the most adventurous jurisdictions make out by adopting these codes. This wait-and-see attitude is not new since historically many jurisdictions have made it a policy not to adopt the latest codes until others have adopted them and worked out the bugs.

Although it might seem that the design and development community has accomplished a great deal in moving toward a more uniform code system in the United States, it is important to examine the impact of the adoption process on uniformity. Some states have adopted a statewide building and fire code with a mini-max provision. Mini-max simply means that local jurisdictions cannot add additional code requirements to the state codes nor can they adopt codes that are not as stringent as the state code. However, it is still more common that most individual jurisdictions, especially the larger cities, enjoy a greater amount of autonomy.

It is not uncommon for a particular region of the country to have a group of individual cities that have all adopted the same model code, but have modified the code to suit their particular constituencies. The bottom line is that fire officials need to work with elected officials to gain the political support required to make changes in existing codes or to adopt new code editions. As an example of a more stringent code requirement adopted to achieve higher levels of fire safety than that required by the minimum code, some communities might require automatic fire sprinklers in single family residences. Whereas some communities may require sprinklers where water is scarce, others set maximum size (sq. ft or sq. m) requirements or provide some other trigger before sprinklers are required. These are but some examples. Other examples might include spacing of fire hydrants, water main sizes, detection requirements, and on and on. The bottom line is that architects and developers must still study and be familiar with a myriad of local code amendments that make the codes different among jurisdictions, even among adjacent jurisdictions. Having said this, arguably the system is becoming more user friendly, codes are becoming more consistent, and the fundamental principles and assumptions used to explain why a code provision is adopted is more typically scientifically based.

Code Implementation

Implementing codes is a tremendous challenge for any community regardless of size. Larger communities tend to have more staff but they also have more buildings and types of occupancies to regulate. Training fire inspectors becomes one of the critical needs, and using fire companies becomes an internal policy issue. More progressive fire departments hire specialists to work with high-hazard occupancies and to consult with architects and engineers during the design and plans review stage. These specialists also provide a good resource for fire inspectors to use when they encounter difficult, unusual, or unfamiliar problems. Many of these specialists are engaged in the development process, which gives them a better insight into why a code provision is written the way it is or simply a deeper understanding of the intent of the code provision in question.

Planning and Zoning Ordinances.   The term ordinance is usually used to describe the provisions set forth by local jurisdictions, as opposed to laws set by the federal government or statutes set by state government. The building code adopted by a city, for example, is typically called an ordinance, as is the fire code, or a city requirement to keep dry vegetation cleaned from vacant property. Planning and zoning ordinances regulate what type of occupancies can be built, where they can be built, how they can be built, and in some cases how they can be designed. These ordinances are important to the over-all fire safety regulatory environment because they establish setbacks from streets, distances allowable between certain types of buildings and occupancy uses, as well as the mix of property that can be built. For example, most cities would not allow dynamite manufacturing within the city limits or the building of a residence next to a petroleum plant. Malls and shopping centers can only be built in certain areas and high-rise buildings are often restricted to certain geographical areas of the city.

Planning and zoning committees are typically appointed by elected officials to administer these ordinances, including appeals and requests for exceptions to them. Often the fire code or building code is a critical factor in whether the exception is granted. It is imperative that the fire marshal participate in the adoption processes as well as the administration of appeals and requests for exceptions as much as is necessary so the fire code is not lost in the detail of the community's general plan and its zoning ordinance. Fire marshals are often asked to testify before the zoning commission on issues that they believe affect the fire requirements in the fire code, but it is also important for fire marshals to review new ordinances and revisions to existing ordinances as a matter of routine. Usually one phone call is all that it takes to get put on the list of reviewers.

Regulations.   Regulations are essentially a set of rules that detail how a law, statute, or ordinance will be enforced by a regulatory agency. For example, a state statute may require that state licensed day care centers meet all applicable state and local fire codes and give the authority to enforce this law to the state fire marshal. The state fire marshal must then write a regulation that details what agency(s) should carry this out, how they should do it, and outlines the consequences of noncompliance. Although regulations typically require public input before they are enacted by the state agency (in this case the state fire marshal), the state agency usually has a great deal of autonomy. The process tends to be similar at the local level. For example the city fire code is an ordinance and then the department might write a regulation for applying some specific section where the regulation provides greater detail for owners, contractors, and inspectors to follow.

Standards.   Standards are different than codes and regulations in that they specify in great detail how something should be installed, what materials should be used, where the standard is applicable, and sometimes, more importantly, where the standard is not applicable. The NFPA develops and publishes more than 300 codes and standards (NFPA Web site, www.nfpa.org, 2003). The standard for the installation of automatic fire sprinklers in commercial occupancies is NFPA 13, *Standard for the Installation of Sprinkler Systems* [5], which provides all of the detailed requirements for the installation of automatic fire sprinklers, including types of acceptable pipe, bracing, testing, and inspections to name only a few areas. Codes then include where these standards are applicable (or not). For example the fire code informs the user about where automatic fire sprinklers are required and the standard provides the detail for installation.

▶ ## FIRE AND LIFE SAFETY EDUCATION

Fire prevention mostly focused on inspections from the 1920s through the 1940s and then in 1947 at the fire prevention conference called by President Truman, it was made clear that public education initiatives were needed to reduce or mitigate the nation's fire problem. Several specific recommendations were suggested in the committee reports ranging from education of the business community to education of children in schools [2]. One of the recommendations called for revising curricula in colleges of engineering and architecture, and engineering schools. Many of the recommendations have been implemented in a variety of formats with varying degrees of success.

Public Education Organizations and Departments

Public fire education seemed to come into its own right in the 1980s with many fire departments creating public education sections within fire prevention while some departments, state and local, even created separate departments (at the state level) or separated divisions within a local fire department. These changes have had the effect of elevating the status and support for public fire education within the fire service. Support is typically translated into resources, which in turn provide money and staff. Even so, the resources available at the state and local level for these activities pales in amount compared to resources provided for suppression and enforcement programs.

Because of these realities, public fire educators have been much more aggressive, and successful, in garnering support from the community through collaborative and partnering efforts. As they reached out to the community, they created associations and organizations that focus on various aspects of the community's fire problem. The membership of these organizations typically reflects the demographic makeup of the community, where they often have more nonfire department members than they have fire department members. Most of these organizations are nonprofit (not-for-profit) organizations that sponsor fund-raisers (golf tournaments are big fund-raisers) to supplement the small amount of public funding they receive. Bringing together the private sector with the public sector, citizen groups, and special interest groups, public fire education has not only captured much-needed funding, it has also developed soul mates to help carry the message.

Public Education Programs

Nationally, cooking equipment remains the leading cause of structural fires and fire injuries, smoking is the leading cause of fire deaths, and intentionally set fires generally cause the largest share of property damage. According to the FBI's report, "Crime in the United States," roughly half of the people arrested for arson were under 18. In 1999, electrical distribution equipment caused the largest share of property damage in home structure fires. Heating equipment is also a significant factor in the country's structure fires [10]. The elderly, disabled, and the young continue to be at greatest risk. Locally, fire investigators play a key role in identifying the specific elements of the fire experience in their community. Contemporary fire departments use all of these data to build effective and targeted public fire education programs. *Risk Watch* targets children in elementary schools. Although it was developed by a consortium through the leadership of the NFPA, based on national data, it is delivered at the local level by fire departments and schools.

One fire department developed a special education program for building contractors, which focused on safe fire practices during construction, especially related to the use of open flames around exposed and unprotected building material. Plumbers' torches and roofers' kettles were responsible for a rash of fires that burned several apartment complexes when they were in the framing phase of construction. As a result of this special educational program, these fires were essentially eliminated. Other examples include programs designed to teach the elderly what to look for and how to be more fire safe. Typical inspection programs focus on commercial occupancies even though roughly 80 percent of the fires in the United States are residential fires with more than 50 percent of them occurring in one- and two-family dwellings. Because privately owned and occupied residential properties are outside the scope of most fire codes, home inspections are voluntary, if conducted at all. Postfire public education programs are an especially effective method of getting the fire safety message into the home, and the message can be targeted to the local fire experience.

Learn Not to Burn.   The development of educational components was soon followed by the introduction of the NFPA *Learn Not to Burn* program in 1979, then by *Risk Watch* in 1998. Both of these programs focus on children and are typically offered as part of the elementary school program in cooperation with the schools. The *LNTB* program teaches 22 key fire safety behaviors. Intended for use at three elementary school levels, *LNTB* is designed so that it can be integrated into a teacher's regular classroom curriculum with a multitude of educational resources, as shown in Figure 7.3. By

▶ **Figure 7.3**   Some of the *Learn Not to Burn*® Resource Materials.

integrating fire safety into normal elementary school subjects that range from math and science to the humanities and history, children can learn how fire safety is a part of everyday life (see the NFPA Web site, www.nfpa. org/Education/LNTB.asp). The *LNTB* program is also offered in Spanish and French for international students.

LNTB Preschool Program. The educational community has known for a long time that children are able to grasp many concepts at an early age. Since children under 6 years old are at high risk for burn injuries, the NFPA developed a program for grades preschool-K. This program takes eight key fire safety behaviors children can develop to avoid fire hazards and prevent burn injuries. Each lesson includes information for the teacher, cassettes with music, learning activities that are fun and designed specifically for these youngsters, letters for the parents, and an activity sheet. The knowledge children learn at this early age is then further developed and reinforced in the *LNTB* and *Risk Watch* programs as they matriculate through school. See Figure 7.4 for an example of some of the materials used in this program.

Risk Watch. *Risk Watch* was developed by the NFPA with the help of nationally recognized experts from the fields of prevention and safety and expands on the *LNTB* experience by addressing injuries related to eight topic areas. They include motor vehicle safety; fire and burn prevention; choking,

▶ **Figure 7.4** *Learn Not to Burn® Preschool Program* Resource Materials.

suffocation, and strangulation prevention; poisoning prevention; falls prevention; firearms injury prevention; bike and pedestrian safety; and water safety (see NFPA Web site, www.nfpa.org/riskwatch/home.html). Like the *LNTB* program, *Risk Watch* is a school-based curriculum but is broader in scope and targets five grade levels instead of three. The grade levels are Pre-K/Kindergarten, Grades 1–2, Grades 3–4, Grades 5–6, and Grades 7–8. Injuries in these topic areas are commonly encountered by today's contemporary fire departments. Through the delivery of EMS and the aggressive expansion into public fire education programs, fire departments found that their programs were not comprehensive enough to address the full range of injuries and accidents they encountered. *Risk Watch*, as shown in Figure 7.5, fills that gap and, by expanding the range of grade levels, it provides for involvement in safety and prevention programs through the eighth grade. Some jurisdictions are expanding their reach into high school as well in an attempt to make sure that key safety behaviors become life skills carried throughout a person's personal and professional life.

Business Education Programs.   Educational programs have also been created for businesses. Using a format familiar to business people, these pro-

▶ **Figure 7.5**   *Risk Watch*® Program Being Used in a School.

grams use a seminar or conference structure to deliver the message. Often offered at a local hotel conference site, the seminars include speakers, workshops, Q&A sessions, displays, and lunch. This approach provides a safe environment for businesses to ask questions about fire safety issues without feeling threatened by a harsh bureaucracy; but that is how business customers often feel when a fire inspector is on their premises. Workshops provide additional opportunities for fire safety specialists to answer more pointed questions. Speakers are not always from the regulatory community. Businesspersons, who have been successful in complying with code requirements, especially if they had seemingly adverse conditions to overcome, can be especially helpful as speakers and workshop leaders.

▶ SUMMARY

Fire prevention programs continue to be fundamental to a community and its overall strategy to provide effective fire and life safety. The notion of fire prevention is not a new concept, but the approach taken to fire prevention in the United States at the beginning of the 21st century, is considerably different than the approach used at the turn of the 20th century. Over the brief span of one hundred years, science and engineering have found a legitimate place in the application of academic theory and knowledge to the practical problems of fire prevention and fire and life safety education. Three historical events have helped shape the direction fire prevention is taking in the United States: the President's Conference on Fire Prevention in 1947, the Wingspread Conferences, held every 10 years from 1966 through 1996, and the National Commission on Fire Prevention and Control report, *America Burning*. These historical events have provided the national drive to evolve fire prevention and life safety education in the United States to what it is today.

Over the years fire prevention bureaus and programs were developed in response to a wider range of fire and life safety programs beyond those narrowly defined as fire problems. Programs that address playing with guns and gunshot wounds, water safety and drowning, to car seat safety and buckle up campaigns are but a few of the ways the fire service has expanded its approach to prevention and education. Contemporary fire and life safety programs reach across the community and involve the delivery of diverse programs to diverse populations.

Fire prevention is typically a function of state and local fire departments (i.e., city, town, or county) but has global implications. Fire departments provide services to their communities using international standards, design techniques used in other countries from all over the world, and technologies that are developed worldwide. While fire inspections are still a major consideration of fire prevention programs, contemporary fire departments are involved in a variety of fire prevention and fire and life safety educational programs, not the least of which are the NFPA *Learn Not to Burn, Learn Not to Burn Preschool,* and *Risk Watch* programs.

Group Activity

Research the following topics to determine which was the most significant in the history of preventing fire loss, and present your findings in class:

- Code development
- Fire prevention organizations
- Public safety education

Review Questions

1. Who was the first president to call a conference on the United States fire problem and when was the conference?
2. What are three of the recommendations that came out of the conference in question 1?
3. What are the three E's?
4. What did Wingspread IV, 1996, emphasize?
5. What two organizations were created as a result of *America Burning?*
6. What types of programs emerged more prominently in the 1970s as many jurisdictions became concerned with injuries and the loss of life from a variety of sources beyond just the traditional fire injuries and deaths?
7. Working with the community on nonenforcement _____ _____ activities often builds the confidence in the community that the fire department needs to manage an effective code enforcement program.
8. What kind of designs set the general goals, objectives, and intended outcomes that have to be met in a building regarding fire and life safety features?

9. What do planning and zoning ordinances regulate?

10. Regulations are essentially a set of rules that detail how a law, statute, or ordinance will be enforced by whom?

11. Standards are different than codes and regulations in that they specify in great detail how something should be installed, what materials should be used, where the standard is _____, and, sometimes more importantly, where the standard is _____ _____.

Related Web Sites

Federal Emergency Management Agency (FEMA)
www.fema.gov

Home Fire Sprinkler Coalition
www.homfiresprinkler.org

International Association of Arson Investigators, Inc. (IAAI)
www.firearson.com

International Association of Fire Chiefs (IAFC)
www.iafc.org

International Association of Fire Fighters (IAFF)
www.iaff.org

International Code Council (ICC)
www.iccsafe.org

National Association of State Fire Marshals (NASFA)
www.firemarshals.org

National Fire Protection Association (NFPA)
www.nfpa.org
www.nfpa.org/education/index.asp

SOS FIRE: Youth Intervention Programs
http://sosfires.com/welcome.html

United States Fire Administration (USFA)
www.usfa.fema.gov
www.usfa.fema.gov/public/index/shtm

8 Controlling Fire Loss through Active Fire Protection Systems

Richard Pehrson
Futrell Fire Consult and Design

Learning Objectives

After completing this chapter, the reader should be able to do the following:

- Identify and explain the operation of various types of detection devices
- List the components of a rate of rise heat detector
- Discuss the principles of ionization and photoelectric detectors
- Describe the components of a water supply system
- Describe the value of fire suppression systems in protecting life and property
- Identify the components of a sprinkler system
- List and explain the components of sprinkler heads
- Discuss the possible uses of total flooding systems and other fire suppression systems
- Identify and explain common automatic fire suppression agents and equipment
- Identify and explain common fire suppression agents and equipment used for manual application

Even when a building is located in an area protected by a fire department, the fire and life safety goals expected by the public and mandated through building and fire codes often require added protection to be built into the structure. Passive fire protection systems were introduced in Chapter 4 as part of the discussion on building construction and examples include a concrete fire wall or structural fire protection materials sprayed on steel beams. This type of fire protection is termed *passive* because it performs its intended duty without the need for outside intervention (turning it on) or mechanical support (electricity or water supply). Although it can contain a fire and prevent the spread of smoke and flame, passive fire protection cannot warn sleeping occupants about a smoldering sofa in their living room or suppress a growing fire before it becomes dangerous.

Active fire protection systems, however, are able to sound an alarm of fire or apply a suppression agent when a fire is still small, but to carry out their functions they require the input of energy or water for example. This action could also include mechanical movement like turning on a fire pump or closing a smoke-tight door. Active fire protection systems are installed for the express purpose of positively changing the course and resulting outcome of a fire in a building. Fire protection systems can be broadly classified into the following four different groups:

1. *Automatic fire detection and alarm systems.* Fires predictably produce products of combustion including smoke, high temperatures, and radiant energy that can be identified by sensors placed throughout a building. Once smoke levels exceeding a certain threshold are measured inside a detector, a signal can be sent to an alarm system that notifies occupants in the building and/or the fire department. Although fire detection and alarm systems can reduce the time it takes for people to leave a building or how long it takes the fire department to arrive, they do not slow the growth of the fire or reduce the production of smoke, thus the fire hazard is still present.

2. *Automatic fire suppression systems.* Automatic fire suppression systems apply a fire-suppressing chemical such as water, carbon dioxide, dry chemical, or foam to a fire without the need for human intervention, resulting in a reduction in the hazard to occupants, structures, and contents. Water flow in a pipe due to the opening of a sprinkler can be monitored and used to signal an alarm of fire, thus automatic fire suppression systems both warn of a fire and control it, thereby reducing the hazard.

3. *Manual fire suppression systems.* In large or tall buildings, it is difficult and time consuming for the fire department to pull hose lines up stairs or across floors. To assist building occupants and fire fighters in quickly applying water to a fire, standpipes (vertical or horizontal pipelines) can be connected to a water supply to provide water where hose lines off fire trucks cannot easily reach. Because both the fire department must be present and valves opened, standpipes do not automatically control a fire. Although fire extinguishers and fire hoses are used against unwanted fires, they require a willing human operator to be present, so are another example of manual fire suppression.

4. *Automatic smoke control/exhaust systems.* When a fire occurs, certain vulnerable populations such as hospital patients or prisoners may not be able to respond to a fire threat by exiting the building without assistance. Given that smoke is the leading cause of fire fatalities, automatic systems that remove smoke from buildings or control its spread provide added protection when a fire is controlled by sprinklers. The design of smoke control systems is a complex subject and is only touched on briefly.

▶ AUTOMATIC FIRE DETECTION AND ALARM SYSTEMS

Value of Fire Alarm Systems

Uncontrolled fire in a building is not a static event, but instead is characterized by fast-moving flame and smoke movement that, if left unchecked, will eventually consume an entire building. To protect people from fire, it is necessary that they (1) be aware of the situation either through observing fire cues like smoke or the sounding of an alarm (notification); (2) properly respond by making a decision to leave the building (reaction time); and (3) initiate travel through the exit system and continue to travel until a point of safety is reached (egress or evacuation time). The total time required for a person to reach a point of safety from the start of the fire is then the sum of the three components:

Time required to reach safety (evacuation time) =
notification time + reaction time + egress time

The problem is that while fire is rapidly consuming more and more fuel, smoke and hot gases are spreading throughout the building, making travel to a point of safety more difficult. Eventually the fire and products of combustion can make travel to a point of safety impossible, at which point the occupant becomes trapped, and his or her life is in danger. The time available from the start of a fire until the path(s) out of the building become impassable is known as the Available Safe Egress Time (ASET). When this available time for safe egress is longer than the time required to evacuate the building, people that decide to leave the building should be successful:

Time is available to exit the building when: ASET > Evacuation Time
Fire is growing faster than people can get out and are at risk when
ASET < Evacuation Time

What solutions are available to correct the situation when ASET is less than evacuation time, so that people have the opportunity to escape? The period it takes a person to walk out of a building during a fire (egress time) depends on how far they have to travel and how quickly the person walks. One of the primary functions of building codes such as *NFPA 5000*™, *Building Construction and Safety Code*™ [1], and life safety codes like NFPA 101®, *Life Safety Code*® [2], is to specify minimum requirements for building design to reduce the egress time by specifying maximum travel distances, locations of enclosed stairways, and numbers of exits in a building. Reaction time is more difficult to address because it deals directly with human behavior (deciding to leave) and physiological reactions (how long it takes the person to awake when sleeping). Some of the greatest gains in reducing the evacuation time can come through decreasing the notification time with fire alarm systems.

Especially when sleeping, building occupants may not become aware of a fire or emergency situation until it is too late—their path out is already blocked. The role of fire alarm systems is to (1) recognize that there is a fire—*detection*; (2) sound the alarm to let occupants know that there is a problem—*notification;* and (3) carry out preplanned actions such as notifying the fire department, closing doors, or activating smoke control equipment. Although many alarm systems are installed for life safety reasons, notifying the fire department early in the development of a fire allows them to arrive sooner, thus being confronted with a smaller fire that is easier to extinguish. Both improved property protection and continuity of business operations (protecting the building owner's mission) are provided. However, it is important to recognize that alarm systems only provide earlier notification to give additional time to leave the burning building, but they are unable to stop or slow the progression of the fire. Should a person decide to stay in the building or be unable to reach a point of safety due to a disability, alarm systems provide less benefit, although they can help notify the fire department earlier.

Detection and alarm systems are traditionally required by building, fire, and life safety codes in certain high-risk occupancies like hospitals, high-rise buildings, places of assembly, and where people sleep. These codes identify *when* an alarm system is required and what type of systems and level of coverage is to be provided (extent of coverage, how occupants are notified, and when alarms are monitored outside the building). In the United States, *how* alarm systems are designed and placed into buildings is addressed through an installation standard, *NFPA 72*®, *National Fire Alarm Code*® [3].

Fire alarm systems can be as simple as a single battery-powered smoke detector on up to complex multifunction systems that combine security, fire, communications, and data that share equipment to protect large building complexes or high-rise structures. Fortunately, many of the component pieces that make up alarm systems are similar. How the detection and notification is carried out and to what extent protection is provided accounts for many of the differences between systems. We first look at the devices that detect the presence of unwanted fire, followed by appliances used to notify occupants of the fire through visual or audible means, and finally the equipment used to control the alarm system and carry out intended functions.

Automatic Fire Detectors

Fires give off a number of telltale signs that they exist; two of the more obvious are smoke and heat. Less familiar, but still very real, are radiant energy and other combustion gases (CO_2, CO, etc.). What products of combustion are given off is highly fuel and scenario dependent, making it important that the correct automatic fire detector is selected for the hazard. A wide range of detectors is available simply because no one detector is adequate for all scenarios. These common fire signatures or products of combustion are used individually or in combination to automatically detect the existence of a fire.

Smoke Detection.   Although often used to describe all of the products of combustion, smoke is technically an aerosol comprised of the solids that form by the coagulation or sticking together of carbon, tars, and other particulates resulting from combustion, the exact composition of which depends on factors such as fuel, amount of oxygen available, and at what point in the course of a fire the smoke is produced. A volume of smoke tends to age with time, resulting from the collisions of particles, tending to increase their size as smaller particles stick together. All of these factors impact the ability of a smoke detector to identify a fire, which directly influences how long it takes for the detector to operate. A smoke detector that includes a horn or sounding device to warn of a fire is not only a smoke detector, but it is now called a *smoke alarm* and is most often found as part of a household warning system.

Most smoke detectors operate on the *ionization principle* and are *spot type detectors* (operate only at one point in a space) that respond to changes in the flow of electrical current between two plates when smoke is present inside the detector, as shown in Figure 8.1. Figure 8.1(a) shows that the positively charged air molecules drift toward a negatively charged plate, while negatively charged air moves to the positively charged plate (opposite charges attract). A small current flow thus occurs during normal operation. When smoke is introduced into the detector, the charged air particles attach to the smoke, re-

AUTOMATIC FIRE DETECTORS

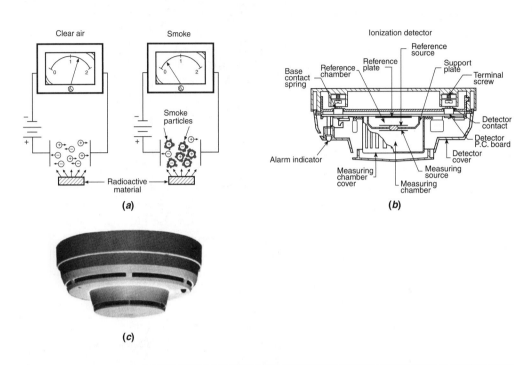

Figure 8.1 Ionization Spot-type Smoke Detector. (a) The operation principle of ionization where a small source of radiation induces a charge, or ionizes, air by changing the number of electrons on a molecule, (b) cross section of an ionization detector, and (c) an exterior view. (*Photo in (c) is provided courtesy of Fire Safety Divisions of Siemens Building Technologies, Inc.*) (*Source: Fire Protection Handbook, NFPA, 2003, Figure 9.2.9.*)

ducing their ability to move to one of the charged plates. As the current level drops below a set level, an alarm signal is sent. The greater the number of small smoke aerosol particles, the more effectively an ionization detector responds, thus it tends to respond well to a rapidly growing flaming fire.

Photoelectric smoke detectors are also known as light-scattering detectors based on the detection technology used to indicate the presence of smoke. When smoke passes a focused beam of light, the smoke particles are able to redirect the light in all different directions, known as scattering. A light-sensitive (photosensitive) receiver in the detector is able to register the scattering that is characteristic of smoke in the detector, as shown in Figure 8.2. A light source directs a narrow beam of light across the inside chamber of the detector and when only air is present, the beam is absorbed by a dark

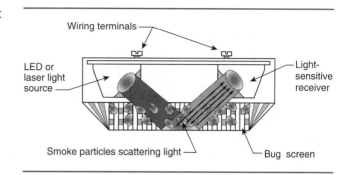

Figure 8.2 Photoelectric Smoke Detector. *(Source: National Fire Alarm Code Handbook, NFPA, 2002, Exhibit 3.34.)*

surface on the far side of the detector. When smoke particles enter the detector, a photosensitive portion of the detector is able to register the light scattered by the smoke, sending an alarm. A long-lasting light-emitting diode (LED) is typically used for the light source, and because larger particles deflect more light, photoelectric detectors tend to respond better to slow smoldering fires that produce a large dark smoke aerosol.

Smoke also blocks a light beam, leading to a newer technology of detecting fires, called *projected beam* or *light obscuration detectors.* An infrared (IR) radiation beam invisible to the human eye is emitted by one detector and picked up by the photosensitive surface of a second detector in line with the first. Because the beam is projected outside the detectors, they are able to protect large spaces such as atria, assembly halls, or arenas where they are placed on opposite walls and project their beam across the open space.

Heat Detection. Heat or high temperature is one of the more reliable indicators of a potential fire, although a common problem is that some fires may burn slowly or smolder for long periods of time, giving off insufficient heat to be detected before hazardous conditions are created for humans. As discussed in Chapter 3, the hot gases above a fire tend to rise until they contact a surface such as a ceiling, at which point they spread out along the ceiling. Heat detectors placed near the ceiling are in an ideal location to intercept the hot gas stream and detect the temperature change (increase) characteristic of fires through several different technologies (only a few of the many available are discussed here). Spot detectors sense heat only at a single location and so are placed throughout the protected space at regular intervals. Decreasing spot detector spacing or temperature ratings is a strategy to allow for a faster response, thus shorter time to detection.

Fixed-temperature heat detectors send a fire alarm signal when the temperature of the detector exceeds a predetermined level. Heat detection technology can include fusible element and bimetallic spot detectors, along with *continuous line detection* (a wire that detects fires over many locations at the same time), as shown in Figure 8.3. A small mass of metal is used to hold back a spring and on heating to the activation temperature, the metal melts, releasing the spring. Once free, the spring pushes an electrical contact closed and completes the circuit, indicating a fire. Due to the damage caused by melting, these detectors do not restore to a normal condition when cooled, therefore they must be replaced after activation.

Rate-of-rise heat detectors are designed to react faster to fire not only by responding as a fixed-temperature detector, but also by sensing the rate of change in temperature characteristic of rapidly growing fires. As shown in Figure 8.4, heating the metal detector also heats the air inside the detector, A, resulting in expansion. With slow heating, such as when a furnace turns on, the expanding air is able to escape through the small vent, B. With more

▶ **Figure 8.3** Fixed-Temperature Heat Detectors. (*a*) Diagram of a spot-type fixed temperature fusible-element heat detector and (*b*) Photo of a typical fixed-temperature (nonrestorable) heat detector. *(Source: (a) Fire Protection Handbook, NFPA, 2003, Figure 9.2.1; (b) Kidde-Fenwal, Ashland, MA; photo courtesy of Mammouth Fire Alarms, Inc., Lowell, MA.)*

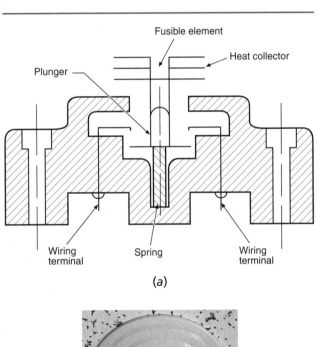

(a)

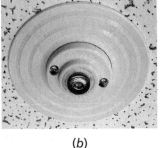

(b)

▶ **Figure 8.4**   Rate-of-Rise
Spot-Type Heat Detector.
*(Source: Fire Protection
Handbook, NFPA, 2003,
Figure 9.2.6.)*

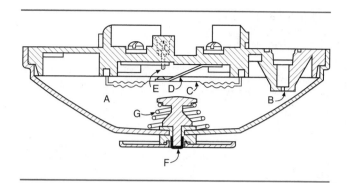

rapid heating during a fire, the thin diaphragm, C, is pushed upward, completing the electrical contacts, D and E, resulting in alarm. Should a slowly
developing fire occur, thermal element F and spring G operate as a fixed-
temperature detector. These detectors are typically designed to respond to
temperature changes greater than 12°F to 15°F (7°C to 8°C) that occur in
less than 1 minute and are available in both spot and line configurations.

Air-Sampling Fire Detectors.   Light scattering or obscuration is used in the
detection chamber of these detectors, but smoke is drawn through the detector from a series of small pipes or tubes located throughout the protected
area.

Flame Detection.   We are familiar with radiant energy in the heat energy
from the sun. When exposed to this energy for too long, our skin burns due
to overexposure. The spectrum of radiant energy, or electromagnetic energy, can be divided into three broad classes: Infrared (wavelengths less than
the visible spectrum), the visible spectrum composed of the colors that humans can detect, and the ultraviolet spectrum (wavelengths larger than visible). Flames give off differing amounts of radiant energy, again depending
on the fuel, burning conditions, and flame temperature. A flame detector
such as the one shown in Figure 8.5 can usually be used both indoors and
outdoors to identify open flaming, especially from fuels that lead to rapid
fire development like flammable liquids. Spark and ember detectors are also
available and can detect solid-state combustion such as a glowing piece of
coal traveling down a conveyor system.

Flame detection systems require that the correct type of detector be selected to match the wavelength of radiation from the fuel of interest, while
at the same time considering obstructions such as vehicles, fog, or rain that
may obstruct the line of sight between the detector and fire. Because lightning and arc welding produce ultraviolet (UV) radiation, false alarms are a

▶ **Figure 8.5** Sample Flame
Detector. *(Source: National
Fire Alarm Code Handbook,
NFPA, 2002, Exhibit 3.10;
Courtesy of Vibro-Meter, Inc.,
Manchester, NH.)*

problem. Gas welding and solar radiation can produce unintended activation with IR detectors. Combination detectors reduce the number of false signals by requiring both a source of IR and UV for alarm, of which fire is one of the few natural sources.

Gas Detectors. A number of different detection technologies are often lumped together and are known as gas detectors. Where flammable or explosive concentrations of gas are generated or could form, a *combustible gas detector* serves to alert personnel not only to the danger of a high concentration of gas that could be toxic or asphyxiate, but to the fact that the concentration of fuel in air may burn with only the introduction of an ignition source. Especially for large gas clouds or inside buildings, concentrations of combustible gases tend to be not uniformly mixed, so, as a safety feature, many combustible gas detectors are designed to alarm at a gas concentration below what will actually burn, such as 10 percent or 25 percent of the lower explosive limit (LEL). Similar detection technology can also be used to detect the many gaseous products of combustion characteristic of fire and are known as *fire gas detectors.*

Alarm Notification Devices

As was already discussed, a primary function of alarm systems is to alert people to emergency conditions that may occur to provide additional time for escape. Audio and visual means are typically used to inform building occupants of a fire, while voice alerting systems can provide additional information and event context to occupants compared to a simple fire signal. For the notification of a fire to be successful, the signal must be observed or perceived by an occupant, who must then recognize the signal and interpret its meaning. Should the intended receiver not hear the alarm or not identify its significance, the alarm system has failed.

Audible notification appliances are designed to sound a recognizable tone or pattern distinctive enough to be recognized over other sounds in the area (such as from machinery, a doorbell, background music). *NFPA 72* [3] mandates a three tone temporal pattern consisting of a repeated sequence of three short 0.5-second sounds followed by a 1.5-second pause with no sound, as the fire signal of choice for audibility. Even if a clear signal is sent, if it is too quiet to be recognized, it will go unheeded. Smoke detectors in the home typically use a sounding horn built into the detector, while most other commercial systems instead use remote sounding horns, located throughout the building to provide a sufficient sound pressure level above ambient noise levels. Sounders or horns are placed throughout the area to be notified, where louder horns allow wider spacing. *NFPA 72* places upper limits on the sound levels permitted to be generated to prevent hearing damage.

Due to hearing loss or other disabilities, not all people are able to perceive sound or recognize a fire alarm depending only on sound levels. Most building, fire, and life safety codes now require that public areas of buildings also include visual notification appliances, usually in the form of strobe lights that activate in conjunction with the audible appliances, as shown in Figure 8.6.

▶ **Figure 8.6**   Visual Strobe Notification Appliance. *(Source: Gentex Corp.)*

Other Functions and Devices

Alarm systems do not only detect the presence of fire, but can be used for a number of other important functions including the following:

- Manual alarm activation points (manual pull stations) allow building occupants to initiate the alarm system through a mechanical action such as pulling a lever or flipping a switch, as shown in Figure 8.7. *NFPA 72* [3] requires that they be located near exit doors, thus in the direction of exit travel, to increase the chance that they are used during a fire.

- Fire sprinkler system water flow detectors can observe water flowing in the sprinkler piping, an indication of a possible fire, and send an alarm signal.

▶ **Figure 8.7** Typical Manual Fire Alarm Initiation Boxes. *(Source: Top, Simplex-Grinnell, Westminster, MA; Bottom, Edwards Systems Technology, Cheshire, CT.)*

- Control valve supervisory signal initiating devices allow the alarm system to monitor the position of sprinkler system and water supply control valves to reduce the likelihood that the valve will be unintentionally closed.

- Fire sprinkler system supervisory devices also allow the alarm system to monitor critical fire protection equipment to continuously verify readiness. This monitoring includes low temperature or water level sensors for water tanks, no power or fuel supply available to a fire pump, along with low air pressure in a dry sprinkler system.

Detector Location

A number of design principles govern the design and layout of detection systems, primary of which is that the products of combustion must reach a detector. As a result, spot detectors are usually best located on the ceiling for reasons already discussed. When it is not possible to locate detectors on the ceiling, *NFPA 72* [3] permits spot-type detectors to be located on certain wall areas as shown in Figure 8.8, although a myriad of additional rules govern detector location to consult for acceptable detector response.

▶ **Figure 8.8**   Acceptable Spot-Type Detector Locations According to *NFPA 72*. *(Source: Fire Protection Handbook, NFPA, 2003, Figure 9.2.15.)*

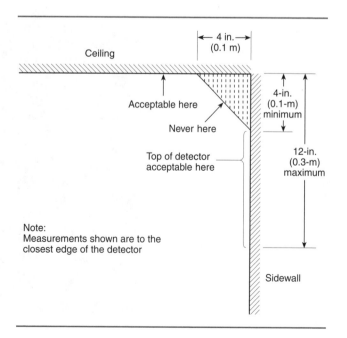

Fire Alarm Control Unit

Conceptually, alarm systems observe a change in conditions and respond appropriately in a prearranged manner. The *control panel* or *fire alarm control unit (FACU)* can be thought of as the brains or nerve center of a fire alarm system and waits for one of three different input signals from devices located throughout the protected structure, as shown in Figure 8.9.

Alarm Signal.   Once a fire is observed by an automatic fire detector or sprinkler waterflow switch, the signal is sent to the alarm panel and causes it to immediately initiate appropriate functions including sounding of alarms and off-site notification of the alarm monitoring company or fire department.

Supervisory Signal.   Supervisory signals represent a condition observed by an input device that is not correct or normal, but is not a fire. A supervisory

▶ **Figure 8.9**   Cover of a Fire Alarm Control Unit. (Top right) Types of signals accepted by the panel, (middle right) notification that results upon receipt of an alarm signal, and (bottom right) options for testing, silencing, or acknowledging an alarm signal. *(Source: Fire Control Instruments, Inc., Westwood, MA.)*

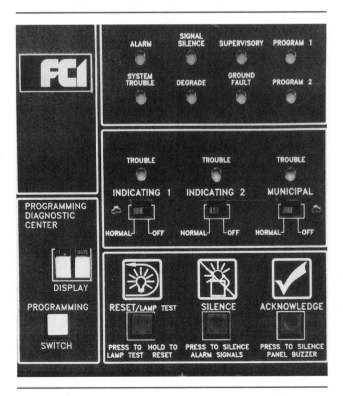

signal should be sent both when the condition is noticed and when it is corrected or returned to normal. These signals are the following:

- Monitoring sprinkler water supply valves to make sure they are not closed
- Observing water levels in a water tank or pressure levels at a fire pump providing water to the sprinkler system
- Detecting that a security guard has not reported to an assigned location at the designated time

Trouble Signal.   Trouble signals indicate a problem with alarm wiring or transmission means or a problem with the system power supply. Signals can be sent from the input device to the panel through wires, radio signals, or other wireless technology. All three types of signals are usually retransmitted from the control unit to the alarm monitoring location or fire department. The fire department is typically notified for alarm signals only, while the building owner, maintenance staff, or other responsible person is contacted to correct supervisory and trouble signals.

Most alarm systems receive their primary power for operation from the building wiring supplied by the electric utility. In some remote areas, a generator may supply the power. Either way, the alarm system should be located on a dedicated branch circuit protected to prevent inadvertent or intentional disconnection. People continue to occupy buildings even when the power is out, so *NFPA 72* [3] requires that alarm systems be provided with a backup or standby source of power, usually in the form of batteries and is required to automatically supply power to the system in the event primary power is unavailable. Batteries are usually sized to supply power to the alarm system for at least 24 hours, followed by 5 minutes of system operation in a full-alarm condition.

Depending on what is protected and how the system functions, *NFPA 72* [3] identifies a number of different fire alarm system types. Although there is some overlap between classifications, it is helpful to classify detection and alarm systems as the following:

Household Fire Alarm Systems

When used to protect one- and two-family homes and the sleeping rooms of apartments and hotels, household fire alarm systems are usually the simplest alarm configuration—consisting only of smoke alarms, possibly interconnected so that they all sound an alarm when one detects a fire. Although the least complicated, household fire alarm systems, or more specifically residential smoke alarms, have had a major impact on reducing the number of fire fatalities in the United States over the past 20 to 30 years. Fire death rates in the United States and Canada are among the highest of any industrialized democracy outside the former Soviet block.

Although household warning systems that are part of a home security system are common, most households are protected with single-station or multiple-station smoke alarms. A single-station smoke alarm contains an integral horn, alarm, or strobe light that sounds when a fire is detected. Given that only the one alarm will sound initially, occupants may not be aware of a fire growing in the basement when they are asleep on the main level of a home. Multiple station smoke alarms are interconnected with wire or by radio signals such that when one unit goes into alarm, all linked units will sound the alarm, providing notification throughout the protected area. Household smoke alarms typically receive power from batteries, building wiring (also called hard-wired), or a combination of both to provide primary and backup power, as shown in Figure 8.10.

Although installation requirements vary according to local building and fire codes, *NFPA 72* mandates that smoke alarms be provided outside each sleeping area and on every level of existing homes, including the basement. For new construction, smoke alarms are also required inside each sleeping room. Heat detectors may also be installed as part of a household fire alarm system or interconnected in combination with smoke alarms (heat detectors are not considered life safety devices because they tend to operate too slowly

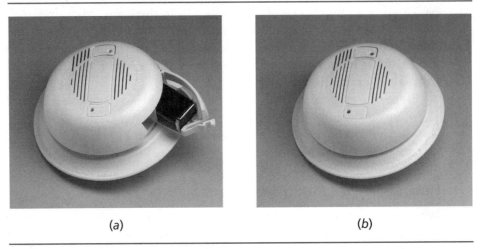

(a)   (b)

▶ **Figure 8.10** Single-Station Ionization Smoke Detectors Where (a) Is Powered with a 9-Volt Battery, While (b) Is Connected to the Building Wiring for Primary Power and Contains a 9-Volt Battery to Provide Backup Power. (© *BRK Brands, Inc. First Alert is a registered trademark of First Alert Trust.*)

and only after smoke alarms for most common residential fire scenarios). National Fire Protection Association data show that smoke alarms are installed in well over 90 percent of the homes in the United States, but at the same time, one-third of these detectors are not functioning due to missing batteries, lack of maintenance, or disconnection from the power supply. To reduce the number of nuisance alarms that lead to disconnecting the power supply to smoke alarms, some communities now require the installation of photoelectric smoke alarms instead of ionization detectors for residential occupancies because they tend to produce fewer nuisance alarms from typical causes (cooking, smoking, and steam from the bathroom).

Protected Premises Fire Alarm Systems

Fire alarm systems are primarily designed to protect both occupants and building, and when provided with detection and notification equipment, meet the definition of *protected premises*, as shown in Figure 8.11. Alarm-initiating devices are the inputs to the system indicating an alarm of fire and

▶ **Figure 8.11** Possible Layout for a Local Protected Premises Detection and Alarm System. *(Source: Fire Control Instruments, Inc., Westwood, MA.)*

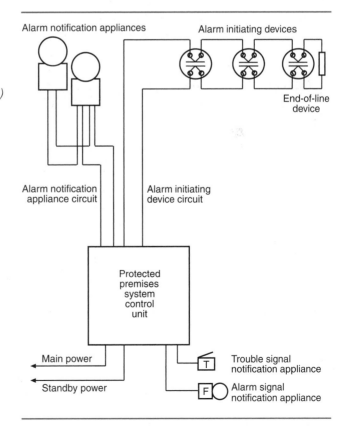

include automatic fire detectors, manual pull stations, and sprinkler water-flow switches. Such signals are carried to the control unit over the alarm initiating device circuit (IDC). Once an alarm signal is received, audible and/or visual notification devices are activated over the notification appliance circuit (NAC). Other functions that improve the level of safety include activation of suppression systems, closing of doors in fire walls, and startup of smoke exhaust fans.

Newer alarm system technology for addressable communication allows for data to be carried both ways between a device and the alarm panel. Communications over such signaling line circuits can include the ability of the control unit to "talk" with a detector to determine if it is becoming dirty, is missing, or requires service. Valuable new alarm verification features can be included to permit the FACU to query the detector over a time period to determine if the signal came from a nonfire source (such as the starting of a fan in a dusty room), instead of a fire (where the signal might show different characteristics such as a slower increase).

With a traditional protected premises fire alarm system, the only alarm signals occur in the protected building—the fire department or other off-site location is not notified of the alarm condition. A significant delay in fire department response can occur when occupants are not educated to call the fire department or during periods when the building is not occupied. A common protected premises fire alarm installation would be found in a small apartment building where system smoke detectors are located in the hallways and other common areas and ring a buildingwide alarm upon activation.

Supervising Station Alarm Systems

Many building, fire, and life safety codes now require that all alarm systems, except for the protection of households, include a way of notifying a constantly attended point to allow notification of the fire department (during an alarm) or other responsible parties (for supervisory or trouble signals). To provide this function, the protect premises alarm system includes a way to communicate with the world outside the building, and when so provided is called a *supervising station alarm system*. The type of supervision provided to monitor for alarm signals depends on the destination of the signal once sent. A clear advantage is that someone is able to watch over a property at all times, even when no one is physically in the building, reducing the chances that a fire can grow without being discovered. Once the alarm signal is received by the monitoring service, it is required to immediately notify the fire department, which provides both for faster arrival on scene and hopefully a smaller, more manageable incident.

Central Station Supervisory Service.   Private companies provide around-the-clock off-site monitoring for alarm signals by trained personnel at a constantly attended location, as shown in Figure 8.12. Based on the type of signal received, the central station notifies either the fire department or a responsible party as outlined in *NFPA 72* [3]. In the case of alarm signals, they are to be forwarded (retransmitted) to the fire department immediately on receipt. A central station monitoring company should also provide additional services, including a runner to respond to the building to verify conditions, repair technicians to restore equipment to operation based on supervisory or trouble signals, and recordkeeping of all signals and actions taken.

Proprietary Supervising Station.   For large campuses or companies with buildings throughout the country, alarm signals may instead be received and retransmitted at a constantly attended location operated by the owner of the protected buildings. It may be more cost-effective for the owners of many buildings to provide their own monitoring (often combined with site security), than to pay a central station to provide the service for each building.

Remote Supervising Station.   Some buildings may not require full central station service or warrant a proprietary station operation. Remote supervising stations provide a minimum level of service, basically limited to alarm retransmission.

▶ **Figure 8.12**   Central Station Monitoring Location. *(Source: SimplexGrinnell, Westminster, MA.)*

Emergency Voice/Alarm Communication Systems

In hospitals, high-rises, and large assembly occupancies, building, fire, and life safety codes recognize that it may not be possible to immediately evacuate everyone from the building at one time and thus mandate additional protection features to address the fact that exiting may involve only a portion of the building population, known as *staged evacuation*. Conflicting signals given to occupants may result in egress delays or occupants taking incorrect actions to save themselves. Emergency voice alarm communication systems are integrated with the fire alarm systems described previously to allow spoken emergency messages to be directed to occupants in a portion of the building as an alternative to activation of buildingwide alarm horns (Figure 8.13). A typical arrangement for high-rise buildings is for the alarm notification system to activate on the fire floor and floor(s) above/below, followed by a prerecorded voice message directing occupants to the nearest exit or place of safety. A trained employee, building security, or the fire department may elect to interrupt the prerecorded message to provide updated information relevant to the current situation. Especially in these large buildings, occupants may be directed to relocate to an area of safety or refuge to wait until the sprinkler system and fire department control the fire. When approved, some emergency voice alarm systems are combined with the building paging system used for daily communications.

▶ **Figure 8.13** Emergency Voice Alarm Communication System Used by the Fire Department to Provide Critical Information about the Specific Incident to Building Occupants Impacted by the Fire. *(Source: SimplexGrinnell, Westminster, MA.)*

AUTOMATIC FIRE SUPPRESSION SYSTEMS

Fire Suppression Agents

Although probably the first material used to control a fire, water is not the only substance available that can be discharged through fire extinguishers, fire fighters' hose lines, or fixed systems to stop the spread of fire. No one fire suppression agent is appropriate for all fires. Some agents may be ineffective on certain types of fires, whereas others can spread the fire or cause an explosion, if applied incorrectly or to the wrong fuel. Selection of an agent for the intended use is a critical first step that should be done well before a fire actually occurs.

A number of other chemical agents are available to suppress fires in specialized applications such as where there is the potential for water damage, where fuels cannot be controlled with water, or where it is difficult to store a sufficient quantity of water, as on airplanes or subway trains. This subject area is undergoing rapid change due to environmental and toxicity issues—agents that were commonplace 20 years ago are now banned from manufacture. It is not only the environment that needs protecting. Agents may cause harm to occupants and fire fighters with and without a fire. What might be a safe agent by itself may become toxic or corrosive as it is broken down during suppression. Prudent fire department operations during any nonwater-based suppression system activation, with and without fire, include complete turnout gear and self-contained breathing apparatus until the space is well ventilated.

Water.   For a number of reasons, water is the most common extinguishing agent used to control or suppress unwanted fires. It is universally available in large quantities at low cost, is easily stored in tanks or ponds, and can be transported readily through pipes and discharge nozzles. Although it is heavy (8.3 lbs/gal), fire department trucks and tankers can carry enough of it to control fires involving everything from cars to small buildings.

One of the reasons why water is such a good agent for suppressing fires in ordinary combustibles such as wood or paper is due to its ability to absorb large amounts of heat energy. When applied directly to the surface of a fire, for example, water will take in heat from the fire enough to raise the temperature of the water up to its boiling point. Water absorbs even more energy while undergoing a change in phase from liquid to a gas (i.e., steam). We know that water boils at 212°F (100°C) and that flames from most fuels burn at temperatures much higher than this, so the steam at 212°F (100°C) that is created by the vaporization of water will keep absorbing more energy until it reaches the temperature of the surrounding hot fire gases.

Taking this much energy away from a fire cools the fuel, hopefully below its ignition temperature, or the temperature necessary for continued burning. Because most solid fuels burn in the gas phase, they too absorb energy to vaporize then burn. By absorbing energy and changing from a solid into a gas (called *pyrolysis*), the fuel is able to participate in combustion—just one of the many chemical reactions that happen around us all the time. The more energy the water is able to take away, the less energy is available to liberate (pyrolyze) additional fuel, reducing the fire intensity further. When enough water is applied to a fire to absorb the energy created, there is no longer sufficient fuel vapor produced above the surface, and the fire weakens. If enough water can be applied to the fuel or flame to absorb all of the energy produced, the fire is not only controlled, but can be considered extinguished or suppressed. (See Chapter 3.)

In a room or compartment, a fire can be controlled indirectly by applying water to the hot gases that accumulate near the ceiling. By absorbing heat during a rapid expansion into steam, the water spray can reduce the radiation heat transfer to the fuel from the flame and upper layer of gases, reducing the rate of pyrolysis, leading to control of the fire. At the same time, liquid water, when converted to steam, expands and can displace both the fuel gases and oxygen that the fire needs to continue burning.

Class A, B, C, D, and K are commonly used to classify the type of hand extinguisher to be used on a particular fire. While a bucket of water is effective on a fire involving wood, paper, or cloth (Class A fire), most people are correctly taught not to use water to put out a fire involving flammable or combustible liquids such as cooking oil on a stove (Class K fire). By disturbing the surface of the burning liquid, water can spread or intensify the fire, without gaining control. The following are fire scenarios where water is usually not a recommended agent of choice:

- Many flammable and combustible liquids (Class B fires), especially those with a lower flashpoint or that will float on water may not be controlled with plain water.

- Because the impurities in water cause it to be an unwanted conductor of electricity, water should not be applied to energized electrical equipment (Class C fires). When water is broken down into a stream of drops from a fire fighter's fog nozzle, some protection against electric shock is provided at lower voltages when sufficient distance is maintained between the nozzle and energized equipment.

- High-value equipment, such as computers or electronics, or other materials could be easily damaged by water.

- Some metals (Class D fires), such as magnesium, lithium, or metallic sodium can be highly reactive with water or burn at such a high temperature that the water molecules (H_2O) can be split into the component hydrogen and oxygen, with explosive results.

- Certain hazardous materials too may react when water is applied or may cause a dangerous or toxic runoff when carried by extinguishing water.

Water Mist Suppression.   Water mist nozzles create a spray of finely divided water that can be used like a sprinkler system to control a fire, but through different suppression mechanisms and with less water. In addition to cooling and blocking heat transfer back to the fuel, a fine mist of water readily vaporizes, displacing the air (oxygen) the fire needs to continue burning. Fixed water mist systems have seen extensive use protecting hazards that are sensitive to water damage such as ship equipment rooms. Portable water mist systems, including extinguishers and backpack units are now available and offer the advantage of greatly reducing the amount of water necessary to control a given fire versus ordinary fire department hose line operation.

Halon (Halogenated Hydrocarbons).   By taking a hydrocarbon molecule comprised of carbon and hydrogen atoms, removing hydrogen atom(s), and replacing them with fluorine, chlorine, bromine, or iodine (halogen atoms), a halogenated hydrocarbon is created. While used as everything from refrigerants to aerosol propellants, many halons are not only noncombustible and possess low toxicity, they are also good fire extinguishing agents because of their ability to interrupt the chemical reactions necessary for combustion. Given their capacity to deplete stratospheric ozone (which is important for blocking harmful portions of ultraviolet (UV) light from the sun), the production of halons was banned in 1994 by the Montreal Protocol. Halon 1211 (bromochlorodifluoromethane) was a common agent for fire extinguishers and local applications protecting equipment, while Halon 1301 (bromotrifluoromethane) worked well when used to completely fill a space (total flooding) such as computer rooms or machine spaces.

Clean Agents (Halon Replacements).   Research for halon replacements has led to a host of new environmentally friendly (i.e., "clean") suppression agents available commercially for use in fire extinguishers and fixed suppression systems. *Halocarbon-based clean agents* are also composed of carbon, hydrogen, fluorine, chlorine, bromine, or iodine atoms, but are selected specifically to reduce their ozone depletion potential (ODP), global warming potential (GWP), and atmospheric lifetime, all measurements of the potential harm the agent could do when released into the atmosphere. Advantages of halocarbon clean agents are that they leave little residue, are nonconductive to electricity, and perform reasonably well as total flooding agents. While more costly, other disadvantages compared to halon 1301 include less effective suppression and increased production of toxic and corrosive by-products during suppression.

Some manufacturers have instead developed clean agent replacements for halons based on *inert gases* or mixtures of gases (such as nitrogen and argon). Although less expensive, because they operate by displacing oxygen, a large quantity of agent is necessary, and there is the potential for asphyxiation in occupied areas. Because of the wide variations between available products, selection of the best clean agent for a given application becomes a difficult task.

Carbon Dioxide (CO_2). For a long time, CO_2 based systems were the choice for protecting many hazards where water was not acceptable, given that CO_2 is nonconductive, cheap, easy to store in large quantities when refrigerated, and not environmentally unfriendly. While fewer CO_2 systems were installed when halon was common, they are gaining in popularity again because of the difficulties associated with selecting an acceptable environmentally friendly clean agent. Unfortunately, the problems with CO_2 systems are still applicable: potential for asphyxiation by displacing oxygen, requirements for large storage volumes, and ineffectiveness on fuels that do not require oxygen from the air to burn. As with the inert clean agents, CO_2 suppresses a fire by displacing enough air (oxygen) from a space that a fire cannot continue to burn (it also provides a minor amount of cooling).

Dry Chemical. By finely grinding effective solid suppression agents into a powder, dry chemicals can be applied through fixed nozzles or extinguishers onto a fire. Because of the large surface area of each solid particle, the fire is cooled somewhat through heat absorbed, vaporizing the agent and blocking of heat transfer back to the fuel, but more important is the ability of the vaporized agent to break the chemical chain reaction necessary for combustion. Dry chemical consists of sodium bicarbonate, potassium bicarbonate, urea-potassium bicarbonate (all appropriate for Class B and C fires), while monoammonium phosphate is multipurpose because it is also effective on Class A fires. To control fires in deep burning solids such as wood chips, multipurpose dry chemical forms a crust over the surface to prevent oxygen from reaching the burning fuel. Dry chemicals are nonconductive to electricity, do not pose a significant hazard to the environment, and cause few problems in humans when they are exposed to them. Disadvantages to consider include the fact that little protection is provided once the solid dry chemical settles from the air, cleanup is difficult, and the agent is corrosive, resulting in damage to sensitive equipment or electronics. Dry chemicals are applied by fire extinguishers, hose lines, and fixed systems of nozzles.

Note that solid materials called *dry powders* are available for hand application to burning metals (Class D fires) and should not be confused with the dry chemicals just discussed for flammable liquids and energized equipment (Class B and C fires). They are not interchangeable and will be ineffective, if not dangerous, if applied on the wrong type of fire. (See Chapter 3.)

Wet Chemical.   Because dry chemicals provide limited protection once they are no longer suspended in air, a flammable or combustible liquid fire is susceptible to reignition shortly after suppression. This shortcoming of dry chemicals has been recognized with fixed suppression systems used to protect deep fat fryers in commercial kitchens. Wet chemical agents were developed to control such fires, and at the same time protect against redevelopment after extinguishment. When discharged on a fire, wet chemical agents mix with grease and fats to form a foam or soapy mixture (saponification) on the surface and block the production of fuel vapor and cool, while at the same time excluding oxygen from reaching the fuel. Wet chemicals are available in both handheld extinguishers and fixed systems for the protection of cooking equipment.

Foam.   In the case of flammable liquids, water can be used more effectively if it is mixed with small amounts (1 percent to 6 percent) of foam concentrate to form a foam-water suppression agent, often called class B foam. By changing the properties of plain water, the foam-water is able to float on top of the burning liquid, reducing the heat transfer back to the fuel necessary for continued vaporization of the fuel and subsequent burning, while at the same time cooling the fuel and blocking oxygen. Aqueous film-forming foam (AFFF) technology can even develop a foam blanket that seals the surface of the fuel, preventing vaporization and limiting the chance of a flash fire occurring. Care must be used, however, in selecting foam or other water additives due to environmental concerns about water runoff, and toxicity to plants, fish, and wildlife.

Depending on how much air is mixed with the foam-water solution (called *aerating*), mixtures with different volumes, densities, and firefighting capabilities are formed. This rate of expansion has a large impact on the foam through its effectiveness to suppress a fire, prevent reignition, and useful period (retention or hold time). The ratio of the final foam mixture volume (water + foam concentrate + air) to the volume of the foam water solution before mixing with air (water + foam) is known as the *expansion ratio*, and can be divided as follows:

- Low-expansion foam has a ratio of 1:1 to 20:1.
- Medium-expansion foam has a ratio of 20:1 to 200:1.
- High-expansion foam has a ratio greater than 200:1.

Many fire departments use a 3 percent concentrate foam for flammable liquid fires. When 3 gal of the concentrate are mixed with 97 gal of plain water, 100 gal of foam-water solution result. If applied to a fire through a medium expansion foam nozzle that adds air to the mixture at a ratio of 30:1, then $(100) \times (30) = 3{,}000$ gallons of aerated foam are made, as shown in Figure 8.14. By passing the foam-water solution at high velocity along openings in the side of the nozzle, air is drawn into the nozzle (aspirated or entrained) at a high rate and mixed to form a large volume of finished foam. High-expansion foams can produce tremendous amounts of expanded foam in a short period and are commonly used to protect aircraft hangers or underground structures where they exclude oxygen, while providing limited cooling.

Improved fire suppression technology has led to the development of safe chemicals that, when added in small quantities (a few percent or less), improve the ability of water to extinguish common Class A fuels such as wood and paper. Such Class A fire-fighting additives/foams can include wetting agents that reduce the surface tension of water and allow it to better soak into materials, up to long-lasting foam mixtures applied to the outside of

▶ **Figure 8.14** Aspirating Nozzle Used by Fire Department to Generate Medium Expansion Foam. *(Source: NFPA 11A, 1999, Figure A-1-9.5(a).)*

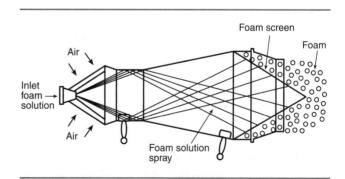

buildings threatened by wildfires. NFPA standards on class A foams include NFPA 1150, *Standard on Fire-Fighting Foam Chemicals for Class A Fuels in Rural, Suburban, and Vegetated Areas* [4], and NFPA 1145, *Guide for the Use of Class A Foams in Manual Structural Fire Fighting* [5].

Now that we have described the different agents that can be used to control or suppress unwanted fire, we look at the systems that have been developed to automatically apply a suppression agent to a fire. Automatic water-based fire sprinklers are the most common type of system installed and are discussed first. Later we discuss portable fire extinguishers that use many of these agents.

▶ WATER-BASED SUPPRESSION SYSTEMS

The first fire sprinkler systems were installed to protect valuable warehouses and manufacturing buildings, which were often constructed out of or stored combustible materials. Thus early on the insurance industry saw the benefits of sprinklers in controlling property loss due to fire. This section looks at the fire protection systems using water to control, and in some cases suppress, unwanted fires. We start with fire sprinklers, the equipment that comprise these systems, and touch on how they are designed and installed. We then look at alternative systems that may use water, or some other chemical, to control combustion.

How Fire Sprinklers Control a Fire

Depending on the application, fire sprinklers are designed to create a spray pattern of water drops that are ideally suited for the environment in which they will fight fire. A sprinkler that produces mostly small drops may see them deflected by the high momentum of the fire gases rising above a growing fire or rapidly vaporized into steam. If few of these drops reach the fuel, it will be difficult to control the fire. This may be fine if a sufficient number of sprinklers can activate around the fire, wetting all of the fuel that has yet to ignite. Figure 8.15 shows an example of fire sprinklers in a warehouse controlling the growth of a fire until the original burning fuel is consumed or the fire department is able to complete manual extinguishment. Sprinklers that open around the fire cool temperatures to prevent the failure of structural steel supporting the building and sprinkler pipe, and at the same time also avoid sprinkler activations far from the fire. Although the initial fire may not immediately be extinguished, it is unable to continue growing, while the fuels that are already burning are consumed.

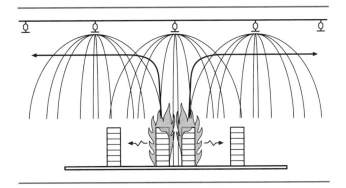

► **Figure 8.15** Example of Fire Sprinklers Controlling a Fire. *(Source: Fire Protection Handbook, NFPA, 2003, Figure 10.9.8.)*

Such sprinklers are operating in what is called *control mode*, meaning they are not designed to completely put out every fire (although they often do), but instead limit the fire to what is already burning, and no more, until the fire department can arrive for final extinguishment. With a control mode sprinkler system, it must be expected that there is water damage to materials around the area of fire origin, in addition to smoke damage in a possibly larger area. When these conditions occur, it should not be considered a failure of the sprinkler system. As long as it operated and controlled the fire until the fire department could carry out final extinguishment (i.e., mop-up and overhaul), the operation would be considered a success.

In the 1980s, research showed that a sprinkler could be developed that would produce enough large drops directed down with sufficient momentum to disrupt the fire plume and flame, allowing most of the water discharged by the sprinkler to reach the seat of the fire. In effect, fire scientists developed a sprinkler that would increase the amount of water able to travel through the fire plume to reach the fuel, which allowed high challenge fires like rack storage of plastic goods in warehouses piled to 40 ft to be protected with sprinklers located only on the ceiling of the building. This was the birth of the Early Suppression Fast Response (ESFR) sprinkler system.

Water Supplies for Sprinkler Systems

Water Sources. Water is available from streams, rivers, lakes, and other natural sources, and these common sources are used by fire departments in areas without fire hydrants on a city water system. Because nature can be unreliable and natural sources of water may be unavailable during droughts or freezing, aboveground tanks, water towers, or underground reservoirs can be used to store water for fighting fire until it is needed in areas without an available public water system.

Fire sprinkler systems are required by NFPA 13, *Standard for the Installation of Sprinkler Systems* [6], to have at least one automatic water supply to provide the necessary water without human intervention or action. Most developed areas have water provided for domestic, commercial, manufacturing, and fire protection through a central waterworks supply and distribution system. Sizing of the underground pipes supplying water to an area or building is usually based on fire protection use, which can be thousands of gallons per minute (gpm), versus other uses that flow at closer to tens or hundreds of gallons per minute. These distribution systems tend to be highly reliable given that pumps are usually in constant operation and faults in the system caused by closed valves, broken pipes, or failed pumps are quickly recognized by the users.

For some buildings, a single automatic water supply may not be enough. Due to the high value of a large manufacturing plant, storage facility, or complex of buildings, an increase in reliability may be provided through multiple water supplies from different sources. Life safety, difficult firefighting, and in some locations the potential for earthquakes results in building and fire codes sometimes requiring two different water supplies for high-rise buildings.

Design and Installation of Components.   A common design could include fire pumps to boost the pressure from the city supply to overcome elevation combined with fixed pressurized storage tanks in the building to provide water when the pumps are not operational. Figure 8.16 shows a highly protected risk (HPR) where the sprinkler system is supported by two different automatic water supplies, public water works, plus a tank and pump.

Water tanks, whether elevated or on the ground, are usually installed according to NFPA 22, *Standard for Water Tanks for Private Fire Protection* [7], which contains guidance on tank sizing and design, construction materials, foundations, stability during earthquakes, and protection from freezing. Due to the complexities of properly installing a reliable pump for fire protection, NFPA 20, *Standard for the Installation of Stationary Pumps for Fire Protection* [8], contains design, installation, and maintenance provisions for electric, diesel, and steam-driven fire pumps. Although wells are able to provide a sufficient flow rate of water for domestic use, providing the necessary volume of water for fire protection makes wells cost prohibitive except in unusual circumstances.

To improve the reliability of sprinkler systems even further, NFPA 13 also requires that systems be provided with a secondary water supply, which need not be automatic, to allow water to be provided to the system during times when the primary supply is out of service or during a fire to allow the fire

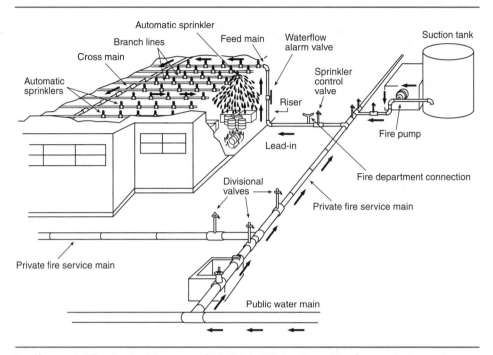

▶ **Figure 8.16** Typical Layout of Multiple Water Supplies for an Automatic Sprinkler System. *(Source: Automatic Sprinkler Systems Handbook, 2002, Exhibit 3.11. Adapted from FM Global Systems.)*

department to augment or support the primary supply. Fire fighters are trained that one of their early tasks when arriving at a fire in a sprinklered building is to provide a secondary source of water to the sprinkler system through the fire department connection (FDC) located at an accessible location outside the building.

Evaluation of Water Supplies

The strength of a water supply is measured in terms of the flow rate in gallons per minute and pressure in pounds per square inch. If a tank resting on the ground is full of water, there is little pressure available to push the water down a pipe or hose and out a nozzle, so we say the pressure is inadequate. Conversely, if a pump is operating and increasing the pressure in a pipe and if the connection between the water source and pump is obstructed so that little water is able to pass, the flow rate is inadequate. To test a fire protection water supply system involves measuring both the pressure and flow rate at the location of interest, as shown in Figure 8.17.

▶ **Figure 8.17** Test of a Municipal Water Supply System. *(Source: NFPA 13, 2002, Figure A.15.2.1.)*

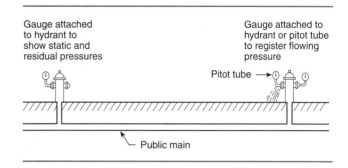

Gauge attached to hydrant to show static and residual pressures

Gauge attached to hydrant or pitot tube to register flowing pressure

Pitot tube →

Public main

Using NFPA 291, *Recommended Practice for Fire Flow Testing and Marking of Hydrants*, as a guide, these steps are followed during a flow test:

1. A pressure gauge is attached to the fire hydrant closest to, or in front of, the building to be protected. The hydrant is opened. After the air trapped in the hydrant is released, the pressure on the gauge is recorded. This *static pressure* represents the available pressure in the system under normal domestic and commercial water usage (i.e., no fire protection water flow).

2. A second hydrant, the flow hydrant, is selected near the first hydrant, and if possible, should be further away from the source of water on mains or piping fed from only one direction (called *dead ends*).

3. One or possibly two caps on the flow hydrant are removed and the hydrant opened to allow water to flow.

4. A pitot tube (also called a *pitot gauge*) is used to measure the flow rate of water out of the flow hydrant, as shown in Figure 8.18. A pitot gauge measures the pressure for the flow out the hydrant (pressure of the stream in the direction of travel, called *velocity pressure*) and can be converted to a flow rate in gpm.

5. At the same time the flow is measured in step 4, the pressure is recorded on the first hydrant. This is the *residual pressure* and represents the remaining pressure in the water system with fire protection water flowing.

Using special logarithmic (log 1.85) graph paper, the flow test results can be plotted, as shown in Figure 8.19 for use in evaluating the water supply. Curve A in Figure 8.19 presents the results from a rather weak municipal water supply where the static pressure is 35 psi (with 0 gpm flowing), and the residual pressure is 25 psi (with 1200 gpm flowing). To protect a warehouse or tall building, this system may have sufficient volume available (1200 gpm at 25 psi or 1550 gpm at 20 psi by reading off the curve), but not enough pressure to overcome friction and elevation in pushing water through pipes and out the

▶ **Figure 8.18** Use of a
Pitot Gauge to Measure
Hydrant Flow Rates.
*(Source: Fire Protection
Handbook, NFPA, 2003,
Figure 10.5.9.)*

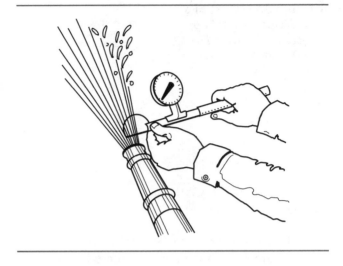

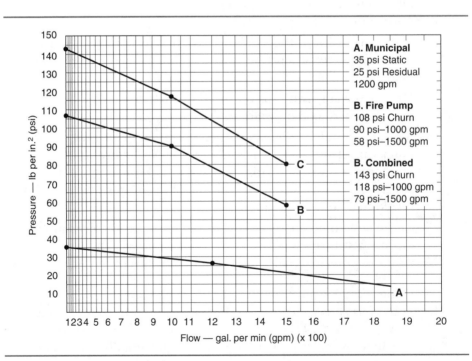

▶ **Figure 8.19** Plot of Municipal Water Supply Test, Fire Pump Performance,
and Combined Results. *(Source: Automatic Sprinkler Systems Handbook, NFPA,
2002, Exhibit 3.4.)*

sprinkler. For important properties, water supply flow tests should be repeated on a regular basis to identify changes in the water system, closed valves, or other conditions that would prevent the fire protection systems from operating properly.

A fire pump can be specified that will increase the pressure available for firefighting and the sprinkler system. A fire pump cannot create water, so it alone will not improve the flow rate, only the pressure available at each flow rate. If the fire pump will provide an increase in pressure of 108 psi with 0 gpm flowing (called *churn*), and 90 psi at 1000 gpm, curve B shows the characteristics of the selected pump that would be provided by the manufacturer. A fire pump taking its supply from this municipal water system results in the combined supply, Curve C, found by adding the results of Curves A and B at a number of points. Curve C represents the total available water supply using a pump on the municipal supply. A sprinkler system with a required flow rate and pressure falling anywhere under Curve C would be considered acceptable, because the combined municipal supply and pump would have both the pressure and flow rate capabilities to meet the system's demand. More information on fixed fire pumps is available from the *Fire Pump Handbook* [10].

The testing of water supplies is an important skill that unfortunately is often neglected. Fire fighters can use these same tools and techniques to verify the performance of pumps located on their vehicles, while at the same time they can measure the expected flow rates from hydrants in their response area to determine the availability of water as part of fire preplanning. Finally, the fire service, fire inspectors, and other code officials are often asked to witness such water supply tests when a new building or sprinkler system is constructed.

Sprinkler System Components and Operation

Before looking at how sprinkler systems are designed, it is necessary to first discuss how they work and look at the special equipment used to distribute water over a fire, send an alarm, and test the system.

Wet Pipe Sprinkler Systems.   The simplest and most common protection is provided by a wet pipe sprinkler system using pipes that are under pressure and filled with water at all times to connect the sprinkler distribution nozzles, or sprinkler heads, to the water supply. Returning to Figure 8.16, the lead-in piping carries water from the supply or supplies into the building, where a series of pipes are located at the ceiling throughout the protected area. Sprinklers are attached directly to branch line piping that typically ranges between 1 in. and $2\frac{1}{2}$ in. in diameter. When viewed from above, this layout looks like the branches of a tree in its simplest form. Sprinkler piping is typically constructed of steel, although copper is also used, as well as with plastic pipe in lower hazard occupancies like apartments, hotels, and offices.

On a wet pipe sprinkler system, all of the sprinklers are maintained closed with a metal link or glass bulb until the heat of a fire opens a small number of sprinklers directly over or around the fire. Contrary to popular belief, on a wet pipe sprinkler system all of the sprinkler heads do not open at once, nor do they operate when a smoke detector activates. As will be discussed shortly, each sprinkler must be heated by the energy from the fire before that single sprinkler will open, releasing water.

From Figure 8.16, the small fire in the building has released enough heat to activate only the one sprinkler directly above the fuel, and in many cases only the first sprinkler is necessary to control or extinguish the fire. To notify the fire department and possibly the building occupants, a water flow alarm is shown on the piping connecting the water supply to the sprinkler pipe. This vertical section of pipe, or *riser*, contains the control equipment necessary for proper operation of a sprinkler system and is detailed in Figure 8.20.

▶ **Figure 8.20** Wet Pipe Sprinkler Riser Connecting the Water Supply to the Sprinkler Piping Located at the Ceiling. *(Source: Automatic Sprinkler Systems Handbook, NFPA, 2002, Exhibit 8.29.)*

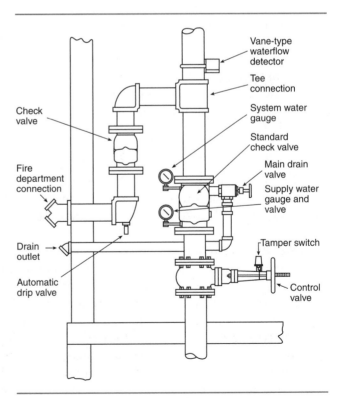

Starting at the bottom of this wet pipe sprinkler riser, a connection is made with the water supply, and immediately a valve is provided to allow the sprinkler system to be shut down for maintenance, construction, or repairs. To keep water from flowing or draining back into the water supply, a check valve is provided next to isolate the sprinkler system, allowing water to travel in only one direction—toward the fire. Pressure gauges are provided on both sides of the check valve: the lower one on the supply side measures the pressure in the water source, while the upper one measures the system pressure above the check valve. Because most water supplies fluctuate in pressure over a given day or season, the water captured above the check valve tends to be at a higher pressure until a sprinkler head opens, releasing water and allowing the flow to pass the check valve. This feature reduces, but does not eliminate, the number of false alarms generated by water pressure fluctuations.

Because the check valve allows water to flow into the sprinkler system, but not out, a drain is required on the riser by NFPA 13. A fire department connection should be provided to allow the fire department to supplement the primary water supply in cases where either the main system valve is shut or if the supply is insufficient due to more sprinklers opening than the system can supply.

Finally, a means to transmit an alarm once water begins flowing is required to notify the fire department and usually building occupants. In this case, Figure 8.20 shows a vane-type water flow detector comprised of a thin plastic disk slightly smaller than the inside diameter of the riser pipe that is attached to an electronic switch and that will detect the movement of water inside the pipe resulting from the opening of even a single sprinkler. Instead of a separate check valve and water flow detector, some wet systems use a special alarm check valve that combines both features. As long as the valve is open and the water supply is working, the simple design of wet pipe sprinkler systems makes them highly reliable. A closed valve is the most common reason why sprinkler systems fail; that is why many fire codes require that valves be locked open to prevent tampering.

Dry-Pipe Sprinkler Systems.   In cold climates or unheated buildings, sprinkler pipes filled with water can freeze causing water damage, or worse, rendering the sprinkler system inoperable due to ice blockage in the pipe. An obvious solution is not to have water in the sprinkler piping where it can freeze (drop below 40°F according to NFPA 13), thus the need for dry-pipe sprinkler systems. As shown in Figure 8.21, the sprinkler piping is filled with air under pressure (unheated area), while a dry-pipe valve holds back the water (located in a heated enclosure). When the heat of a fire opens a nearby sprinkler, the air is released, reducing the pressure in the pipe until

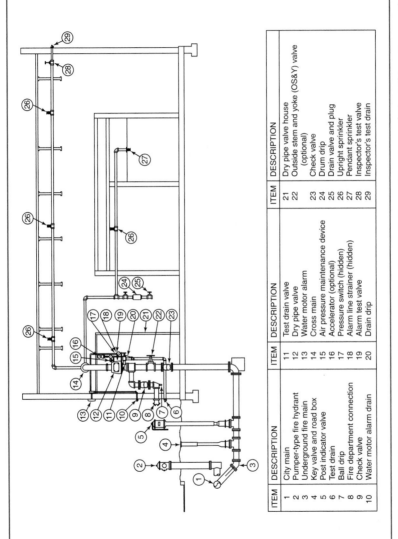

ITEM	DESCRIPTION
1	City main
2	Pumper-type fire hydrant
3	Underground fire main
4	Key valve and road box
5	Post indicator valve
6	Test drain
7	Ball drip
8	Fire department connection
9	Check valve
10	Water motor alarm drain

ITEM	DESCRIPTION
11	Test drain valve
12	Dry pipe valve
13	Water motor alarm
14	Cross main
15	Air pressure maintenance device
16	Accelerator (optional)
17	Pressure switch (hidden)
18	Alarm line strainer (hidden)
19	Alarm test valve
20	Drain drip

ITEM	DESCRIPTION
21	Dry pipe valve house
22	Outside stem and yoke (OS&Y) valve (optional)
23	Check valve
24	Drum drip
25	Drain valve and plug
26	Upright sprinkler
27	Pendant sprinkler
28	Inspector's test valve
29	Inspector's test drain

▲ **Figure 8.21** Dry-Pipe Sprinkler System Protecting an Unheated Building. *(Courtesy of The Viking Corporation.)*

the pressure on the top of the dry valve is no longer able to hold back the water.

One problem with these systems is that water is not immediately discharged on the fire, causing a delay that can allow the fire to grow larger during the period while air is expelled from the sprinkler and the entire pipe system fills with water. NFPA 13 typically mandates that a dry system discharge water within 60 seconds of the first sprinkler opening. To verify this requirement, an inspector's test connection is located on the most remote section of piping and can be opened to allow measurement of the time for water to reach the most distant portion of the sprinkler system.

In an effort to reduce the amount of time it takes for water to be discharged from a dry system, a number of solutions are available. A dry system can include a device called an *accelerator* to detect the reduction in pressure representative of an opened sprinkler head and direct the air pressure in the piping under the dry valve, forcing it to open sooner. Another device is an *exhauster*, a special type of valve that senses a reduction in pressure and opens to release the air in the piping, again allowing the dry-pipe valve to open sooner. Finally, most dry-pipe valves are designed using a check valve with a larger top surface area (air side) versus the bottom (water side). Such a *differential dry-pipe valve* allows a lower air pressure on the order of 30 to 40 psi to hold back water at pressures of 150 psi or more (see Figure 8.22).

▶ **Figure 8.22**   Differential Dry-Pipe Sprinkler System. *(Source: Fire Protection Handbook, NFPA, 2003, Figure 10.11.9.)*

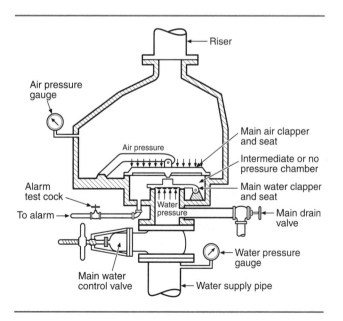

For small areas subjected to freezing conditions, another possible option is to use antifreeze solutions mixed with water to reduce the freezing temperature of water in order to prevent damage due to the expansion of water when it freezes. Instead of maintaining air in the piping, a wet sprinkler system is used with antifreeze placed in the piping in the areas subject to freezing. Unfortunately, many antifreeze chemicals are corrosive and toxic, so are not permitted in sprinkler systems connected to domestic or city water systems. Even when using low toxicity or nontoxic chemicals, most water utilities will require expensive backflow prevention equipment to keep the antifreeze mixture from migrating back to the public water system.

Preaction Sprinkler Systems. Preaction sprinkler systems are a solution to a problem associated with wet and dry pipe sprinkler systems protecting hazards that have a high potential for water damage. If a pipe is broken on a wet or dry system, an unintended discharge of water will occur. A preaction valve is similar to a dry valve, except that a mechanical lever instead of air pressure is used to keep the valve that holds back water closed. To allow water into the sprinkler piping, a smoke or heat detector must be triggered in response to a fire and send a signal to the fire alarm panel, which in turn releases the valve allowing water into the piping. A sprinkler must still open due to heat for water to be discharged. Damage to a pipe or sprinkler head alone does not allow water to flow, unless a detector also activates. Preaction systems are commonly used in areas such as computer rooms or museums where the consequences of water damage is great or in large freezer warehouses where the maintenance associated with a false activation of a dry system is costly.

Deluge Sprinkler Systems. A deluge system takes a preaction system one step further and protects buildings such as aircraft hangers or flammable liquid storage areas where a rapidly developing fire is expected. Sprinklers are located on branch lines as with the other types of systems, but all of the sprinklers are open, that is, the link holding back the water on each sprinkler head has been removed. When the detection system activates sending a signal to the valve to open, water is allowed into the piping and immediately flows from all sprinklers at the same time. Contrary to what is portrayed by Hollywood, a deluge system is the only common fire sprinkler system in use today that discharges water from all sprinklers at once based on the activation of a smoke or heat detector. Granted, usually the effect is dramatic on TV, but it is false nonetheless. Deluge systems are not found in schools, hotels, or highrise buildings, only in unique industrial settings with high hazards.

Sprinklers. Fire sprinklers are the nozzles attached to pipes carrying water over the area to be protected. The vast majority of sprinklers installed in the United States today are considered standard spray sprinklers that produce

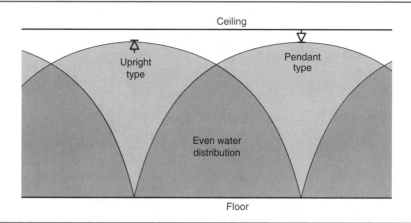

▶ **Figure 8.23**   Standard Spray Upright (SSU) and Pendant (SSP) Sprinkler Water Distribution Pattern. *(Source: Automatic Sprinkler Systems Handbook, NFPA, 2002, Exhibit 3.22.)*

an umbrella-like pattern by directing the discharge of water down toward the floor as illustrated in Figure 8.23.

Notice how the discharge from one sprinkler reaches all the way to the next sprinkler. This overlap is an intentional safety feature that allows adjoining sprinklers to provide protection, should a sprinkler not activate or operate improperly. To ensure that a proper discharge pattern is developed, NFPA 13 requires a minimum pressure of 7 psi on all sprinklers, although many extended coverage or special application sprinklers specify a higher minimum pressure identified by the manufacturer.

At their simplest, sprinklers are small nozzles with a plug or valve that opens when heated to a predetermined temperature, thus allowing water to discharge. Once heated to activation, the sprinkler cannot be turned off and must be replaced (there were a few on-off type sprinklers manufactured in the past, but few are sold today).

Fusible sprinklers, as shown in Figure 8.24, use a small amount of metal alloy (i.e., mixture) in the form of a disk or ball as part of a mechanical element designed to hold back the force of air or water until activation. Certain metals, including tin, lead, cadmium, and bismuth, each with a unique melting point, form a *eutectic alloy* when mixed and are characterized by a melting point less than any of the metals in the mixture. By adjusting the ratio of each metal, sprinklers can be manufactured with a wide range of activation temperatures. To prevent false activation, NFPA 13 specifies a maximum allowable ceiling temperature usually 25°F (14°C) less than the operating temperature and also stipulates distances sprinklers must be maintained away from heat-producing appliances, fireplaces, open flames, and skylights.

▶ **Figure 8.24** Fusible Disk Standard Spray Upright Sprinkler. *(Source: Fire Protection Handbook, NFPA, 2002, Figure 10.10.3.)*

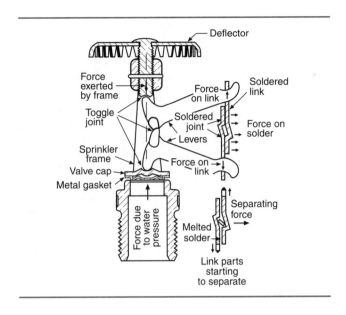

Let us assume that an ordinary temperature rated sprinkler was manufactured with an activation temperature of 165°F (74°C) and is subjected to the hot gases near the ceiling from a growing fire. The link starts at room (ambient) temperature before the fire and is now heated closer and closer to its activation temperature. Because the metal link has a small amount of mass that takes time to heat, the temperature of the fire gases around the sprinkler may actually be greater than the activation temperature, but the sprinkler will not have opened (known as *thermal lag*). When the link reaches 165°F (74°C), the eutectic alloy metal in the link is softening, but it still does not activate until the solid metal undergoes a phase change and turns into a liquid. Analogous to melting ice, the metal must absorb additional heat energy (called the *latent heat of fusion*), before it liquefies, releasing the pieces in the link and allowing the two levers to move apart and fall away from the sprinkler. With no force to counteract the pressure of the water, the cap or plug holding the sprinkler closed falls away. Water contacts the top of this upright sprinkler (i.e., the deflector) and is directed down toward the floor.

Instead of using a metal link and levers to hold water back, *glass bulb* (or *frangible bulb*) sprinklers use a small very high quality glass bulb that is nearly full with liquid except for a small bubble of air. See Figure 8.25, and note the difference in the shape of the deflectors between the pendant and upright sprinklers. As the sprinkler bulb and liquid are heated during a fire, the liquid expands, increasing the pressure in the bulb, thus decreasing the size of the air bubble. Eventually the bulb is heated enough so that the liquid expansion has compressed the bubble to the point that it disappears or is ab-

▶ **Figure 8.25**   Viking Model M Glass Bulb Standard Spray Pendant (*left*) and Upright Sprinkler (*right*). *(Courtesy of The Viking Corporation.)*

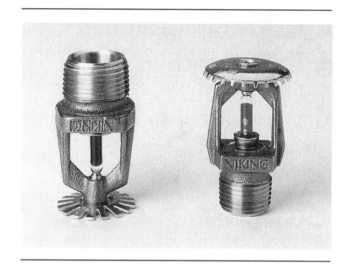

sorbed into the liquid. With only liquid in the bulb, the pressure inside rises quickly, and any additional expansion in the liquid due to heating breaks the bulb. By controlling the size of the bubble and the properties of the liquid, sprinklers with different operating temperatures can be constructed.

Design and Installation of Sprinkler Systems

The installation of fire sprinkler systems is governed by rules designed to promote uniformity between installations, while ensuring reliability and the construction of systems that will perform properly. In the United States, the guidance followed is NFPA 13. The current edition was published in 2002, but new versions are available every 3 years and contain the guidance or "rules" used by designers and installers to provide a code-complying system. These systems are usually designed by trained individuals working for installation contractors or engineering firms and in some cases may be required by state or local law to be licensed or certified in some way. Installation of the pipe and sprinkler heads is then traditionally undertaken by plumbers, pipe fitters, sprinkler fitters, or others, and again depending on the location may require special training or licensure. Although fire fighters may not design and install sprinkler systems, they need to have a basic understanding of these subjects to guide them when they are conducting prefire plans, inspecting buildings on the fire company level, or responding to a system activation.

Layout of Sprinkler Piping.   Figure 8.16 on page 302 shows a sprinkler system. Each of the branch lines containing sprinklers is supplied with water through a cross main from only one direction, forming a dead end. When branch lines are located on both sides of the cross main, the system tends to look like a center trunk of a tree with branches coming out from the center, leading to the name *tree system* for this type of sprinkler layout. When laid out in this fashion, tree systems are simple to design, but since water is supplied in only one direction, a single blockage or plug in the pipe can prevent water from reaching all sprinklers downstream of the block. When the branch lines are supplied by a cross main on both ends, water can travel in multiple paths leading to what are called gridded sprinkler systems. Because water can travel in multiple directions, pipe sizes can be smaller, resulting in more economical designs.

Sprinkler Design and Occupancy Hazard Classification. Intuitively one would expect that more water is necessary to control a fire in a hardware store filled with goods than an art museum containing relatively few artworks on the walls. NFPA 13 solves the problem of determining how much water is necessary for each type of building use through *occupancy hazard classifications*. Although there are an infinite number of ways buildings can be used, building areas are divided into one of the following five different hazard classifications:

1. *Light Hazard.* Occupancies where the quantity and combustibility of contents is low, resulting in fires where a low rate of heat release is expected and includes theaters, restaurant seating areas, museums, offices, and residential buildings.

2. *Ordinary Hazard Group I.* Occupancies where the combustibility of contents is low, but the quantity of goods stored do not exceed 8 ft in height, and moderate rates of heat release are expected. This group includes manufacturing of noncombustible materials such as glass or dairy products, canneries, and automobile parking garages.

3. *Ordinary Hazard Group II.* Occupancies where the combustibility and quantity stored is moderate to high, stockpiles do not exceed 12 ft in height, and moderate to high rates of heat release are expected. This classification includes manufacturing of combustible materials such as alcoholic beverages, paper and wood products, plus printing, repair garages, and most stores and shops (mercantile).

4. *Extra Hazard Group I.* Occupancies where dust, lint, or other materials are present that could result in rapidly developing fires with high rates of heat release. This group includes aircraft hangers, plywood manufacturing, upholstering with plastic foams, and saw mills.

5. *Extra Hazard Group II.* Occupancies where moderate to substantial amounts of flammable or combustible liquids are stored or used. Extra

Hazard Group II occupancies can also include lower hazard occupancies where the shielding of combustibles is extensive, preventing sprinkler water from reaching the hazard. This classification includes flammable liquid spraying or coating, plastics processing, solvent cleaning, and the manufacture of homes or modular buildings (due to the roofs doing what they should—preventing water from passing below).

Sprinkler Area of Operation.   Although NFPA 13 intends for a high level of safety to be provided by sprinklers installed according to the standard, it is not expected that every sprinkler in a building will operate during a fire. Fire incident statistics support the idea that few sprinklers actually open during a real fire when the sprinkler system is operating properly. For all of the sprinklers in a building to operate, generally a very large fire that has grown out of control has occurred and is an example of a system failure, not a successful operation. The exception is deluge sprinkler systems, where all of the sprinkler heads are intentionally open before the fire starts.

The number of sprinklers considered necessary to open is specified by NFPA through the concept of the *hydraulically most demanding area of sprinkler operation.* Whether a building is one story with an area of say 100,000 ft^2 or multiple stories with 25,000 ft^2 on every floor, the standard requires the designer to plan for the same small area of sprinklers to open. Thus, irrespective of the building area, NFPA 13 only requires the sprinkler system to be able to provide enough water (flow rate and pressure) to a limited area of sprinklers, and this area is usually much less than the entire area of a single floor. Figure 8.26 shows the method NFPA 13 uses to relate the floor area of sprinkler operation in ft^2 that the system will be able to supply with the required amount of water necessary to protect each of the five different occupancy classifications. As an example, for an extra hazard group I occupancy, the designer may select any area of sprinkler operation along the extra hazard group 1 curve, which varies between 2500 and 5000 ft^2. The potential exists for all sprinklers protecting that floor area to operate during a fire, thus the pipe sizing must be able to support this number of sprinklers opening.

The amount of water provided by the system is also specified in Figure 8.26 through the sprinkler *discharge density* in gpm/ft^2 and represents the amount of water sprinklers will discharge over the coverage area of the sprinklers. For a light hazard occupancy with a design area of 1500 ft^2, all of the sprinklers in this area must each flow enough water to cover the floor area protected with 0.1 gpm/ft^2. This amount is 0.1 gallons of water over each square foot of area, each minute of sprinkler operation. If sprinklers in this area are spaced 15 ft apart, they each protect an area of 15 ft × 15 ft = 225 ft^2. To provide the required density, each sprinkler in the design area must be

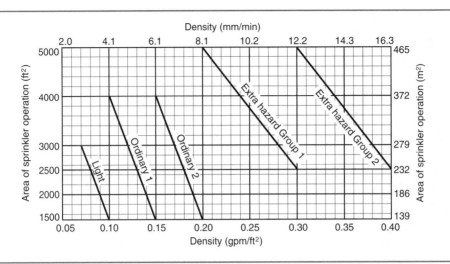

▶ **Figure 8.26**   Hydraulic Calculation Area/Density Curves. *(Source: NFPA 13, 2002, Figure 11.2.3.1.5.)*

able to flow a minimum of 0.1 gpm/ft^2 × 225 ft^2 = 22.5 gpm. The total flow required would be the density × the area (the flow from each sprinkler × the number of sprinklers in the design area). The curves in Figure 8.26 allow the designer to select any one point on the curve and use that *area/density* combination for the sprinkler design—the sprinkler system need not meet all points on a curve.

Sprinkler Systems for Residential Buildings.   To this point, the discussion of fire sprinkler systems has been based on installation according to NFPA 13. When protected with sprinklers installed according to NFPA 13, the intent of the installations is to provide a high level of safety against fire in terms of property protection, life safety, and business continuity. The standard accomplishes this goal through specification that sprinklers be installed throughout a building (including combustible concealed spaces such as attics), safety factors during design, and redundancies (both an automatic and secondary water supply are required).

More than many other industrialized countries, the United States has suffered a high rate of fire fatalities where people live and sleep—their homes, apartments, dormitories, and hotels. The frustration with many fire service leaders is that even with an immediate fire department response, people still die from smoke and fire exposure before fire fighters arrive. To encourage the installation of fire sprinkler systems that would be designed primarily for life safety, additional standards were developed to modify the installation rules in NFPA 13 with the goal of providing a reasonable level of safety at a reduced cost. NFPA 13R, *Standard for the Installation of Sprinkler Systems in Residential Occupancies up to and Including Four Stories in Height* [11], and NFPA 13D, *Standard for the Installation of Sprinkler Systems in One- and Two-Family Dwellings and Manufactured Homes* [12], are two such documents. It turns out that it is easier (i.e., cheaper) to design and provide systems for life safety than for property or business conservation.

Fire data gathered over many years from fires that were attended by the fire department show that a large number of the fatal fires in the residential environment have an area of origin involving the living room, kitchen, or bedrooms. Few fatal fires begin in attic spaces or bathrooms, although these fires can still do significant property damage. Thus NFPA 13D and 13R concentrate protection in those areas that do the most good or will protect the most people. To achieve a high level of life safety, the goal of such residential sprinkler systems is to prevent the phenomenon of flashover, a period during the development of a fire when the room of fire origin becomes fully involved in fire and large amounts of smoke and hot fire gases can be transported throughout a building, blocking escape routes. Although there may be a significant amount of damage to the building or contents for fires starting in unprotected areas such as an attic, as long as everyone makes it out of the building the system operation would be considered a success. (See Chapter 3.)

The installation of residential sprinklers is accelerating in the United States as more building and fire codes require their installation, prices are reduced, and people become more familiar with this lifesaving technology. A new trend has been the installation of residential sprinklers in new developments and areas undergoing rapid housing development. It is recognized that new buildings that are sprinklered require fewer fire department services, so growing communities can often delay establishing new fire stations or provide them at wider spacing.

▶ ALTERNATIVE SUPPRESSION SYSTEMS

The more common nonwater-based suppression agents and their extinguishing mechanisms were discussed earlier. While some are used for manual application via fire extinguishers, many are best suited for automatic fire suppression systems used to protect critical equipment, operations, and high fire hazard areas. Even in fully sprinklered buildings, alternate suppression systems are valuable, either providing additional protection for a high hazard (flammable liquid storage room) or as protection for critical computer equipment to control a fire at a size much smaller than necessary to activate a sprinkler and without the water damage. Because they require proper operation of detection, control, agent storage, and distribution systems, alternate agent systems have the potential for a lower reliability than sprinklers due to increased mechanical components and limited agent storage. Therefore it is not recommended that they be considered a replacement or substitute for sprinklers, but instead should be used to complement sprinkler protection, especially in those areas where water is not the agent of choice.

Dry chemical is discharged as a finely divided solid suspended in air. Water mist is discharged from a nozzle as very small drops of water, while most other agents are discharged as a gas (but may be stored as a liquid when under pressure). This fact drives when these systems can be used and where they provide the best protection.

When a hazard can be enclosed (completely or nearly completely) with construction that will contain the agent, a *total flooding* system is usually installed to completely fill the space with agent, allowing it to disperse throughout the enclosure and any equipment and is most common with gaseous agents, as shown in Figure 8.27. Detectors are provided throughout the protected space to sense the presence of a fire and send a signal to the control unit. Suppression agent is maintained in cylinders through closed valves connected to the control unit that will open during a fire. To distribute the agent, the cylinders are connected to piping and open nozzles throughout

▶ **Figure 8.27** Total Flooding System Used to Protect a Computer Room and the Under-Floor (Raised Floor) Area Containing Cables and Electrical Equipment. *(Source: Fire Protection Handbook, NFPA, 2003, Figure 11.1.2.)*

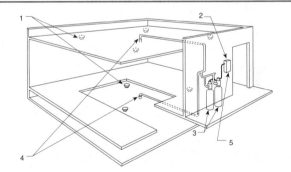

1. Automatic fire detectors installed both in room proper and in underfloor area.
2. Control panel connected between fire detectors and cylinder release valves.
3. Storage containers for room proper and underfloor area.
4. Discharge nozzles installed both in room proper and in underfloor area.
5. Control panel might also sound alarms, close doors, and shut off power to the area.

the space. To reduce false activations, the detectors may be zoned to require two separate detectors to activate before a signal of fire is sent to the agent distribution control unit. If doors, windows, ductwork, or other openings cannot be closed prior to agent discharge, it is necessary to provide additional agent to account for leakage.

Some agents, such as dry chemical, are better suited for local (flooding) application protecting well-defined hazards like a small combustible liquid dip tank where the agent can be distributed directly on the surface of the fuel, without the need for containment by the building. If used to protect an entire room, dry chemical tends to rapidly settle out of the air, limiting the protection. Local application is also appropriate for agents that are discharged mostly as a liquid, because they too would be difficult to apply throughout a room, as shown in Figure 8.28. Because the protected equipment may be located in the middle of a large plant, total flooding the entire building is not practical, so agent is applied only over the portions of the equipment with the highest hazard.

High-speed explosion suppression systems use water or other agents distributed through nozzles within 100 milliseconds of when a fire is detected. These systems are used to protect against damage caused by dust explosions, or when manufacturing pyrotechnics, explosives or some hazardous chemicals, for example. High-speed water spray systems are addressed in NFPA 15, *Standard for Water Spray Fixed Systems for Fire Protection* [13], while systems specifically designed to suppress an explosion and resulting pressure increase are covered in NFPA 69, *Standard on Explosion Prevention Systems* [14]. It is important to recognize that these systems only have the capability to control explosions with a flame front traveling less than the speed of sound (deflagration) and not against explosions where the flame front travels at a

▶ **Figure 8.28**   Local Application System Protecting Portions of a Printing Press Where Combustible Inks Are Applied. *(Source: Fire Protection Handbook, NFPA, 2003, Figure 11.1.5.)*

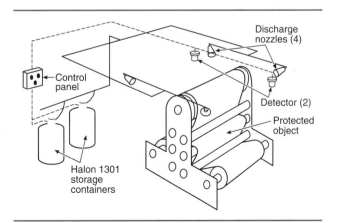

speed greater than that of sound (detonation). High-speed detection and control equipment is an integral part of these systems to provide a signal that a fire condition exists and to initiate agent discharge while minimizing false alarms.

▶ MANUAL SUPPRESSION—STANDPIPE SYSTEMS AND FIRE EXTINGUISHERS

Manual fire suppression, especially the stretching of hose lines through tall or large area buildings, requires significant human resources and time. To assist responding fire fighters in these buildings, most building, fire, and life safety codes require the installation of standpipes in high-rise buildings, warehouses, and manufacturing plants. As a backup to support fire department operations, they are even necessary in sprinklered buildings, although both systems often share common piping. A good source of information for fire fighters to learn about operations in buildings with sprinklers and standpipes is NFPA 13E, *Recommended Practice for Fire Department Operations in Properties Protected by Sprinkler and Standpipe Systems* [15]. As with any mechanical system, standpipes can fail during a fire, leading to conditions in which fire fighters lose their water supply while attacking a fire, so it is common for preincident plans to include notification of building engineers so that they can be dispatched to the scene to monitor the operation of critical fire protection systems.

The following three main levels of service are outlined in NFPA 14, *Standard for the Installation of Standpipe and Hose Systems* [16], the document used to design and install standpipes:

- Class I standpipe systems contain a $2\frac{1}{2}$-in. hose line connection designed to supply firefighting water for use by the fire department or on-site fire brigade. By providing a large flow rate at pressures sufficient to operate fire department fog nozzles (65 to 100 psi), this standpipe could support a number of fire department hose lines used concurrently to control a fire in an unsprinklered building or during sprinkler system failure. Each connection can be sized to provide up to 250 gpm of water.

- Class II standpipe systems are designed to be used by building occupants or the fire department during final extinguishment of a fire and include a $1\frac{1}{2}$-in. hose line connection and usually a supply of hose. Building, fire, and life safety codes often require Class II standpipe systems in warehouses, stages, prisons, and auditoriums used for exhibits.

- Class III standpipe systems include the necessary equipment and design to serve both as a Class I and Class II standpipe.

Standpipes may also be found in underground structures such as tunnels or transit centers, shopping malls, open or unheated parking garages, and rooftop helipads. As with sprinkler systems, the piping for a standpipe can either be wet or dry. To be considered an *automatic standpipe*, there must be a connection to a primary source of water that requires no intervention for proper operation. A *manual standpipe* either requires the fire department to connect hose lines outside to the standpipe as a water source or the opening of a valve at the water supply connection to allow water into the standpipe.

The building, fire, and life safety codes requiring the standpipe installation often indicate the number of standpipe connections required, type of fire hose thread, and where connections should be located. For many buildings, standpipe connections are placed in the enclosed exit stairwells at every floor level (to protect the pipe and fire fighters), near major exits, or so that all portions of the protected area can be reached with 100 ft of hose plus 20 to 30 ft nozzle discharge distance. Each point of connection is provided with a valve that the fire department can operate to discharge water, plus a threaded $2\frac{1}{2}$ in. or $1\frac{1}{2}$ in. discharge opening for connection of the hose.

Especially with high-rise buildings, it is common to see combined standpipe and sprinkler systems supported by the same water supply and sharing the same vertical distribution piping located in an enclosed exit stairway, as shown in Figure 8.29. This assembly is located on each floor and provides the

▶ **Figure 8.29** Combined Sprinkler System and Standpipe Riser, Typical of What Is Installed in a High-Rise Building. *(Source: NFPA 14, 2003, Figure A.7.10.1.3.1(a).)*

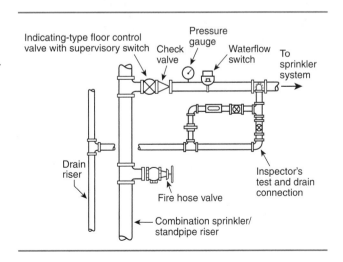

valves and controls for both systems. Notice that a separate sprinkler system is created on each floor, complete with a flow switch to allow alarm indication by floor, a drain, control valve, and inspector's test connection. The term _riser_ was used to describe the system control piping that connects the water supply to the sprinkler distribution pipe located at the ceiling. The term _riser_ is also used for the vertical piping of a standpipe system from the lowest level of a building to the top floor. A combined sprinkler and standpipe riser includes the necessary valves, flow alarms, and pressure gauges to serve both functions.

The last type of equipment that can be used to control or suppress fire is the portable fire extinguisher that is designed to be used by building occupants, specially trained individuals, or fire fighters. Many of the fire suppression agents discussed earlier in this chapter can be stored in the cylinder of a fire extinguisher, along with a pressurizing gas such as nitrogen for those agents such as water or dry chemical that will not flow out of the extinguisher under their own pressure when the valve is opened. As with alternate agent suppression systems, it is important that the correct portable fire extinguisher be selected to protect the hazard, especially with flammable or combustible liquids, energized electrical equipment, and combustible metals. Common fire extinguishers are shown in Figure 8.30 through Figure 8.32. Because the carbon dioxide changes phase from a gas to a liquid when under pressure, the type of extinguisher shown in Figure 8.31 must be weighed to determine if it is full, because a pressure gauge will not give an accurate reading of how much agent is in the liquid form.

▶ **Figure 8.30** (a) Water-Based Fire Extinguisher That Uses Stored Pressure of Air, Nitrogen, or Other Gas to Discharge the Agent from the Container with (b) Close Up of Head Assembly. _(Source: Badger Fire Protection.)_

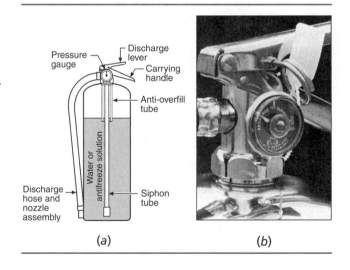

(a)

(b)

▶ **Figure 8.31**   Carbon Dioxide Fire Extinguisher. *(Source: Fire Protection Handbook, NFPA, 2003, Figure 11.6.7.)*

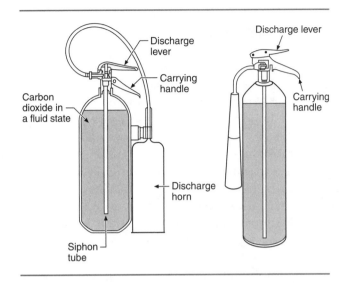

▶ **Figure 8.32**   Dry Chemical Fire Extinguisher That Uses a High-Pressure Gas in a Separate Container to Pressurize the Cylinder Containing the Agent When the Puncturing Level on the Side Is Pressed Firmly.   *(Source: Ansul Incorporated.)*

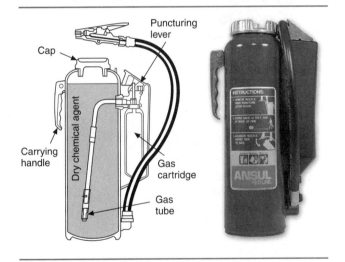

▶ SMOKE CONTROL AND EXHAUST SYSTEMS

The final type of fire protection system discussed is less common than suppression and detection systems and is more complicated and often misunderstood. Although detection systems sound the alarm and suppression systems discharge agent on a fire, what can be done to minimize the impact of smoke? Even sprinkler-controlled fires can produce tremendous volumes of cool smoke sufficient to block escape routes. Systems are available that can safely direct smoke away from occupied areas, supplement fire walls and fire doors to limit spread, or remove the smoke from a space to provide time for occupant escape or fire department operations.

It is common in high-rise buildings that egress routes such as the exit stairs are maintained at a higher (positive) pressure compared to the rest of the building, thus limiting smoke spread into the stairs. The fire floor can also be exhausted, while floors above and below are pressurized to limit smoke migration to other areas of the building, as shown in Figure 8.33. The pressurization method for smoke control can limit the spread of smoke through small openings typically found in buildings. Buildings and floors can be divided separately (zoned) to create a "pressure sandwich" where the fire area is maintained at a lower pressure than nonfire areas. This strategy is especially useful in hospitals and nursing homes to protect patients who are unable to move and are protected with a defend-in-place strategy.

Although the pressurization method supplements fire-resistive construction such as fire walls and doors, when an opening can not be protected with doors, a high-velocity flow of air can be directed through the opening to limit smoke spread to nonfire areas.

▶ **Figure 8.33**   (a) Pressure Difference across a Barrier of a Smoke Control System Preventing Smoke Infiltration to the High-Pressure Side of the Barrier and (b) Diagram of a Smoke Control Zone Limited to Part of a Building's Floor.
(Source: The SFPE Handbook of Fire Protection Engineering, 3rd ed., Figure 4-12.5 and Figure 4-12.18(e). ©Society of Protection Engineers. Used with permission.)

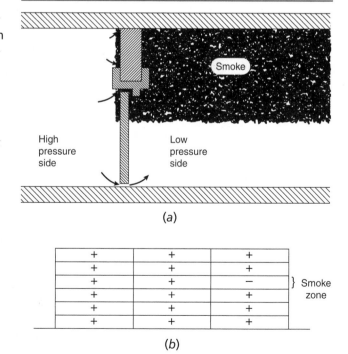

▶ **Figure 8.34**   Smoke Exhaust System Operating in a Large Volume Space (atrium), Thus Increasing the Time Available for Occupants in Upper Levels to Leave the Building. *(Source: Fire Protection Handbook, NFPA, 2003, Figure 12.6.10.)*

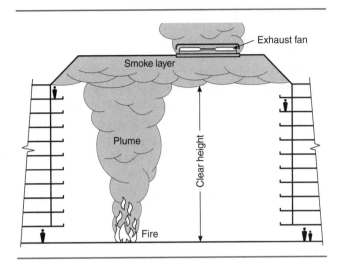

One of the more effective ways to deal with smoke is to not just contain it, but to exhaust it to a safe location outside the building. Especially common in large spaces like sporting arenas and atria, smoke exhaust systems are able to both limit the spread of smoke to unaffected areas and delay the descent of smoke in the space, thus providing additional time for exiting, as illustrated in Figure 8.34. Smoke exhaust can also be used in warehouse spaces and big-box stores to remove smoke during a sprinkler-controlled fire to facilitate fire department operations/suppression and to limit smoke damage.

▶ **SUMMARY**

Although fire fighters are not usually involved in the design and installation, they certainly are the first to respond when an unwanted fire sets the operation of an automatic fire protection system in motion. Safe fire department operations in buildings necessitate a thorough knowledge of how these systems should respond and what the consequences are when they do not work properly. The incident commander faces a difficult decision during a sprinkler-controlled fire, which is when to shut off the water supply to the sprinklers, for if done too early, the fire can redevelop and extend beyond the initial fire area before the water supply is re-established.

Special suppression systems that use agents other than water to control a fire are expected to be tailored to the hazard they protect, thus are usually able to rapidly knock down fires in most cases. Even though the fire is out, anyone entering an environment containing a suppression agent, including during a false activation, must be trained and protected with self-contained breathing apparatus until the area can be well ventilated. Fire fighters can make a dramatic difference, both good and bad, in the operation of smoke control systems because the control is usually provided with manual overrides located at the fire department command center. Unfortunately, all too often fire fighters arrive at a building fire where automatic fire protection systems are operating properly, and they proceed to shut everything down. Especially with smoke control systems, turning them off while they are operating properly only increases the exposure to smoke and hot gases for building occupants and fire fighters. Proper training and education on this equipment can only come before an incident and not during the course of battle.

Group Activity

Which of the protection devices do you believe is most critical to have operating in a facility such as a nursing home? Why? Be prepared to defend your answer.

Review Questions

1. Into what four groups are fire protection systems broadly classified?
2. The total time required for a person to reach a point of safety from the start of the fire is the sum of what three components ?
3. What is the role of a fire alarm system?
4. Most smoke alarms operate on what principle?
5. _____ _____ placed near the ceiling are in an ideal location to intercept the hot gas stream and detect the temperature change (increase) characteristic of fires.
6. How do rate-of-rise heat detectors react faster than fixed temperature heat detectors to fire?
7. What type of systems require that the correct type of detector be selected to match the wavelength of radiation from the fuel of interest?
8. What kinds of appliances are designed to sound a tone or pattern distinctive enough to be recognized over other sounds in the area?

9. What signifies a problem with alarm wiring or transmission means or a problem with the system power supply?

10. What type of stations provide a minimum level of service, basically limited to alarm retransmission?

11. What is the most common extinguishing agent used to control or suppress unwanted fires?

12. By finely grinding effective solid suppression agents into a powder, what can be applied through fixed nozzles or extinguishers onto a fire?

13. The simplest and most common protection is provided by what type of sprinkler system that uses pipes under pressure and filled with water at all times to connect the sprinkler distribution nozzles, or sprinkler heads, to the water supply?

14. One of the more effective ways to deal with smoke is to not just contain it, but to instead do what?

Suggested Readings

NFPA 10, *Standard for Portable Fire Extinguishers*, 2002 edition.

NFPA 11, *Standard for Low-, Medium-, and High-Expansion Foam*, 2002 edition.

NFPA 12, *Standard on Carbon Dioxide Extinguishing Systems*, 2000 edition.

NFPA 12A, *Standard on Halon 1301 Fire Extinguishing Systems*, 2004 edition.

NFPA 17, *Standard for Dry Chemical Extinguishing Systems*, 2002 edition.

NFPA 17A, *Standard for Wet Chemical Extinguishing Systems*, 2002 edition.

NFPA 69, *Standard on Explosion Prevention Systems*, 2002 edition.

NFPA 750, *Standard on Water Mist Fire Protection Systems*, 2000 edition.

NFPA 2001, *Standard on Clean Agent Fire Extinguishing Systems*, 2004 edition.

9

Fire Investigation

Richard Custer
Arup Fire USA

Learning Objectives

After completing this chapter, the reader should be able to do the following:

- Describe the role of fire investigation in reducing fire loss
- Explain the difference between fire investigation and arson investigation
- Describe the investigative process
- Describe how the observations and actions of fire fighters can assist fire investigation.

Fire investigation is the process of determining the origin, cause, and development of a fire [1]. It involves the study and analysis of physical evidence at the fire scene and in the laboratory and consideration of witness statements and documentary evidence to determine the location of the fire's origin and its cause. Determination of cause includes identifying the energy source and circumstances of ignition and the first fuels involved in the ignition. Fire investigation also involves identifying the factors relating to the spread and extent of the fire and the resulting injuries and property damage.

There are many reasons for investigating fires, ranging from identification of hazardous practices, dangerous materials and products, and the effectiveness of building and fire code provisions to the apprehension and conviction of arsonists. The primary objective, however is to reduce fire losses, and preventing fires in the first place is the best way to start. Successful fire safety education programs and fire prevention codes are based on data from careful and accurate investigation of fires. Behavior patterns that result in fires or injuries or can be shown to prevent or reduce injuries, once identified, can be the basis of training and educational programs such as the National Fire Protection Association's (NFPA) *Learn Not to Burn®* curriculum and *Risk Watch®*.

Fire protection standards and building codes are developed to limit fire losses through the use of systems such as sprinklers and smoke control and construction provisions such as egress systems to aid in escape and fire-rated construction to contain fire. Investigation of fires can provide real-world evaluation of the effectiveness of these standards and codes. Historically, major fires resulting in large life losses or property damage have led to changes to codes such as those made to the automatic sprinkler requirements for assembly occupancies in the State of Rhode Island Building Code as a result of The Station nightclub fire in which 100 lives were lost in 2003.

▶ FIRE INVESTIGATION ORGANIZATIONS

Many organizations carry out fire investigations. These organizations form basically two main categories, public and private. A third category is a mixture of both public and private interests and is involved in the development of codes and standards.

Public Organizations

Public organizations are governmental and include federal, state and local fire and law enforcement organizations. Fire departments are generally required by local or state laws to investigate the causes of fires in their jurisdictions. Although this investigation is often done by the fire department, local or state police authorities may have the responsibility or be called in when a crime is suspected.

In the United States in incidents in which federal laws might have been broken (such as suspected arson, bombing, or possible terrorist activities), federal agencies including the Bureau of Alcohol Tobacco and Firearms (ATF) in the U.S. Department of the Treasury or the Federal Bureau of Investigation (FBI) in the U.S. Department of Justice may be called to assist or have jurisdiction. In the case of aircraft or rail incidents, the National Transportation Safety Board (NTSB) may have primary investigative responsibility.

Other federal agencies that may become involved in fire or explosion incidents include the Chemical Safety Board (CSB), the Mine Safety and Health Administration (MSHA), the Occupational Safety and Health Administration (OSHA), and the Consumer Product Safety Commission (CPSC).

Private Organizations

Private organizations are largely insurance companies and the investigative services and consultants that provide services to the insurance industry and attorneys involved in fire-related litigation. Although both the public and private organizations share interest in the origin and causes of fires, the private organizations are often focused on the reasons for the extent of the damages or injuries in order to identify conditions, actions, and defective designs or products that may have contributed to the loss.

Those organizations involved in development and effectiveness of codes and standards such as NFPA are focused on identifying both failures and successes of the code and standards provisions. Public organizations may be involved with private organizations where large losses or high numbers of deaths and injuries lead to the consideration of changes to standards or building codes. Whereas some specific fires may be closely investigated by these organizations, regional or national trend data over time is also an important input to the codes and standards process.

▶ INVESTIGATION PROCESS

The following sections provide an overview of the investigation process and identify the important areas of observation and actions for first responders that can assist in the investigative process.

Fire Incident Data

The results of investigations can be used to provide a broad picture of causes, losses, and performance of fire safety features of buildings. Tables 9.1 through 9.3 provide examples of national summary data for fire causes, fire deaths and injuries, and fire spread factors. See Chapter 2 for additional information. Data can also be analyzed at the state or local level and provide insight to regional or local fire prevention or building code issues.

There is an important difference between *fire investigation* and *arson investigation*. Fire investigation is the process of determining cause of the fire. If the fire investigation concludes that cause of the fire is incendiary, then the crime of arson has been committed and an arson investigation is undertaken. Because arson is criminal in nature, arson investigation is often undertaken by those with police or law enforcement powers.

▶ **Table 9.1**   Major Causes of U.S. Structure Fires Reported to Fire Departments, 1994–1998 annual average.

Cause	Fires	Civilian Deaths	Civilian Injuries	Direct Property Damage (in $ Millions)
Defined by Heat Source				
Smoking material (i.e., lighted tobacco product)	35,400	990	2,540	430
Electric-powered equipment	208,900	970	7,670	2,484
Open flame (e.g., match, lighter, torch, candle)	106,000	820	5,550	1,563
Fueled equipment	119,300	640	3,580	1,149
Defined by Equipment Involved				
Heating equipment	75,700	490	1,910	839
Electrical distribution system	58,100	350	1,670	1,028
Cooking equipment	116,300	330	5,120	560
Defined by First Ignited Item				
Upholstered furniture	15,000	640	1,670	272
Mattress or bedding	29,700	540	2,890	373
Structural member or framing	48,700	320	880	1,111
Defined by Material in First Ignited Item				
Fabric, textile, or fur	89,700	1,570	6,490	1,077
Wood or paper	174,100	920	3,580	2,947
Flammable or combustible liquid	30,000	300	2,160	505
Defined by Behavior or Event				
Mechanical or electrical failure	158,300	670	3,700	2,078
Incendiary or suspicious causes	94,500	640	2,530	1,680
Abandoning something (e.g., cigarette)	35,300	610	1,960	374
Child playing	23,500	300	2,170	284

Note: Unknowns have been allocated proportionally.

Source: National estimates based on NFIRS and NFPA survey data.

▶ **Table 9.2** Locations of Fatal Victims at Ignition in Structure Fires, 1994–1998, Annual Averages.

Victim Location Relative to Fire	Structures		Homes		Structures Other Than Homes	
Intimate with ignition	611	16.3%	548	15.7%	68	27.7%
Not intimate, but in same room	921	24.6%	850	24.3%	72	29.4%
Not in room, but on same floor	1,149	30.7%	1,097	31.4%	48	19.4%
Not on same floor, but in building	987	26.4%	936	26.8%	48	19.7%
Outside building of origin	36	1.0%	31	0.9%	6	2.4%
Unclassified and other known	40	1.1%	36	1.0%	3	1.4%
Total	3,744	100.0%	3,498	100.0%	246	100.0%

Note: Unknowns have been allocated proportionally.

Source: National estimates based on NFIRS and NFPA survey data.

▶ **Table 9.3** Sizes of Home Fires, 1994–1998.

Extent of Flame Damage	Fires (%)	Civilian Deaths (%)	Civilian Injuries (%)	Property Damage (%)
Confined to room	72.7	19.6	57.2	21.4
Beyond room, confined to floor	5.3	12.1	11.3	10.9
Beyond floor	22.2	68.3	31.5	67.6
Total	100.0	100.0	100.0	100.0

Source: National estimates based on NFIRS and NFPA survey data.

Fire Causes

Although fire investigations are done for a wide range of objectives, finding the place where the fire started (the *origin*) and identifying the circumstances that brought the ignition source and the first fuel ignited together (the *cause*) is usually the primary task. NFPA 921, *Guide for Fire and Explosion Investigations*, defines four fire causes: accidental, natural, incendiary, and undetermined [1].

Accidental Causes.   Accidental fires are those in which deliberate human action was not involved in the ignition or spread of the fire. Fires caused by failure of an electrical or heat producing appliance, sparks from open burning, fire following a vehicle accident, storing combustibles too close to heat sources, or careless use of smoking materials are examples of accidental fires.

Natural Causes.   Natural fires are those not involving human action or intervention for ignition or spread. Examples include fires following earthquakes, lightning, or resulting from electrical wires brought down by windstorms.

Incendiary Causes.   Incendiary fires are those that are deliberately ignited under circumstances in which the person(s) involved knows that the fire should not be ignited.

Undetermined Causes.   Fire cause is classified as undetermined when the cause cannot be proven. The fire may also still be under investigation and the cause determined later, should evidence become available.

Type of Loss

Fire investigation involves finding the cause of the losses in the areas of damage to property, injury, or death and contribution of human fault to the circumstances of the cause. Investigation of the cause of damage to property considers the factors responsible for the spread of the fire and extent of the loss and include the adequacy or appropriateness of fire protection, building code features, and materials or products involved.

Investigation of the causes of injury or death deals with the factors of life safety components such as adequacy of detection and alarm systems, egress systems, building construction, smoke management, and the role of products producing smoke or toxic combustion products. The investigation of the human fault considers the role that human actions played in the cause or spread of fire or in injury or loss of life.

Arson Investigation

Arson is defined in NFPA 921 as the crime of maliciously and intentionally, or recklessly, starting a fire or causing an explosion [1]. The precise legal definition varies among jurisdictions as defined by statutes and judicial decisions. The issue of arson might arise in civil litigation between parties in insurance disputes where there is insufficient evidence for criminal action but an insurance claim is in question due to an allegation that the insured set or caused the fire to be set.

Arson investigation begins after the determination that there was an incendiary cause and involves the use of fire scene evidence, witnesses, and documentary evidence to identify and charge individuals with the crime of arson. Documentary evidence is evidence in written form and includes materials such as business, personal, insurance, and public records such as those from the building, health, police, and fire departments.

▶ CONDUCTING THE INVESTIGATION

Scientific Method

The investigative process is built around the collection and analysis of evidence (data), development of alternative theories (hypotheses) of how the fire might have started and spread, and testing the alternative theories against the evidence collected and test results. This process is often referred to as the *scientific method* and its use in fire investigation is detailed in NFPA 921 [1]. Figure 9.1 outlines the basic steps in the scientific method as applied

▶ **Figure 9.1** Use of the Scientific Method. *(Source: NFPA 921, 2004, Figure 4.3.)*

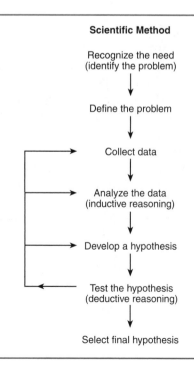

Scientific Method

Recognize the need
(identify the problem)

↓

Define the problem

↓

Collect data

↓

Analyze the data
(inductive reasoning)

↓

Develop a hypothesis

↓

Test the hypothesis
(deductive reasoning)

↓

Select final hypothesis

to fire investigation. In general, the theory that is most strongly supported by the evidence and requires the least complex explanation or sequence of events is more likely the cause.

Fire Scene Analysis

The purpose of the fire scene analysis is to collect data that will be used to assist in developing cause theories. The data consists of information regarding burn patterns, fuel sources, ignition sources, and evidence of fire spread. The data is documented by interviews with building owners, occupants, passers-by, and first responders including the fire and police departments. Data collection also includes documentation of visual evidence as accomplished by still photos, video, and preparation of sketches and diagrams. Data can also be developed through preservation and testing of physical evidence collected from the scene. The physical evidence may be fire debris collected for analysis to determine its composition, flammability for its role in fire ignition and spread, or as an accelerant in an incendiary fire. Documentary evidence such as business or personal records can also be identified and collected at the scene. Figure 9.2 shows a sample form used to gather information about evidence collected at the scene.

General Procedure.   The scene investigation usually follows a procedure of examination beginning with the exterior of the building or the perimeter surrounding the fire and proceeding to the interior of the building or vehicle or, in the case of wildland fires, the burned areas of vegetation. The exterior investigation provides initial information about the extent of fire spread and may, for example, show which floors or rooms were involved to the extent that damage to the exterior of the building can be noted. Possible evidence such as containers for ignitable liquids or areas where an exterior fire may have entered a building can also be discovered during the exterior examination. The interior examination generally proceeds from the area of least damage to the area of greatest damage.

Origin Determination.   Interior examination is intended to assist in locating the area of origin (area where the fire started) based on the assumption the fire burned longest and created the greatest damage in the area of origin. When the investigation involves fires of limited damage or spread, this assumption is generally valid and locating the area or the actual point of origin can be relatively simple. As the degree of damage and spread becomes greater, the burn patterns become more difficult to interpret due to involvement of more flammable fuels, changes in ventilation for the fire, and

EVIDENCE FORM

Date of Incidents _____/_____/_____   Storage Location: _____   Case No. _____

ITEM NO.   DESCRIPTION   LOCATION

_____   _____   _____   ❑ DESTROYED   ❑ RELEASED

_____   _____   _____   ❑ DESTROYED   ❑ RELEASED

_____   _____   _____   ❑ DESTROYED   ❑ RELEASED

_____   _____   _____   ❑ DESTROYED   ❑ RELEASED

_____   _____   _____   ❑ DESTROYED   ❑ RELEASED

_____   _____   _____   ❑ DESTROYED   ❑ RELEASED

_____   _____   _____   ❑ DESTROYED   ❑ RELEASED

_____   _____   _____   ❑ DESTROYED   ❑ RELEASED

_____   _____   _____   ❑ DESTROYED   ❑ RELEASED

_____   _____   _____   ❑ DESTROYED   ❑ RELEASED

_____   _____   _____   ❑ DESTROYED   ❑ RELEASED

How was evidence received?   Date received: _____/_____/_____   Date stored: _____/_____/_____

Removed from scene by investigator.
Received by investigator from: _____
                                                         Name, company, or dept.

Received via: ❑ UPS   ❑ FEDEX   ❑ AIRBORNE   ❑ USMAIL   ❑ IN PERSON

Freight _____   **Other** _____
                          Name of company   Describe

_____   _____
              Received by   Case investigator

LOCATION EVIDENCE REMOVED

Owner _____   Address 2 _____

Company _____   City _____

Address 1 _____   State _____ Zip _____ Phone _____

▶ **Figure 9.2** Sample Evidence Collection Form. *(Source: Field Guide for Fire Investigators, NFPA, 2003, Figure 2.7.)*

INTERNAL EXAMINATION

Investigator	Date Pulled	Date Examined	Date Returned

EVIDENCE DESTRUCTION

Authorized by   Date

Investigator's authorization   Date

Destroyed by   Date

EVIDENCE RELEASE

Signature of person receiving evidence

Person receiving evidence (please print)   Date

Company name

Address

City   State   Zip code

Authorized by   Date

Investigator's authorization   Date

Released VIA _____

Remarks _____
_____
_____
_____
_____
_____
_____

EXAMINATION BY OTHERS

Name   Date of examination

Company

Address

City   State   Zip code

Phone

Authorized by

Investigator's authorization   Date

Name   Date of examination

Company

Address

City   State   Zip code

Phone

Authorized by

Investigator's authorization   Date

Name   Date of examination

Company

Address

City   State   Zip code

Phone

Authorized by

Investigator's authorization   Date

▶ **Figure 9.2**   *(Continued)*

suppression activities as the fire progresses. In some cases where a building or vehicle is entirely destroyed, witness statements and first responder observations can be essential to identifying the area of origin and in some cases finding the origin may not be possible.

Cause Determination.   Determination of fire cause involves examination of the evidence found within the area of origin. Possible ignition sources and potential first fuels ignited are identified from the evidence or by interviewing persons who may provide prefire details of the conditions in the area of origin. However, identifying a fuel and an ignition source does not necessarily identify the cause. It is also necessary to determine whether the ignition source is capable of igniting the target fuel and the sequence of events that brought the ignition source to the fuel for the time needed to ignite it. Thus a process of elimination may be involved in establishing a most likely cause.

The decision regarding whether the cause of a fire is incendiary may be based on physical evidence at the scene such as finding flammable liquid containers, timing mechanisms, or unexpected fuels or fuel arrangements at the point of origin with no rational explanation for their presence. The results of laboratory tests can reveal the existence of liquid accelerants or chemical compounds associated with incendiary fires in the area of origin for which there is no explanation.

Evidence Examination and Testing

Evidence examination and testing is classified as either nondestructive or destructive. *Nondestructive examination* or testing is of the type that will not change the conditions, appearance, or physical properties of the material or item being examined or tested. Once changes have occurred, the evidentiary value of the item may be compromised. It is general practice to avoid any destructive testing at the scene. It should be noted that simply turning a valve or moving a switch can be destructive and thus should be avoided. There may be items of evidence such as sprinkler valves, detection devices, and door hardware that are not related to the origin or cause of the fire but may be essential to understanding the issues of fire spread, extent of damage, and injury or death.

▶ FIRST RESPONDER

Role of the First Responder

Where do fire service personnel who are not charged with or specifically trained for fire investigation fit into this process? As first responders, fire fighters and EMS personnel are in a position to make possibly the earliest observations of initial conditions as they arrive at a fire or explosion incident and to note important information about the changing conditions during and after the rescue and fire suppression activities. After the incident, fire fighters might be asked by their departments to prepare witness statements describing their observations. Both public and private investigators often conduct interviews or hold debriefing sessions with the responding fire fighters to obtain information.

Fire investigation covers a wide range of interests and objectives. The first responder's observations are important not only to the public investigation or origin and cause but to the additional investigations often by the private sector, of the reasons for the fire spread and the performance of building code requirements and fire safety features.

The fire fighter's role in these activities is to provide factual personal observations and answer questions and not to offer opinions. The basis for the information should be what was seen, heard, or smelled. One of the important activities for the investigator is to develop an understanding of the order in which the fire events occurred, which is accomplished by using a timeline to organize the information [1]. Figure 9.3 shows an example of a timeline. First responders may be asked when a certain event such as "flames coming from the second floor" happened. It would be rare that the answer "18:30 hours" could be given. Here it is the sequence that would be useful. One answer might be "just as Engine-14 was arriving on the scene" or "just moments after the order to leave the building was given (recall sounded)."

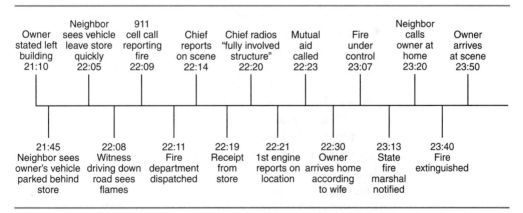

▶ **Figure 9.3**   Illustration of a Timeline. *(Source: NFPA 921, 2004, Figure 20.2.5.4.)*

Observations and Actions

Clearly the primary duties of first responders at fire incidents is rescue of endangered persons, suppression of the fire, protection of exposures, and operational safety. The secondary role of observing the conditions and progress of the fire must not interfere with those primary duties. This section identifies useful observations and actions for first responders that can be helpful to all aspects of fire investigation. For ease of organization, this information is presented according to the enroute, arrival, rescue and suppression, overhaul and postfire phases of the incident.

Enroute. While enroute to the scene, responders should review the nature of the call as stated in the dispatch and recall the type of occupancy involved and any information about the location such as from preplan activities or past inspections. They should recall the presence of flammable products, hazardous processes, the construction features, and number of occupants.

First responders should note the amount of smoke visible and where, enroute, it was first smelled and seen. They should also note the direction of smoke movement as an indication of wind direction. Often the wind direction at the scene will be different from that at the local airport or other reporting point remote from the fire. While the color of smoke may not indicate what is burning, it may be a clue as to the ventilation or burning conditions at the time. A change in smoke color may also be of value when correlated with fire spread or fire suppression activity data. If an unusual smoke color is seen, it should be noted and reported. The presence of a red-brown smoke color, for example, could indicate a chemical reaction that may be helpful to investigators in understanding origin, cause, or spread factors.

On Arrival. When arriving at the scene, first responders should look for the location(s) of visible flames coming out of the building and what floor levels are involved. Are there any exterior fires? They should note whether flames have penetrated the roof and identify locations where smoke is seen but not flames. Responders should observe whether portions of the building have collapsed and note any unusual odors and the color of the flames. While the orange-yellow color of flames may not give much of a clue as to what is burning, other colors such as bright white, violet, or green may indicate the presence of specific combustible materials or chemical compounds.

First responders should listen for sounds of residential smoke alarms, building fire or evacuation alarms, or operation of the water motor gongs associated with sprinkler system operation. When there are water motor gongs operating, water is often flowing from below the gong. This can be useful in determining which sprinkler system is operating in large facilities where there may be many separate sprinkler systems. Responders should note the unit numbers or company apparatus in their vicinity when they arrive as this may help establish the relative sequence of events for later development of timelines. Figure 9.4 shows a sample form that can be used to record notes during this process.

During Rescue and Firefighting Operations.   When entering buildings for rescue or an interior attack, first responders should note which doors are already open and which had to be forced. They should listen for alarms sounding and note whether sprinkler systems are operating at their location. Where was smoke first encountered and what was the visibility? Where was flame first seen? Responders should tell investigators if flames were seen coming from a specific location or piece of equipment in a room or space. The location of evidence of materials that appear to have been used to spread fire within the building and evidence of unsuccessful attempts to start fires such as timing mechanisms or ignition devices associated with fuel packages should be reported. Should an interior attack have to be abandoned, interior evidence may be difficult or impossible to locate without such observations. Figure 9.5 shows an example of a form that can be used to record some of the investigation details.

When fire fighters are carrying out an exterior attack, events such as fire moving from room to room on a particular floor or spreading between floors should be noted. Other events to note include sudden localized increase in burning rate or smoke production, outbreak of fire in locations remote from the initial fire, or change in color of flame or smoke. Responders should try to keep track of the order in which events occurred. One way to help with this is to link the event with fireground activities such as the arrival or departure of apparatus (including rescue and EMS units), advancing or withdrawal of attack lines, charging of monitor nozzles, deck guns, or nozzles on aerial ladders or platforms. These events may be recorded on emergency communication tapes. The timing of major events at a fire such as collapsing of portions of a building may be captured on video by news media or witnesses.

One area of observation of importance to investigators is the reaction of the fire to various firefighting activities such as use of attack lines, opening roofs for ventilation, and breaching walls for access. Did the fire diminish or flare up when hose lines were used? Did smoke or flame come from openings resulting from ventilation activities or breaches in the building?

During Overhaul Operations.   When carrying out overhaul operations, potential evidence such as business and personal records, equipment manuals,

FIRE INCIDENT
FIELD NOTES

Agency:	File No.:

INCIDENT

Location/address						
Property description	Structure	Residential	Commercial	Vehicle	Wildland	Other
Other relevant info						

WEATHER CONDITIONS

Indicate relevant weather information					
	Visibility	Rel. humidity	GPS	Elevation	Lightning
	Temperature	Wind direction	Wind speed	Precipitation	

OWNER

Name		DOB	
d/b/a (if applicable)			
Address			
Telephone	Home	Business	Cellular

OCCUPANT

Name		DOB	
d/b/a (if applicable)			
Permanent address			
Temporary address			
Telephone	Home	Business	Cellular

DISCOVERED BY

Incident discovered by	Name/Address	DOB	
Telephone	Home	Business	Cellular

▶ **Figure 9.4** Fire Incident Field Notes Form. *(Source: Field Guide for Fire Investigators, NFPA, 2003, Figure 2.2.)*

Fire Incident Field Notes Continued
File No.:

REPORTED BY

Incident reported by	Name/Address		DOB
Telephone	Home	Business	Cellular

INVESTIGATION INITIATION

Request date and time	Date of request	Time of request	
Investigation requested by	Agency name	Contact person/telephone no.	
Request received by	Agency name	Contact person/telephone no.	

SCENE INFORMATION

Arrival information	Date		Time		Comments		
Scene secured	No	Yes	Securing agency		Manner of security		
Authority to enter	Contemporaneous to exigency		Consent		Warrant		
			Written	Verbal	Admin.	Crim.	Other
Departure information	Date		Time		Comments		

OTHER AGENCIES INVOLVED

Primary fire department	Dept. or agency name	Incident no.	Contact person/phone
Secondary fire department(s)			
Law enforcement			
Private investigators			

ADDITIONAL REMARKS

▶ **Figure 9.4**   *(Continued)*

STRUCTURE FIRE

Agency	Case number

TYPE OF OCCUPANCY

Residential		Single family		Multifamily		Commercial		Governmental
Church	School			Other:				
Estimated age:		Height (stories):			Length:		Width:	

PROPERTY STATUS

Occupied at time of fire? ❑ Y ❑ N	Unoccupied at time of fire? ❑ Y ❑ N	Vacant at time of fire? ❑ Y ❑ N
Name of person last in structure prior to fire:	Time and date in structure	Exited via which door/egress:
Remarks:		

BUILDING CONSTRUCTION

Foundation Type	Basement		Crawl space		Slab		Other:	
Material	Masonry		Concrete		Stone		Other:	
Exterior Covering	Wood	Brick/stone	Vinyl	Asphalt	Metal	Concrete		Other:
Roof	Asphalt		Wood	Tile		Metal		Other:
Type of Construction	Wood frame	Balloon	Heavy timber	Ordinary	Fire resistive		Noncombustible	Other:

ALARM/PROTECTION/SECURITY

Sprinklers ❑ Y ❑ N	Standpipes ❑ Y ❑ N	Security camera(s) ❑ Y ❑ N
Smoke detectors ❑ Y ❑ N	Hardwired ❑ Y ❑ N	Battery ❑ Y ❑ N
Were batteries in place ❑ Y ❑ N	Location(s):	
Hidden keys ❑ Y ❑ N Where:	Security bars windows? ❑ Y ❑ N	Doors? ❑ Y ❑ N
Remarks:		

CONDITIONS DOORS/WINDOWS

Doors	Locked	Unlocked but closed	Open	
	Forced entry ❑ Y ❑ N	Who forced if known?		
Windows	Secure	Unlocked but closed	Open	Broken
	Broken by first responders ❑ Y ❑ N		Remarks	

▶ **Figure 9.5** Sample Structure Fire Notes Form. *(Source: Field Guide for Fire Investigators, NFPA, 2003, Figure 2.1.)*

FIRE DEPARTMENT OBSERVATIONS

Name of first on scene:	Department

General observations:

Obstacles to extinguishment?	First-in report attached ❑ Y ❑ N

UTILITIES

Electric	On Off None		Overhead Underground	
	Company	Contact	Telephone	
Gas/fuel	On Off None		Natural LP Oil	
	Company	Contact	Telephone	
Water	Company	Contact	Telephone	
Telephone	Company	Contact	Telephone	
Other	Company	Contact	Telephone	

COMMENTS

▶ **Figure 9.5**   *(Continued)*

operational documents, and files may be encountered. These materials should be preserved to the extent possible and identified to investigators. Overhaul is important in making sure that the fire is out and by its very nature results in moving or removal of materials and items or furnishings and equipment, thus disturbing the fire scene. Although it is not reasonable to recall the pre-removal location of every item, an effort should be made to carry out the overhaul in a systematic way that can help identify at least the approximate location of items. One way would be to put all items from a particular location in the same area.

Observations should be made of the fire protection systems and the position of sprinkler valves, fire doors, and other built-in fire safety measures noted. In some cases demolition for safety purposes of salvage operations may destroy this type of evidence before it can be documented by the investigators.

Postfire. Toward the end of the overhaul operations and after the fire is out, investigators will begin trying to reconstruct the scene and determine the origin and cause of the fire. At this time, first responders who have specific information or observations to provide should notify fire officers or investigators. This notification may result in an interview or preparation of a formal witness statement. Typically, the first responders are asked to describe their activities on the fire scene and their observations of fire growth and spread. Often, a floor plan of the building or a site plan of the fire scene is used to help the responders recall their positions and plot the locations of the events they observed.

Later, various other investigators and experts will begin their work. The observations of the first responders may be needed to assist in this work. Similar questions regarding firefighting and fire spread may be asked along with questions regarding operation of fire detection, alarm, and suppression systems. There will be interest in any facts that will help understand spread of fire beyond the area of origin and the performance of the building's fire safety systems and fire protection design.

▶ SUMMARY

Reduction in fire losses is achieved through building codes, fire protection standards, and fire prevention codes and regulations that directly address the causes of fires and the factors that control fire growth and spread. In the development of these codes, standards, and regulations, it is important to assess the success of existing documents and identify areas where changes need to be made. The basis for these assessments and changes is supported in part by statistics resulting from fire origin and cause investigations and from special investigations directed to understanding the reasons for fire growth and spread and injuries and deaths.

An important source of information for these investigations is the observations of the responders who are first on the scene. These individuals can provide early information about the fire as well as continuing observations during rescue, suppression and overhaul activities. These observers can report what they saw, heard, and smelled and what they did during their operations. By recognizing and reporting the early facts, as outlined in this chapter, they can contribute greatly to more consistent and reliable investigations and the associated reductions in fire losses.

Group Activity

Draw a figure to depict the general procedures involved with fire scene investigation. Be sure to include origin determination and cause determination in your work.

Review Questions

1. What is the difference between fire investigation and arson investigation?

2. The investigative process is built around the collection and analysis of evidence (data), development of alternative theories (hypotheses) of how the fire might have started and spread, and testing the alternative theories against the evidence collected. This process is often referred to as what?

3. The scene investigation usually follows a procedure of examination beginning with what part of the building?

4. Determination of fire cause involves examination of the evidence found where?

5. When carrying out overhaul operations, potential _____ such as business and personal records, equipment manuals, operational documents, and files may be encountered.

6. Typically, who is asked to describe their activities on the fire scene and their observations of fire growth and spread?

7. Reduction in _____ _____ is achieved through building codes, fire protection standards, and fire prevention codes and regulations that directly address the causes of fires and the factors that control fire growth and spread.

Suggested Readings

NFPA 921, _Guide for Fire and Explosion Investigations._ Quincy, MA: National Fire Protection Association, 2004.

Ahrens, A., P. Frazier, and J. Heeschen. "Use of Fire Incident Data and Statistics," Section 3, Chapter 3, in _Fire Protection Handbook_, 19th ed. Quincy, MA: National Fire Protection Association, 2003.

Custer, Richard. "Fire Loss Investigation," Section 3, Chapter 1, in _Fire Protection Handbook_, 19th ed. Quincy, MA: National Fire Protection Association, 2003.

DeHaan, John D. _Kirk's Fire Investigation_, 5th ed. Upper Saddle River, NJ: Brady Prentice Hall, 2002.

10 Planning for Emergency Response

William Neville
Fire Service Consultant

Learning Objectives

After completing this chapter, the reader should be able to do the following:

- Explain the basics of the Incident Command System
- Explain the role of the fire service in responding to fires and similar emergencies
- Explain the role of the fire service in responding to nonfire emergencies
- Explain the role of the fire service in emergency planning within the community
- List and describe the major organizations that provide emergency service and illustrate how they interrelate

The goal of this chapter is to introduce the reader to the essentials and certain complexities of planning for fire agency emergency response. Considerable emphasis is given to the management of emergency incidents because that is so essential to effective emergency operations and responder safety and is so often given less-than-appropriate attention. Although this chapter focuses on fire service planning, it is important to note that communities are becoming more aware of the need to plan for the possibility of catastrophic incidents and consequently the need to deploy the resources (equipment and personnel) from a wide variety of organizations. See Chapter 5 for more information on fire department structure and management.

INCIDENT COMMAND SYSTEM

At any emergency incident several management level tasks must be accomplished within a short time frame in order to achieve timely control of the emergency. The situation must be analyzed, usually in a very brief time span, in order to accomplish the following:

- Establish strategic goals
- Prioritize the tasks necessary to achieve control of the situation
- Develop an Incident Action Plan, to include both an operational and a risk management plan
- Request any needed additional resources
- Assign appropriate resources (personnel and equipment) to each task
- Brief responders on their tactical objectives and safety issues

Once assigned a task, it is critical that responding personnel know the following:

- Who their supervisor/subordinate(s) is/are
- How to communicate with that supervisor
- What to communicate to that supervisor
- Updated situation status affecting their tasks, which in turn can impact overall incident scene operations

As the incident progresses, it is vital to effective operations and responder safety that command personnel be updated with information on the following:

- Progress toward control achieved by assigned resources
- Problems encountered by assigned resources that could affect their timely achievement of their assigned task(s) and impact other Incident Action Plans
- The need for additional resources for achievement of assigned tasks
- The availability of committed resources for reassignment

To provide those essential attributes to safe and effective incident control, an organizational system is required that is adaptable to the seemingly unending variable situations—large, small, technical, hazardous, and even humorous—that our nation's emergency response agencies face every day.

Over the more than 200 years of the North American fire service, innumerable incident management schemes have been devised and, to a smaller extent, implemented. In larger organizations it was not unusual to find differing incident management schemes in different elements of the same organization. In all too many instances in our discipline's history, no universally understood or accepted management system was applied. Even today, many fire fighter deaths and injuries can be attributed to the failure to implement an incident scene management system and its components.

One of the more successful attempts, in terms of formal adoption by multiple fire protection agencies, is the FIRESCOPE Incident Command System (ICS), originally developed in Southern California to improve the management of large incidents involving multiagency response. The system has been adopted by the Federal Emergency Management Agency (FEMA), federal fire protection agencies, and numerous state and local agencies.

Currently the state of California requires that all state and local agencies (fire, police, sheriffs, public works, etc.) are to use the FIRESCOPE Incident Command System (ICS) to manage all emergency incidents. Federal fire-fighting agencies and many local fire agencies have been using the system for nearly two decades. Wildland fires, high-rise fires, single-family dwelling fires, petroleum refinery fires, hazardous material spills, structural collapse search and rescue, shipboard fires, and even parades and picnics have been successfully managed using the ICS.

FIRESCOPE and the National Wildfire Coordinating Group (NWCG) have issued many documents on utilization of and training in the ICS. These documents can be purchased on the Internet at firescope.oes.ca.gov or www.nwcg.gov. Another source for further information on incident management is the International Fire Service Training Association, Fire Protection Publications, Oklahoma State University. It publishes a family of documents under the title Incident Management Model Procedures Guides, which can be ordered online at www.ifsta.org. The model presented is titled Incident Management System and is functionally identical to and coordinated with FIRESCOPE's Incident Command System but presented in a simplified format, with the focus on smaller scale incidents.

The National Fire Protection Association (NFPA) has a standard on Incident Management Systems (IMS) titled NFPA 1561, *Standard on Emergency Services Incident Management System* [1]. While it does incorporate the concepts of ICS, it is written from a health and safety perspective for emergency responders. As noted earlier, the lack of training in and utilization of IMS is often cited as a factor in fire fighter fatality investigations. A complimentary standard to NFPA 1561 is NFPA 1600, *Standard on Disaster/Emergency Management and Business Continuity Programs* [2]. Emergency management personnel, as well as members of the private sector, need to understand the utilization of IMS. NFPA 1600 bridges the gap in many areas, but the following two specifically:

1. The relationship between the Emergency Operation Center (EOC) and operational command at the incident scene

2. The need for a comprehensive emergency management plan, based on the risks and hazards in each community

These standards and others have moved forward in addressing national issues or incident management. The Department of Homeland Security is developing and issuing a National Incident Management System (NIMS) as a component of a new National Response Plan (NRP). NIMS is all-risk based, and will be required training for all federal, state, and local responders.

Key Concepts

The key concepts of the ICS include: Incident Commander, Unity of Command, Span of Control, Responder Safety, Common Terminology, Personnel Accountability, Prescribed Communication Channels, Resource Management, and Incident Action Plans.

Incident Commander. Every incident to which an agency responds will have an Incident Commander (IC) who has the responsibility for all management functions at the incident. In those cases where the magnitude or complexity of the incident make it problematical for the Incident Commander to effectively control all management functions, the system design provides a standardized means of delegating certain functions to one of four major organizational elements.

Command at an incident is immediately established by the individual in charge of the first-arriving company or the first-arriving individual of the emergency organization, regardless of rank or function. Transfer of command typically takes place when a more qualified person (by rank, experience, or agency protocols) arrives on scene or in cases of relief of personnel at extended incidents.

Change in the incident situation may make a jurisdictional or agency change in command legally required, for example, if a fire spreads from one jurisdiction to another or a fire scene becomes a crime scene due to discovery of evidence of arson. In addition, it allows for an orderly transfer of the control functions. Such delegation does not, however, relieve IC of overall responsibility. It is basic to the ICS concept, that, at every incident the following management functions are, to some degree, performed:

- Operations: management of those functions directly applicable to the primary mission of the incident, e.g., the assignment of tactical objectives
- Planning/Intelligence: management of those functions involving the collection, evaluation, dissemination, and use of information about the incident situation and status of resources committed to the incident, that is, the direction of the compilation and utilization of incident information. Included in this area is the use of technical specialists

- Logistics: management of those functions relating to the provision of facilities, services, and material required to support all incident activities
- Finance/Administration: management of those functions relating to financial and administrative aspects of incident management

Figure 10.1 illustrates the relationship between the four major elements of the ICS.

Responder Safety.   Responder safety is a primary concern of the IC. The ICS recognizes this by placing an Incident Safety Officer in direct contact with the IC and, in extreme circumstances, allowing the individual to order modifications to plans or operations in the IC's name.

Unity of Command/Chain of Command.   Unity of Command means that every person assigned to the incident has one assigned supervisor, no more, no less. Except in emergency situations, where safety is an issue, direction will be received from a single, designated person. Freelancing or operating independent of supervision is not allowed when operating within the ICS.

Common Terminology.   Creating common terminology was the first challenge faced by the designers of the ICS. The meaning of the various terms used in the system are quite specific and strict adherence to system terminology in both system titles and commonly used resources is critical to a smoothly operating incident management system. What is also critical in common terminology is the use of clear text: No ten codes are used, and commonly used terms mean the same thing to all responding agencies. Management system terminology includes the following:

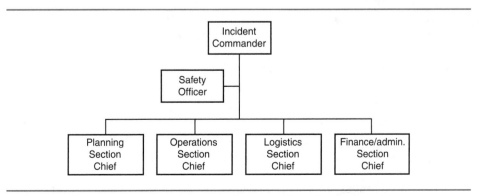

▶ **Figure 10.1**   Four Major Elements of the Incident Command System.

- Commander: Individual in charge of the overall incident
- Chief: Individual in charge of one of the four Sections
- Director: Individual in charge of a Branch
- Supervisor: Individual in charge of a Group or Division
- Leader: Individual in charge of Strike Team or a Task Force

Personnel Accountability. Personnel accountability (essentially, the location and assignment of all incident personnel) is addressed through several different incident command procedures as follows:

- Check-In with Incident Command (or a designated representative) is required of all responding personnel immediately upon arrival at the incident scene and prior to initiating operations.
- Unity of Command ensures that personnel do not get lost in the system. Except in unusual circumstances, personnel reporting to the incident as an entity, remain together as an entity.
- Resource Status is maintained by the Planning Section. Each unit's location, time of assignment, and availability, condition (i.e., available for assignment, actively assigned, out of service) is maintained by an individual or a unit (Resource Unit) designated for that function.

Many acceptable personnel accountability systems are in use. The critical factors here for responder safety are that the system is used at all incidents and it is constantly maintained during the incident.

Prescribed Communication Channels. Incident communication control is essential in large-scale or multi-agency incidents. Ideally, hardware is made available to units according to their assignments so that communications between resources working on the same assignment can easily communicate with their supervisors, but that is not the typical case in multiagency incidents. Incident Command doctrine requires that communication channels be designated taking into account the equipment available at the scene (e.g., this might require that handheld radios be exchanged among working elements; in other situations, resources might be assigned according to their communication capabilities). Another solution is to make use of radio caches available from federal and some state sources. These potential issues should be addressed prior to the emergency. NFPA 1221, _Standard on Emergency Services Communication Systems_ [3], addresses these issues.

Resource Management. In large-scale incidents, management of resources is addressed by grouping resource elements. Combinations of resource elements many be assembled as follows:

- Task Force: A grouping of resource elements (e.g., fire companies), typically for a specific tactical purpose, under the command of a Leader, with compatible communication capability. Task Forces are commonly but not exclusively assembled at the scene of an emergency. An example of a Task Force would be: a truck (ladder) company, two engine companies, and a rescue squad under the command of a Task Force Leader (Battalion Chief) tasked with safely evacuating occupants trapped on upper floors of a high-rise building with fire on the lower floors. After completion of that assignment, the Task Force will normally be disbanded and its elements reassigned.

- Strike Team: A group of resource elements with similar capabilities, compatible communication capability, under the command of a Leader. Strike Teams are commonly but not exclusively assembled at their home agency(s). An example of a Strike Team assembled by a larger fire agency for a mutual-aid response to a distant wildland fire would be: five engine companies, all with 400–600 gal water tanks, 1250–1500 gal per min fire pumps, 1000–1500 ft of 2.5 in. or larger fire hose, and 500 ft of 1.5–1.75 in. fire hose, staffed with four or more fire fighters and under the command of a Strike Team Leader (Battalion Chief). Strike Teams are typically left intact for the duration of an incident.

- Division: A group of resources assigned the responsibility for a specific geographic area within the boundaries of the incident. Divisions are typically formed by the Incident Commander, or in larger incidents, the Operations Chief with concurrence of the Incident Commander. Divisions are commanded by a Division Supervisor. Responsibility for fire control of one floor of a multi-story building would be an example of an assignment given to a Division Supervisor. Divisions may be made up of several single resources (e.g. fire companies), Strike Teams, and/or Task Forces.

- Group: A group of resources assigned responsibility for a specific tactical function at the incident scene. Groups are typically formed by the Incident Commander, or in larger incidents, the Operations Chief with concurrence of the Incident Commander. Responsibility for limiting water damage to lower floors of a building with fire control activity taking place on upper floors, would be an example of an assignment given to a Group Supervisor; Groups may be made up of several single resources (e.g., fire companies), Strike Teams, and/or Task Forces.

- Branches: Utilized when the number of Divisions or Groups become too numerous, too geographically separated, or functionally too diverse for a single person to be coordinated by a single command officer. Communication difficulties may also dictate the use of Branches. Branches are commanded by Branch Directors. A Branch may be made up of several "single resources" (e.g., ambulances), Task Forces, Strike Teams, Divisions and/or Groups. At fire incidents involving extensive emergency medical care, an Emergency Medical Service (EMS) Branch is often formed.

The use of this type of organization limits the span of control for any single officer, reduces radio communication traffic, and increases coordination of resources. This terminology is used in FIRESCOPE and the National Wildfire Coordinating Group ICS documents, publications of the International Association of Fire Service Training, and in National Fire Academy command and control courses.

Figure 10.2 shows an illustration of the Operations Section resource groupings using Branches, Divisions, Groups, Strike Teams, and Task Forces. In the Fire/Rescue Branch, both geographical Divisions and functional Groups organizational elements are used. Many agencies use the term Sector to identify both geographical and functional elements of an incident.

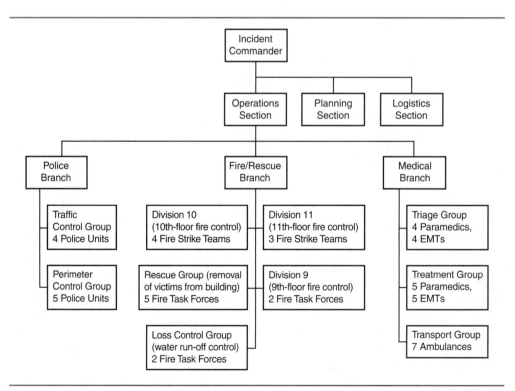

▶ **Figure 10.2**   Operations Section Resource Groupings.

In this illustration, Structural Strike Teams are made up of two engine companies and one truck company, and they are formed prior to arrival at the incident. The Task Forces were formed at the incident scene and are staffed and equipped in accordance with their assignment at the incident. The chain of command from the bottom up is: Single Resource to Task Force/Strike Team to Division/Group to Branch to Section to the Incident Commander. Note that the Police and Medical Branches have delegated responsibilities and specific critical objectives based on function and their subelements are therefore identified as Groups. In the case of the Police and Medical Branches at this incident, there is no need to use Divisions or Groups as the police and EMS resources committed to the incident are limited. Therefore the individual units retain their identity rather than being utilized as part of a Division or Group. Also note the following:

- Operations Section Chief can control a group approximating 200 by interfacing with only three Branch Directors
- The Police Branch Director can exercise control over 18 police officers by interfacing with two Group Supervisors
- The Medical Branch Director can exercise control over 32 medical personnel by interfacing with three Group Supervisors
- The Fire/Rescue Branch Director can control the actions of approximately 150 through interface with three Division Supervisors and two Group Supervisors

Incident Action Plans.   An Incident Action Plan (IAP) is required at every incident to provide supervisory personnel with direction in carrying out their assignment(s). The plan may be written or oral, depending on the situation. It would be impractical to provide a comprehensive written plan at a fast-moving structure fire incident, but at multiday wildland fires, written plans facilitate consistent direction. The use of an Incident Briefing Form is strongly encouraged for faster moving incidents to record incident status and resource assignments.

Figure 10.3 shows an example of an Incident Briefing Form that can be used to track resources ordered and on scene as well as the organizational structure and the situation status. On page 1 of the form, the current command organizational structure should be depicted, making erasures and additions as needed. On all three pages (and any additional pages required), a permanent record of resource ordering and arrival should be maintained. The assignment should be altered as changes in incident deployment are made. On pages 2 and 3 (and any additional pages required), a record of the current situation should be illustrated. On multifloor buildings it may

Incident Briefing Form

Incident location/name	Time of alarm	Incident Commander	Control time	Conclusion time

RESOURCES

Resource ID	Time ordered	Time on scene	Incident assignment and time

COMMAND STRUCTURE (current)

Page 1

▲ **Figure 10.3** Sample Incident Briefing Form.

Incident Briefing Form

RESOURCES

Resource ID	Time ordered	Time on scene	Incident assignment and time

SCENE PLAN VIEW (current)

Page 2

▲ **Figure 10.3**   *(Continued)*

Incident Briefing Form

	RESOURCES			SCENE ELEVATION VIEW (current)
Resource ID	Time ordered	Time on scene	Incident assignment and time	

Page 3

▲ **Figure 10.3** *(Continued)*

be necessary to have several drawings to indicate the lobby area, the involved floors and the staging area. Typically during long-term incidents, the IAP is updated at each operational period.

Modularity.   The ICS is designed to function at any size incident, from the small grass fire in an urban vacant lot to wildland fires involving thousands of acres; from the one-room fire in a residence to high-rise fires involving thousands of square feet of structure. To be effective, it should be used every day at every incident, not just for large-scale incidents. Only those organizational elements required to carry out the IAP are implemented. The duties of any Section not activated are automatically the responsibility of the Incident Commander. At smaller incidents, a single person may carry out the functions for an entire Section, for example, assigning Planning Chief to a single individual. It is possible, but not typical, for the Incident Commander to delegate the responsibilities of more than a single Section to one individual. This typically might be done on an interim basis or at smaller incidents.

Figure 10.4 illustrates an organization used to combat a fire in a five story apartment structure. The first-in engine company officer assumed initial Incident Command. The initial Incident Commander (Engine 1 Officer) directed the second-arriving engine company (Engine 2) to initiate search and rescue to remove any occupants unable to exit the building

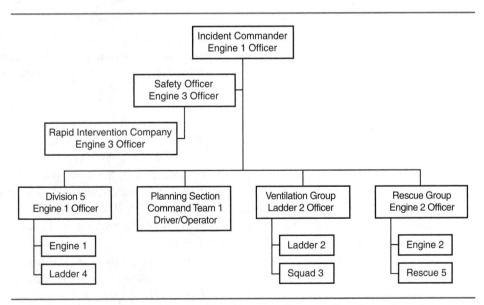

▶ **Figure 10.4**   Organization Used to Combat Fire in a Five-Story Apartment Structure.

under their own power. The initial Incident Commander further directed the first-arriving ladder company (Ladder 2) to initiate removal of smoke and other fire gases. Upon arrival and discussion of the IAP with the Engine 1 Officer, another ranking officer assumed the Incident Commander position and immediately assigned the Engine 5 officer as Division 5 Supervisor to manage all fire fighting efforts on the fifth floor. He then designated the Ladder 2 officer as Ventilation Group Supervisor and Engine 2 officer as Rescue Group Supervisor. The role as Leader of the Rapid Intervention Company (RIC) was assigned to the Engine 3 Officer. There must be an RIC leader while the IC still is responsible for Safety Officer functions. He then assigned his driver/operator to the role of Planning Chief to track the status of the assigned resources and the situation. The second truck company to arrive (Ladder 4) was assigned to assist Division 5 (Engine 1 Officer) on the fifth floor. Upon their arrival, Squad 3 and Rescue 5 were assigned to the Ventilation Group (Ladder 2 Officer) and Rescue Group (Engine 2 Officer) respectively. Also note the following:

- To direct the actions of the approximately 30 personnel working the incident, the Incident Commander need only directly interface with five individuals: RIC Leader, Division 5 Supervisor, Planning Chief, Ventilation Group Supervisor, and the Rescue Group Supervisor.

- The titles are associated with the Incident Command System position being filled rather than the administrative rank of the individual.

- Because the span of control is within reasonable limits (five persons reporting directly to the IC), the IC has assumed the roles of the Logistics Chief and Administrative/Finance Chief and elected not to implement these positions.

Span of Control.   Span of control refers to the number of persons directly reporting to a supervisor. The ISC specifies that an effective span of control is within the range of three to seven. Essentially, the span of control is driven by the complexity of the functions being performed, the ease (or difficulty) of communication and the breadth of the area being supervised. Therefore, smaller spans of control might be expected at hazardous material incidents than at grass fires on a flat plain. Larger spans of control might be expected where communications between all units was very good, than where there was a mix of incompatible radio frequencies making face-to-face communication over a large area fire scene a requirement. The means for controlling an individual's span of control is addressed through a specified organizational structure. Essentially the delegation from the Section Chiefs is done either through functional assignments (specific task assignment, not geographically designated) or geographically (assignment within a geographical area).

APPLICATION EXAMPLES

Most incidents to which urban fire agencies respond are controlled by one or two units (e.g., engine company, or engine company and ambulance). Use of the Incident Command System is necessary in these circumstances to establish its use and provide a mentoring or shadowing component for those operations in different supervisory positions.

Small-Scale Incidents

At a single-company incident (e.g., light brush fire adjacent to a dwelling), the Company Officer is the IC who personally fulfills the following basic management functions:

- Upon arrival, assumes command, identifies location, surveys the scene, and advises Dispatch that his company can control the situation without further dispatch of resources (Incident Command)
- Evaluates the situation and forms his plan of control, e.g., to control the flanks, then move toward the head of the fire to completely surround the fire, then extinguish all hot embers, then create a barrier around the fire by using shovels to remove vegetation around the burn site to a distance of 1.5 times the width of the tallest vegetation standing at the perimeter of the fire area (Planning)
- Directs the tactics of his crew, for example, orders the use of Class A foam through a 1½-in. or 1¾-in. hose line to attack fire on the flank adjacent to the dwelling to reduce the threat to that exposure, the use of another hose line on the opposite flank to control the spread of the fire (Operations)
- Directs the support functions, for example, directs the pump operator to carry additional hose to both flanks to provide sufficient hose to reach the head of the fire (Logistics)

Large-Scale Incidents

At larger (in terms of resources committed) incidents, the ICS becomes more formalized. Typically the need to move equipment from apparatus to the involved portions of the building becomes such that resources have to be assigned to that task, hence the implementation of the Logistics Section and subelements thereof (Units). The Safety Officer role becomes more complex, and often a qualified Safety Officer may be dispatched with any ensuing alarm. The Planning function not only has more resources to track (Resource Unit) but must also document the activities of the assigned units (Documentation Unit) and track an expanding and more complex situation

(Situation Unit). Additionally, in larger scale incidents, the Planning Chief is often asked to analyze the situation and prepare a plan for control of the incident for consideration by the Incident Commander. Therefore an officer is assigned as Planning Chief and often provided additional personnel in the form of a Company to carry out these tasks.

Figure 10.5 illustrates an expanded Incident Command organization for the previously described fire on the fifth floor of a five-story apartment house. The IC has requested a greater alarm after determining that the first alarm assignment will be inadequate to control the situation in a timely fashion. In this agency, the Shift Commander is automatically dispatched to all greater alarms and on his arrival assumes Incident Command after an orderly transfer. Also included in greater alarm assignments are the Department Safety Officer to serve as the Incident Safety Officer and the Department Information Officer to serve as the Information Officer.

The incoming IC reassigns the current IC to Planning/Intelligence and assigns his Aide/Technician to assist the Planning Chief in recording the actions taken by the department (Documentation Unit). The Planning Chief's Aide/Technician then is assigned to the tracking of the situation (Situation Unit). The Apparatus Operator from Engine 9 is detailed to track resources assigned to the unit (Resource Unit).

The second-arriving chief officer is assigned as Operations Chief and she augments the Groups and Division already in place as shown. The Engine 9 Officer is assigned as Staging Manager and sets up an area three or more floors below the fire floor where fire-fighting teams awaiting assignment to active Divisions or Groups are staged. The Operations Chief has established a Medical Group to triage and transport occupants injured by the fire and products of combustion. The Rapid Intervention Crew is reassigned to the Operations Section. A Loss Control Group is formed to mitigate damage from water and smoke.

A Logistics Section is implemented to move supplies (hose, air cylinders, fans, etc.) from the area where all greater alarm apparatus are parked (Base) to the lobby area of the building where the Ground Support Unit will move the equipment to either the fire floor or Staging as directed. Of the approximately 50 personnel assigned to this incident, nearly half are in support roles.

High-rise fires are, in many fire officers' opinions, one of the most difficult emergency incident management challenges that a fire agency can face. In this scenario, a fire has broken out on the 10th floor of a 12-story, non-sprinklered building. Figure 10.6 and Table 10.1 illustrate the organization being used for a high-rise fire. It is by no means illustrative of a worst-case scenario. Many high-rise fires have required the utilization of 200 or more

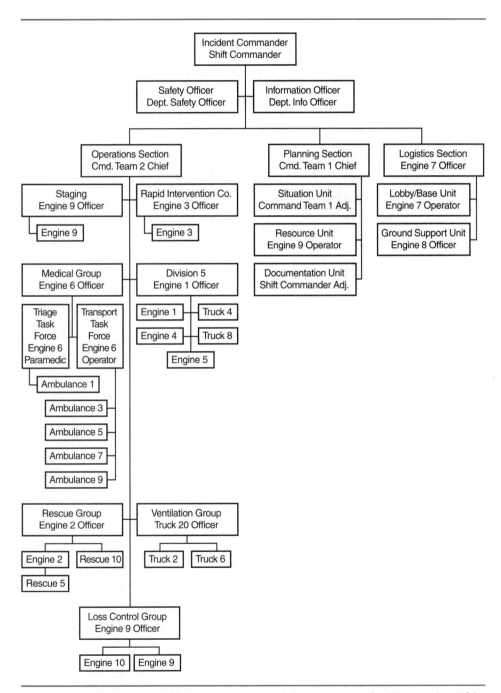

▶ **Figure 10.5** Expanded Incident Command Organization for Fire on the Fifth Floor of a Five Story Apartment Structure.

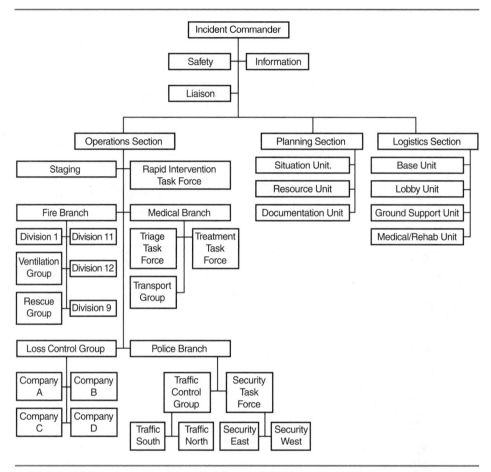

▶ **Figure 10.6** Organization Used for a High-Rise Fire.

fire fighters at the scene in order to control the fire. Figure 10.6 illustrates the management skeleton of the Incident Command organization at a working high-rise fire. Resources in the Operations Section are shown down to the Task Force level only. The Unit level is illustrated in the Planning and Logistics sections.

Also note the following:

- Approximately 25 percent of the total personnel working the scene are in support positions (including those standby resources in Staging and Base). In a long duration fire, that number would increase dramatically as relief would become necessary for the various positions.

▶ **Table 10.1** Number of Staff Used for High-Rise Fire

Type of Staff	Number of Staff
Approximate total staffing for this organization	250
Incident Command Staff	5 (includes Incident Commander aide/technician, safety officer, information officer, and liaison officer)
Operations section	193
Staging	16 (includes 3 companies in reserve)
Rapid intervention task force	10
Fire branch	120
Medical branch	30
Loss control group	17
Planning/intelligence section	6
Logistics	50
Base unit:	24 (including 5 companies standing ready to provide relief for working crews)
Ground support	12
Lobby control	8
Medical/rehab	6

- The IC can control the actions of all personnel on the scene through interfacing with six officers (Operations Chief, Planning/Intelligence Chief, Logistics Chief, Safety Officer, Information Officer, and Liaison Officer).

- The Operations Chief can control the actions of more than 190 personnel through interface with six officers (Fire Branch Director, Medical Branch Director, Loss Control Group Supervisor, Police Branch Director, Staging Manager, and the Rapid Intervention Task Force Leader).

- The Logistics Chief can direct the actions of the 50 personnel under his command through interface with four Unit Leaders (Base Unit Leader, Lobby Control Unit Leader, Ground Support Unit Leader, and the Medical/Rehabilitation Unit Leader)

- The Planning Chief has been assigned a company to assist him in his duties. He will assign them as needed to the three Units implemented for this incident.

- It is important to realize that had the building been protected by an automatic fire sprinkler system, it is relatively certain (with 95+ percent assurance) that the fire would have been controlled by the automatic activation of three or fewer sprinkler heads and utilization of not more than those units initially dispatched. When planning for emergency response, the built-in fire protection features should be considered as well as the unique qualities of the risk.

Multijurisdictional Incidents

Even more complex than the foregoing illustrated high-rise incident is that situation in which the principal responsibility(ies) is (are) shared by more than a single agency. An outgrowth of the ICS is the continuing development of a management concept called *Unified Command*. This concept is designed to facilitate the management of an incident where different jurisdictions (functional, statutory, and/or geographical) are required to work together in order to achieve control of an incident. Experience teaches that an integrated approach to the management of such incidents will provide a safer environment for responders and a more effective control effort. Some examples of situations that may indicate the need for Unified Command are

- An incident that overlaps jurisdictional boundaries (e.g., a wildland fire spreading from one jurisdiction to another, requiring more than a single fire agency response)
- An incident where statutory jurisdiction is shared (e.g., a multivehicle fire on a state highway resulting in several vehicle fires requiring fire agency response, multicasualties requiring the response of an independent emergency medical provider, and the response of the highway patrol to control traffic and investigate the vehicle accident)
- An incident where the primary responsibility shifts in the following manner as the incident progresses (e.g., an aircraft crash with fire and trapped passengers):
  1. In the initial phase, fire control and removal of occupants are the primary function. In this phase the fire agency assumes command. Representatives of the other AHJs assume Deputy roles as they arrive on scene.
  2. In the second phase, almost simultaneously with the first, on-scene triage, treatment, and transport to hospital(s) takes place. Once the fire is controlled and occupants removed, command shifts to the EMS. Because reignition of the remaining fuel is a concern, the fire agency remains on the scene in an active role.
  3. In the third phase, the deceased must be removed to an interim morgue and to the extent possible, identified. Typically the jurisdic-

tion's coroner's office has this responsibility and assumes command after all injured victims have been treated and/or transported. Because reignition of the remaining fuel is a concern, the fire agency remains on the scene in an active role.

4. In the fourth phase, after all the deceased have been removed from the immediate scene, the National Transportation Safety Board (NTSB) assumes command as it has the responsibility to investigate the cause of the incident. If a criminal act is suspected, the appropriate law enforcement agencies will be requested and take their place in the Unified Command organization.

Figure 10.7 illustrates the first phase organizational structure after all responsible agencies have arrived. The Fire Officer assumes Incident Command as he is very likely to be the first on the scene and at this point have the primary duties (i.e., fire suppression and rescue of victims from the aircraft). Representatives from the EMS agency, the coroner's office, and the National Transportation Safety Board assume Deputy positions to the Incident Commander. (In actual practice, the NTSB and the coroner's office may not arrive until some time later, but for illustration sake they are shown as being on scene at this point, as they would assume these positions on their arrival.)

To facilitate communications to field units at the scene, an EMS representative functions as a Deputy to the fire officer functioning as the Operations Chief. Branches are implemented for the Fire and Medical functions to facilitate tactical communications. At this point, the Planning/Intelligence

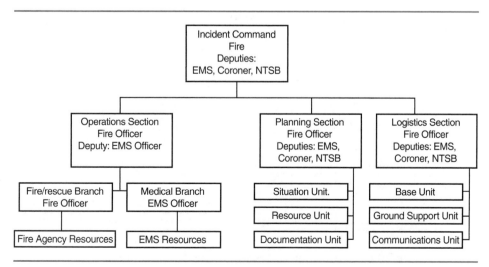

▶ **Figure 10.7** Illustration of First Phase Organizational Structure.

Section is under the management of the fire agency. The EMS representative works closely with the fire Section Chief as planning for the two agencies must be closely coordinated for safe and efficient operations. The coroner's representative will be planning for body identification/relocation once the injured victims have been removed from the aircraft. The NTSB representative will be putting together his investigative team and estimating equipment and facilities needed for the investigation once all victims are removed and the fire completely extinguished. All four agencies involved have logistical concerns and all are represented at the Logistics Section level. Again because the fire agency has the primary function at this time, it has assumed the Logistics Chief position.

Figure 10.8 illustrates the second phase when the primary function at the incident is the triage, treatment and transportation of the injured. Because reignition is a concern, fire resources remain in an active role at the scene. Representatives from the EMS agency, the coroner's office, and the NTSB assume Deputy positions to the Incident Commander. (In actual practice, the NTSB and the coroner's office may not arrive until some time later, but for illustration sake they are shown as being on scene at this point, as they would assume these positions on their arrival.)

To facilitate communications to field units at the scene, an EMS representative functions as a Deputy to the fire officer functioning as the Operations Chief. Branches are implemented for the Fire and Medical functions to facilitate tactical communications.

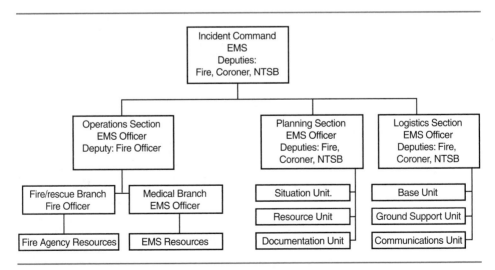

▶ **Figure 10.8** Illustration of Second Phase Organizational Structure.

At this point, the Planning/Intelligence Section is under the management of the EMS agency. The Fire representative works closely with the EMS Section Chief, as planning for the two agencies must be closely coordinated for safe and efficient operations. The coroner's representative will be planning for body identification/relocation once the injured victims have been removed from the aircraft. The NTSB representative will be putting together his investigative team and estimating equipment and facilities needed for the investigation once all victims are removed and the fire completely extinguished. All four agencies involved have logistical concerns and all are represented at the Logistics Section level. Because the EMS agency has the primary function at this time, it has assumed the Logistics Chief position.

In the third phase, shown in Figure 10.9, the coroner's office has the primary responsibility. Therefore the senior coroner's official assumes Incident Command. Because the reignition of the aircraft fuel remaining is still a threat, the fire resources stay on scene in an active role. The senior fire official assumes a Deputy position to the IC (the senior coroner's official). As all live victims have been treated and/or transported at this time, the EMS official no longer has any function and is not represented at the IC post. NTSB is still in position as a Deputy to the Incident Commander for purposes of input regarding recording victims' positions, and other details.

At the Operations position, the coroners office has assumed the Chief's role with the fire agency represented in the Deputy's role. The two active Branches are now fire and the coroner's office. The Planning Section Chief's

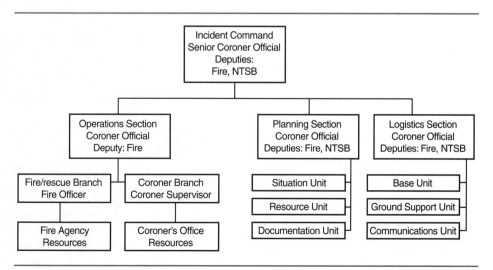

▶ **Figure 10.9**   Illustration of Third Phase Organizational Structure.

role is now assumed by the coroner with fire and NTSB representatives as Deputies. The same relative positions are held in the Logistics Section.

In the fourth phase, shown in Figure 10.10, the NTSB now has the primary function of investigating the cause of the incident. Should there be any suspicion of a deliberate action, the FBI and/or Department of Homeland Security would be called in and participate in the Unified Command. As reignition is still a risk, the fire agency remains on scene as an active participant. In this phase, the Operations Chief, Planning Chief, and Logistics Chief positions will all be held by an NTSB official with a fire official in the Deputy position.

Unified Command is a cooperative effort that facilitates the ability of all agencies with either geographical or functional authority and responsibility to participate in the management of an incident by joint development and utilization of incident goals, objectives, and strategic and tactical plans that are mutually agreed to, without losing agency authority or responsibility.

Agencies that contribute resources to an incident, with no statutory responsibility, are not part of the Unified Command, but may contribute Agency Representatives to effectively represent their interests through the Liaison Officer position on the Command Staff. Successful management of multiagency incidents requires an attitude of genuine cooperation and regular training with credible scenarios involving agencies that may respond to actual incidents. Such training provides participating agencies with clarification of their roles as well as the roles, responsibilities, capabilities, and limitations of other participating agencies.

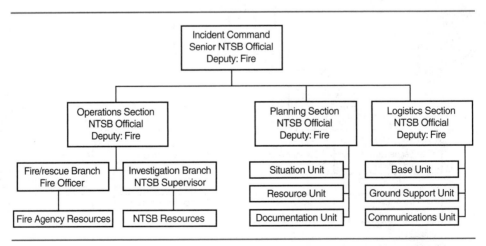

▶ **Figure 10.10**   Illustration of Fourth Phase Organizational Structure.

▶ ## FIRE SERVICE RESPONSE PLANNING

Planning Formats

The fire service has planned for emergencies for generations. Typically called *prefire planning*, it has focused on individual risks, usually buildings. The formats of such planning are nearly as numerous as there are fire departments (NFPA estimates the number of fire agencies in excess of 30,000 in the United States) and include the following:

On-Site.   Some agencies make this planning an on-site exercise involving the resources they believe would be involved in an incident at the location. The participating units lay out hose lines, raise ladders, and deploy personnel as they believe appropriate to the emergency scenario. The exercise may be documented for review. A less demanding approach is to bring the officers of those units to the scene for a walk-through of the risk with discussions of the strategy and tactics deemed feasible.

Tabletop.   Often the effort consists of a tabletop exercise where the risk is depicted on wall charts or chalkboards and officers discuss the various special requirements for controlling the incident described by the instructor.

Building Inventory.   Yet another approach can be called the risk inventory method. In this method, risks are inspected and, using coded symbols, their relevant details (sprinkler connections/controls, utility controls, ingress/egress points, hazardous materials, roof construction, special values, etc.) indicated on drawings, which are distributed to those units deemed appropriate (first-alarm assignment, senior officers, dispatch center, etc.).

Due to the wide scope of emergencies to which today's fire agencies respond, the term *prefire planning* should probably be changed to *preincident planning*. See NFPA 1620, *Recommended Practice for Pre-Incident Planning* [4], for further information on preincident planning. Many fire agencies train on predictable emergencies based on the agency's response experience, environmental circumstances, national fire experience, potential for life safety hazard, importance to the community, and presence of hazardous materials in a particular type occupancy, such as the following:

- Extrication of victims from vehicle accidents
- Fire attack and victim rescue from airplane crashes
- Flood water rescue
- Shipboard fire control

Structure Fires

The amount of resources a community dispatches on receipt of a structure fire varies with the normal worst case situation the agency estimates it will be confronted with on arrival. The goal is to have sufficient resources immediately available to control the incident in a timely fashion, thereby limiting personal injury and property damage to that already suffered before the agency's arrival on the scene. Of course, such guesstimates are not accurate in every case encountered by a fire agency.

Most reported structure fires are controlled by two or fewer fire companies, but that fact should not cause fire planners to reduce the number of units responding, as the immediate availability of a strong fire fighting force can be the difference between limited damage and extensive life and property loss. Agencies should consider automatic aid or automatic response agreements if local resources are inadequate to provide sufficient resources for reasonable initial attack. Mutual aid should be available to any agency that finds itself overwhelmed by an incident or incidents. These aid agreements should be documented, reviewed, and practiced on a regular basis.

Greater Alarms.   The high-rise incident scenario demonstrates how events may proceed in such a fashion that the initially dispatched units are inadequate to deal with the situation, thus the provision for greater alarms. Some fire agencies design the request for additional resources at the scene of an incident to be done in segments, usually termed _alarms._ An IC's request for a second alarm may double the resources dispatched on first alarm, a third alarm may triple the resources dispatched, and so on. Other agencies require that ICs specify what additional resources they require (e.g., two engine companies, three truck companies, four rescue companies, five ambulances, etc.).

Fire Control Functions.   All reported structure fires should receive response of multiple units, even if that requires the use resources from outside the community (often referred to as _mutual aid_). The essential functions that must be addressed for control of a fire within a building are as follows (not in chronological order):

- _Assumption of Command:_ Assuming command of an incident and utilization of an IMS.
- _Search and Rescue:_ Search for, and rescue of, persons trapped in the building; may entail holding occupants at a safe location within the fire building, typically called _protect in place._

- *Confinement:* Stopping fire by controlling all pathways by which it might extend through the structure, including closing doors and applying of sufficient suppressant (usually water) to knock down the fire, that is, reduce flaming.
- *Ventilation:* Removal of smoke, heat, and fire gases from the building.
- *Extinguishment:* Application of sufficient suppressant to reduce the temperature of the burning material below its ignition point.
- *Property Loss Mitigation:* Minimizing the damage to structure and contents from smoke, heat, and water through removal of valuables and/or protecting with water-resistant covers.
- *Overhaul:* Actions initiated after fire extinguishment to ensure that extinguishment has been complete and premises are as safe as is possible.

Protecting in Place.   Although the first priority of fire fighting is to protect persons, in some cases control of fire growth and removal of the resultant smoke, heat, and gases (ventilation and confinement) before extensive search and rescue may be the safest and most effective means of protecting lives, both of fire fighters and building occupants. If all the responders focus on search and rescue, the fire will continue to grow, perhaps cutting off safe routes of exit, and additional smoke, heat, and noxious gases will be generated, endangering fire fighters and trapped occupants.

Indirect Fire Loss.   Some community leaders and even fire officials have indicated that structural fire protection should be diminished because, in comparison to emergency medical service demand, the volume of structure fire incidents is small (30 percent or less in those agencies providing first responder or more advanced forms of emergency medical service).

What can be overlooked is the impact on the community of structure fires, particularly in commercial or industrial occupancies. A typical EMS response affects relatively few people, whereas a commercial or industrial fire can severely affect the community at large. As an example, a commercial/industrial fire can cause permanent job loss in a community. A study funded by the National Fire Prevention and Control Administration (now United States Fire Administration) found that the majority of employers suffering heavy fire loss did not rebuild in the same community. As further examples, in smaller communities, a fire in a school can severely disrupt a community's educational processes, a fire in a theatre can diminish or eliminate that form of entertainment, or fire in a church can make worship difficult. These losses are often referred to as *indirect fire loss.*

Wildland Fires

For the purpose of this discussion, vegetation fires can be roughly divided into two categories, those without structure exposures (termed here as _wildland fires_) and those with structures exposed (termed here as _urban-wildland interface fires_). Federal, state, and county forestry fire-fighting agencies have historically focused on limiting fire spread in wildland fires in order to protect valuable timber, watershed, and other natural values, while urban/suburban fire agencies focused on structural fire fighting to limit the spread of fire within and outside of buildings. The incursion of residences into areas formerly only sparsely populated has brought a new dimension to both forestry and urban/suburban fire services. In Southern California, an area where this wildland-urban interface risk has existed for decades, fire agencies typically dispatch more resources on the initial alarm to a wildland fire than to a fire reported in a large structure.

Aircraft.   A few agencies are so committed to rapid control of these fires that they maintain fleets of fire-fighting helicopters to ensure timely response to areas difficult to reach by road. The helicopters also serve as air ambulances and support other fire operations through roof rescues and transportation of personnel from distant stations to the fire scene. In some areas, the use of National Guard or Army Reserve helicopters is an option for certain, limited uses. In high-risk areas, the leasing of helicopters and flight crews may be an alternative to owning and operating a helicopter fleet. A lease arrangement may, however, constrain the helicopter's use because of the cost of each use. This constraint is significant because the maximum benefit of air attacks is typically in the early stages of these fires, slowing them until ground units can make effective attacks.

Federal, state, and a few local agencies also can call on large fleets of fixed-wing aircraft (contracted or owned) to combat wildland fires. Although used in certain instances to protect structures from approaching wildfire, these aircraft are typically used to create a fire break from which fire fighters can attempt control of the fire. Extreme care must be exercised to avoid collapsing structures or injuring fire fighters or structure occupants with the weight of the suppressant drop. Such drops are far more difficult with fixed-wing aircraft because of their relatively high speed when making drops. Helicopters can virtually come to a standstill when making drops, so their accuracy is significantly better than fixed-wing aircraft. In the past, fixed-wing aircraft also had the advantage of being able to carry much heavier loads of suppressant than helicopters, but fire fighters now have helicopters at their disposal that are capable of carrying more than 2000 gallons of suppressant.

Apparatus Design.   The need to negotiate both paved streets and wildland unpaved roads, to operate in areas without fire hydrants or other water sources for fire protection, and to develop sufficient fire flow to attack structures has caused a change in apparatus specifications for both forestry and urban/suburban fire agencies. Attention must be paid to structural apparatus to ensure adequate road clearance off pavement, sufficient water tank capacity to carry out fire attack without being connected to a water supply, and a sufficient supply of 1¾-in. or 1½-in. hose to stretch lines behind homes in the wildland. Some forestry apparatus is shifting toward greater pump capacity and in some cases equipping the engines with self-contained breathing apparatus (SCBA) for structure firefighting. (See NFPA 1901, *Standard for Automotive Fire Apparatus* [5], and NFPA 1906, *Standard for Wildland Fire Apparatus* [6], for further information.)

Fire Fighter Training.   Some federal wildland fire fighters are trained in structural firefighting and urban/suburban fire fighters are being exposed to wildland firefighting. This additional training is critical for fire fighter safety and improving effectiveness.

Marine Firefighting

Marine firefighting presents special problems for fire agencies because access is often limited to a single approach and distances from fire apparatus to vessels involved with fire can be a significant handicap to rapid attack.

Fireboats.   In some larger ports, the problems are partially addressed through the use of dedicated fireboats. In some ports, several boats are in service. These specialized craft vary in size from less than 20 feet in length to more than 100 feet. Their pumping capacity can be from 500 to more than 20,000 gallons per minute. Because of the cost of the vessel and adequate staffing, several ports serving large ships still do not provide this protection. In one large West Coast port, the fire agency has been able to document the actual savings of having fireboat resources at hand. This fire agency recently was funded to replace 80 percent of its fireboat fleet. (See NFPA 1925, *Standard on Marine Fire-Fighting Vessels* [7] for further information.)

Fireboat Alternatives.   In some cases tugboats equipped with some fire equipment are utilized. These arrangements have several disadvantages over dedicated fireboats. Coordination between the boat crew and fire fighters on board the ship can be difficult as they are responsible to different bosses. Obtaining the tugs for training exercises can be difficult and expensive, thereby limiting training efforts. The tugs may not be immediately available for fire-fighting duties, thereby delaying a timely fire attack. However, the expense of such an arrangement will, in most cases, be dramatically less than maintaining a fireboat company on a 24/7 basis.

U.S. Coast Guard Capabilities.   Some ports have placed heavy reliance on U.S. Coast Guard assets to provide marine fire protection. Coast Guard officials point out that their vessels are not designed as fireboats (though some are equipped with fire pumps). Further, the Coast Guard cannot assure port officials of the availability of their vessels as they are charged with missions requiring them to leave the port on short notice.

Land-Based Fire Fighters.   In recent years, shipboard disasters have generated greater interest in training land-based fire fighters in marine firefighting. NFPA has a technical committee on the subject and has produced standards to assist agencies in preparing for these seldom-occurring but often devastating incidents. For more information see NFPA 1405, _Guide for Land-Based Fire Fighters Who Respond to Marine Vessel Fires_ [8].

Aircraft/Airport Fire Protection

In the context of this document, aircraft/airport fire protection is addressed in the following two categories:

1. Those situations where large commercial passenger aircraft provide regular service. Aircraft rescue and fire-fighting capabilities at these airports are regulated by the Federal Aviation Administration (FAA). The aircraft fire-fighting apparatus are sized in relation to the size aircraft providing the service. The location of deployment of these apparatus is also controlled by FAA response time regulations and NFPA standards.

2. Those airports not regularly served by commercial passenger aircraft. The FAA does not regulate response time or the vehicles used. At many, if not most of these airports, the responders often have other duties such as security, refueling, or structural fire protection. Many of these airports utilize structural fire-fighting apparatus with Class B foam capability.

Whatever situation, a comprehensive response planning capability must be preplanned and exercised in order for it to work when the incident occurs.

Apparatus.   At larger, airline served airports, it is not usually considered feasible to utilize the same apparatus for both airport structure fire protection and aircraft rescue and firefighting (ARFF). See NFPA 414, _Standard for Air-_

craft Rescue and Fire-Fighting Vehicles [9], for further information. The aircraft fire-fighting vehicles tend to be very heavy as they carry large water tanks. The weight can make it difficult for these vehicles to closely approach airport structures as access roads are often built in layers to accommodate simultaneous loading and offloading of aircraft passengers. These layered roadways are essentially bridges and not typically built to accommodate heavy vehicles. Further, the design of most ARFF vehicles does not easily accommodate standard structure fire-fighting equipment.

Personnel.   Fire-fighting crew size is not mandated by the FAA, however, it is specified in NFPA 1710, *Standard for the Organization and Deployment of Fire Suppression Operations, Emergency Medical Operations, and Special Operations to the Public by Career Fire Departments* [10]. Crew size has caused much disharmony between airport fiscal managers and fire officials. Many of the modern aircraft rescue and fire-fighting vehicles can be operated by a single person, but more personnel are needed to accomplish rescue in downed aircraft. In those airports located immediately adjacent to areas serviced by structural fire agencies, fire companies can often provide timely response to augment the aircraft fire-fighting crews. In those airports located remotely from other fire agencies, sufficient staffing should be provided for both operation of the apparatus and search and rescue tasks. Because aircraft firefighting can be very hazardous, recurring proficiency training is strongly recommended by experts in the field.

Reference to NFPA Standards, Guides, and Recommended Practices on the subject are strongly recommended for any fire-fighting agency with ARFF responsibilities. For further information see the following NFPA publications:

- NFPA 402, *Guide for Aircraft Rescue and Fire-Fighting Operations* [11]
- NFPA 403, *Standard for Aircraft Rescue and Fire-Fighting Services at Airports,* [12]
- NFPA 405, *Recommended Practice for the Recurring Proficiency Training of Aircraft Rescue and Fire-Fighting Services* [13]
- NFPA 414, *Standard for Aircraft Rescue and Fire-Fighting Vehicles* [9]
- NFPA 424, *Guide for Airport/Community Emergency Planning* [14]
- NFPA 1003, *Standard for Airport Fire Fighter Professional Qualifications* [15]
- NFPA 1710, *Standard for the Organization and Deployment of Fire Suppression Operations, Emergency Medical Operations, and Special Operations to the Public by Career Fire Departments* [10]

Other Types

There are fire organizations that are maintained solely for a single type of risk. Some large petroleum refineries have a fire response team dedicated to only that facility. (See NFPA 600, *Standard on Industrial Fire Brigades* [16] and NFPA 1081, *Standard for Industrial Fire Brigade Member Professional Qualifications* [17] for further information.) Some military installations have fire organizations that, while established to provide protection for a specific facility, will respond off-site for mutual aid to other fire agencies.

Comprehensive Fire Protection Planning

In the early 1970s, a number of government and academic studies began to broaden the discussion of fire service deployment and thus introduced the creation of public fire protection planning models. These models attempted to establish fire service deployment policies based on locally driven conditions and objectives and focused on developing comprehensive plans that addressed the broader aspect of a fire protection delivery system. This systems approach encouraged communities to consider prevention and mitigation policies as well as suppression. This approach was termed "Community Fire Protection Master Planning" and was promulgated by the then-titled National Fire Prevention and Control Administration (now United States Fire Administration). See Figure 10.11.

An example of this approach was the incorporation of additional fire mitigation technology into local building and fire codes. By using codes that incorporated this new technology, a community could view these built-in building fire safety features as part of the overall local fire defense system,

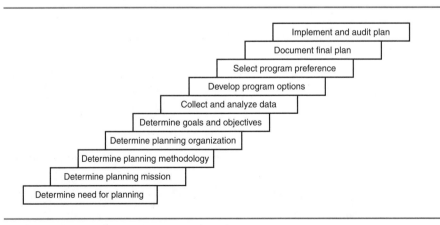

▶ **Figure 10.11** Fire Protection Planning Steps.

which, if applied, could greatly impact the level of fire defenses provided by a municipality. In many communities, however, those issues are still largely perceived through the windshield of a responding fire apparatus.

An important element of the process is the determination of "what there is to burn." This element takes into consideration all those factors affecting the challenge facing the fire agency. Another element of this comprehensive approach to fire protection planning was the effort to incorporate input from community stakeholders as to the level of fire protection desired by those paying for it.

Once the community has selected its desired level of response capability, technology should be utilized to plan fire resource deployment. There are several computer-aided deployment programs available for this purpose. All that is needed is a digital map of the community, determination of speeds that can be safely used on the various streets, and what response time is desired. The computer will indicate response times from any proposed location.

Other Fire Protection Planning Factors

Federal Regulations.   At this writing, one federal regulation weighs heavily on fire protection planning. The Occupational Safety and Health Administration (OSHA) requires, essentially, that before a fire fighter can enter a burning structure, four fire fighters must be on the scene. The fire fighters are to enter the structure in pairs, leaving at least two fire fighters outside to serve as a rescue crew should those fire fighters inside encounter problems requiring additional personnel for their safety. This regulation, 29 CFR 1910.134 [18], is termed "2 in/2 out." There is no OSHA requirement regarding the number of fire fighters to be on each apparatus, therefore apparatus could be staffed with two fire fighters, but they would be required to wait until the arrival of another two fire fighters before entering the building. There are exceptions to the rule to allow fire attack when it is known that there is the possibility of saving life with immediate entry, or when the fire is determined to be in its incipient stage. NFPA 1500, *Standard on Fire Department Occupational Safety and Health Program* [19], also addresses this issue as a safety factor.

NFPA Standards.   Additionally there is an NFPA voluntary standard, NFPA 1710, which essentially requires not less than four fighters be on every responding engine and ladder company [10]. If the community is protected by a mostly volunteer fire organization, NFPA 1720, *Standard for the Organization and Deployment of Fire Suppression Operations, Emergency Medical Operations, and Special Operations to the Public by Volunteer Fire Departments* [20], could be the appropriate document.

(ISO) Grading Schedule. Another factor considered by many fire protection planners is the Insurance Services Office (ISO) Grading Schedule, which is applied to cities of less than 250,000 population. ISO has the responsibility to evaluate and classify a community's fire protection capabilities in most of the nation's states. This classification can be used by insurance companies to determine premiums. The original intent of the schedule was to measure the ability of a community's fire defenses to minimize the risk of substantial losses from inner-city conflagrations. The primary focus of the schedule was to evaluate the community's fire department, water supply, and building construction. In time, fire service leaders began to view the schedule as the primary yardstick to plan for their departments.

Many city and business leaders began to see their community's fire insurance rating as one of several selling points that measured the quality of life within their community. It was not uncommon for community leaders to market their fire insurance rating as a selling point to draw business and industry into the community. Their belief was that lower fire insurance ratings correlated into the sustainability of jobs and tax revenue.

For more than 50 years, applying the schedule as a fire defense planning tool went unimpeded within most cities and towns. However, beginning in the 1960s many within local government as well as the public at large began to question the validity of how and why municipalities provided services to the public. Questions began to rise as to the validity of the schedule and its use as the primary tool for planning of local fire defenses. It had long been argued that the schedule did not reflect the changing urban environment, technological changes in fire protection, or the correlation between the investment in the level of public fire protection, public education efforts, and a community's direct fire losses.

Today the schedule is still used to set fire insurance rates. In 1980 ISO initiated the most comprehensive revisions of the schedule to date. Today's grading schedule reflects the changes in community fire risks and fire service operations. The most significant change is the assessment of fire defenses to suppress large-scale conflagrations to containing fire losses to the building of origin.

▶ SPECIAL OPERATIONS PLANNING

In the past several decades, fire agencies have accepted a broader scope of responsibilities. The breadth and depth of the services provided varies greatly with each agency.

EMS

Perhaps the most common of these broader scope activities is emergency medical service or EMS. It is delivered in several levels, with some agencies providing a combination of all of them, other agencies only the basic level. Before assuming EMS duties, agencies should carefully analyze the training and equipment requirements to be assured that the agency can appropriately fund those requirements.

First Responder.   Usually responsible for cardiopulmonary resuscitation (CPR) and general first aid practices. The use of automatic external defibrillator (AED) is becoming more common in the fire service. This level of service (first responder) typically requires personnel to attend 16 to 40 hours of training. Some jurisdictions require annual recertification in this set of skills.

Basic Life Support (BLS).   Responsible for first responder skills and patient assessment, O_2 therapy, the ability to support advanced life support, activities and in some cases, limited intubation. Typically 100 to 160 hours of training and periodic retesting are required. Personnel trained to certain levels of EMT can be nationally certified.

Advanced Life Support (ALS).   Responsible for assessing and treating immediately life-threatening events occurring to victims of medical and trauma emergencies. EMT-Paramedic (EMT-P) scope of practice includes many life-saving techniques, for example, drug therapy, intubation, defibrillation, and intravenous drug therapies. Typically 1200–1400 hours of initial training and structured reoccurring medically supervised training are required. EMT-P is a national certification.

Emergency Transport.   The transportation of the sick and injured to the appropriate medical facility. This can be done by either BLS or ALS personnel, or a combination of those skill levels. Ambulances can be staffed in many different ways to fit the needs of the local community. Ambulances are operated by local government, hospitals, and private enterprise. The training for personnel usually follows national standards. It is not unusual for all three type of providers to exist in the same community.

Fire Department Ambulance Service.   The primary advantage of fire department–operated ambulances is said to be expanded utilization of fire personnel, the ability to provide all-risk emergency interventions, e.g., both physical rescue and medical care, medical care in a hazardous environment, full firefighting, rapid intervention duty at fire scenes, and other benefits. These functions would typically be outside the normal scope of other EMS providers. The cost of staffing ambulances with fire fighters should be measured against the cost of using private sector personnel.

Rescue Response Teams

Urban Search and Rescue. The Federal Emergency Management Agency (FEMA) sponsors Urban Search and Rescue (USAR) programs in many regions of the nation. The teams are strategically located throughout the country. Many of the teams are made up of a consortium of local agencies located in close proximity to one another. The teams were originally developed and are equipped and trained to rescue persons trapped in collapsed structures. The program is currently administered by the Department of Homeland Security.

Swift Water/Flood Rescue Teams. Fire agencies in those areas where they are called upon to water-related life-threatening incidents are forming Swift Water/Flood Rescue Teams. These teams provide rescue and evacuation services for persons requiring aid due to boat or swimming accidents in swiftly moving rivers and streams and for persons trapped in floods. Training typically requires 40 to 60 hours initially and ongoing skill maintenance exercises. The equipment may include powerboats and other watercraft and can be costly.

Rope-Supported Rescue Teams. Some fire agencies maintain rope-supported rescue teams for rescue of persons trapped on cliffs or other vertical surfaces (bridges, buildings, etc.). Typically requires 60 hours of initial training and ongoing skill maintenance exercises. This endeavor is OSHA regulated.

Trench Rescue Teams. Another OSHA-regulated function that has been undertaken by many fire agencies is trench rescue. Special training (40 to 60 hours) is required for trench rescue teams as the work can be extremely hazardous. Typically, the team's equipment is carried on a vehicle dedicated for that purpose.

Confined Space Rescue. Confined space rescue is yet another OSHA-regulated field involving many fire agencies. The teams are trained and equipped to rescue victims trapped or overcome in areas and equipment not designed for human occupancy, for example, petroleum tanks, railroad tank cars, or underground utility vaults. For further information, see NFPA 1006,

Standard for Rescue Technician Professional Qualifications [21], and NFPA 1670, *Standard on Operations and Training for Technical Search and Rescue Incidents* [22].

Helicopter Operations.   A small but growing number of fire agencies have undertaken helicopter operations to engage in one or more of the following functions: wildland and structure fire control, rescue missions, and air ambulance. In practice, helicopter operations often combine the full scope of a fire agency's operations including the special operations just described. Extensive training of aircrews is essential for safety and effectiveness, particularly in those instances where the helicopter crews are not fire fighters. Some agencies have obtained helicopters from military salvage inventories, but increasingly the value of the aircraft are justifying the purchase of newly manufactured machines. There are private sector organizations that specialize in leasing helicopters to fire agencies, typically those with large wildland protection responsibilities. As was discussed with tugboats being used in place of fireboats, these arrangements can be somewhat limiting in their use and training.

Hazardous Material Response.   In most communities, hazardous material response has been a fire agency with little or no additional funding. Recognizing this lack of funding as well as the hazardous situation encountered at most structure fires, OSHA has required hazardous material training, at least at the operations level for essentially all emergency responders. Before electing to assume more than these basic responsibilities, agencies should review the OSHA training and equipment requirements to determine if funding is available to provide requirements of higher level hazardous materials response. Although federal agencies are referenced in this document, some states will adopt these regulations under state OSHA plans and others will adopt these regulations through adoption of a similar regulation established by the Environmental Protection Agency (EPA) and appropriate state agencies.

The development of both facility response plans and community emergency plans is required by numerous federal laws, including Superfund Amendments and Reauthorization Act (SARA), Title III, "The Emergency Planning and Community Right to Know Act of 1986" in 42 USC 116 [23]. In addition, special dedicated communication capability may be required, special equipment requirements are numerous and costly, and a written personal protective equipment program is required by OSHA. In addition, an ICS must be used and all supervisory personnel must be trained in ICS.

Elements of the program should include personal protective equipment selection and use; storage, maintenance, and inspection procedures; and training considerations. The selection of personal protective clothing should be based on the hazardous materials and conditions present. For more information see the following NFPA publications:

- NFPA 472, *Standard for Professional Competence of Responders to Hazardous Materials Incidents* [24]
- NFPA 473, *Standard for Competencies for EMS Personnel Responding to Hazardous Materials Incidents* [25]
- NFPA 704, *Standard System for the Identification of the Hazards of Materials for Emergency Response* [26]
- NFPA 1991, *Standard on Vapor-Protective Ensembles for Hazardous Materials Emergencies* [27]
- NFPA 1992, *Standard on Liquid Splash–Protective Ensembles and Clothing for Hazardous Materials Emergencies* [28]
- NFPA 1994, *Standard on Protective Ensembles for Chemical/Biological Terrorism Incidents* [29]

Disaster Response Planning

A logical extension to fire service preincident planning is community disaster mitigation planning, the difference being the involvement of all elements of the community's disaster response organization (fire, police, public works, purchasing, Red Cross, water provider, etc.). In many communities, the local fire agency takes the lead role in disaster mitigation planning, in others the police agency, in others an independent emergency planning agency is assigned that responsibility. Successful disaster planning must be closely linked with the previously described unified command and like that planning depends on an attitude of cooperation. The lead agency should see itself as a coordinator, rather than a boss. See Table 10.2.

Risk Analysis.   Disaster planners should first analyze the large-scale risks facing the community. The analysis should not only look at the possibility of the event, but the impact on the community. In prioritizing a risk for planning, a risk that is less likely to occur may be rated higher than another more likely event that will have less negative consequences on the community as a whole.

Consider a community facing two possible large-scale risks, a wildland fire that would threaten 100 single-family residences on the edge of the community, and an earthquake that could make rubble out of the downtown commercial area, including the community hospital and high school. The wildfire has a far greater chance of occurrence than the earthquake, but the earthquake has far greater consequences on the community as a whole.

▶ **Table 10.2** Community Organizations' Roles in Emergency Planning

Organization	Emergency Roles
City manager/mayor	Overall coordination and policy making
Fire department	Fire control, hazardous material control, physical rescue (extrication from vehicles and buildings), flood rescue, emergency medical service
Police department	Crowd control, traffic control, area evacuation, maintenance of order
Public works department	Heavy lift, barricades, diking
Library department	Public information
Health department	Air monitoring
Building department	Building safety surveys
Hospital consortium	Victim care
Social services department	Temporary housing
School district	Busing, food services

While preparations to deal with both risks should be undertaken, it seems reasonable that the greatest share of planning and mitigation resources should be spent on dealing with the earthquake threat.

Training. Most emergency preparedness sources recommend not less than an annual disaster exercise. For the most part, these drills are held at the community's emergency operations center with an "umpire" feeding the "players" bits of a preplanned scenario depicting the selected "disaster" on a timed schedule.

Emergency Operations Center. In larger cities, the emergency operations center (EOC) may be a dedicated facility: in smaller towns a conference room, library, or gymnasium may be used. Important factors to consider when planning an EOC include the following:

- Large enough to accommodate the community's senior leadership
- Access to appropriate communications capability (telephone, public service radios, TV reception, commercial radio, etc.)
- Means to display large-scale maps
- Remote from expected impact areas
- Appropriate acoustics
- Reasonable access by community senior leadership

Although a community may have a preferred EOC location, preparations should be made to establish an EOC in alternate locations as the preferred location may not be available when needed. The purpose of the EOC function includes the following:

- Set community policy on matters pertaining to the situation (e.g., determination of which facility will receive the limited resources on hand, the high school or the hospital)

- Acquisition of resources requested by incident commander(s) (heavy equipment not in community's inventory, staffing from outside the community, etc.)

- Provision of technical information (recommended procedures for a specific hazardous material unfamiliar to responders, rate of water rise in the river during flood conditions, etc.)

- Care for displaced persons (e.g., arrange for temporary housing and food)

- Provide guidance in restoring the community to normal

It is critical for those in the EOC to recognize that they are not managing field operations and be wary of micromanaging. For further information see NFPA 1600 [2].

 ## SUMMARY

Planning for fire service response is a complex issue with many elements that must be considered in order to provide effective and efficient fire protection. The Incident Command System is a critical factor in planning for emergency response in that it provides a guideline for staffing, organization, training, and equipping for emergencies as well as for managing emergency scenes. The system requires the following:

- Someone be in charge from the moment of the arrival of the first emergency responder to the time when the last unit is made available for further response.

- Responder safety be a primary concern of the Incident Commander

- Every person working the scene be responsible to one and only one supervisor

- Common terminology be used to avoid confusion

- The status of all personnel on the scene be continually known to command

- Communication channels be established and adhered to at the emergency scene
- Resources at the scene be organized into effective elements
- A plan of action, written or oral, be established for every emergency incident
- Only those elements of the system that are needed for incident control should be implemented
- The span of control for any supervisor be limited to what he can effectively manage (in general, three to seven subordinates)
- The Incident Commander is responsible for all aspects of incident management, regardless of how basic or complex the organization of the incident.
- Agencies implement a system of shared authority (unified command) at those incidents where more than a single agency is responsible for actions undertaken at the incident.

The goal for structured fire response planning should be to have sufficient resources immediately available on the fireground to control the incident in a timely fashion without unreasonable risk to fire fighters. Achieving this goal may require developing agreements with neighboring agencies to augment the response to some occupancies. These agreements should include provisions for automatic response (at time of alarm) to specified situations and response when specifically requested. All agreements should be documented, reviewed, and exercised regularly.

Group Activity

Imagine that you are the command structure/management team in charge of the following incident:

There has been a report of an explosion at a local high school in your jurisdiction. On arrival, EMS and police officers are greeted by youngsters fleeing the building. One of the youth reported seeing armed men entering the school.

What plans would you put in place? What agencies do you contact for assistance? What resources might you need? What will your command structure look like? (You have 25 to 30 minutes to plan your response to this scenario.)

Review Questions

1. The purpose of _____ _____ _____
   is to provide a flexible, adaptable structure for the management and coordination of emergency incident operations.

2. What term means that every person assigned to the incident has only
   one assigned supervisor?

3. "Span of Control" refers to what?

4. At a single company incident (e.g., light brush fire adjacent to a
   dwelling), the _____ _____ is the Incident
   Commander who personally fulfills all the basic management functions.

5. What is the name for a cooperative effort that facilitates the ability of all
   agencies with either geographical or functional authority and responsibility to participate in the management of an incident by joint development and utilization of incident goals?

6. The difference in the community-wide impact of structure fires (v. EMS)
   is sometimes overlooked. What type of occupancy fires are most likely to
   adversely affect communities?

7. Which organization sponsors Urban Search and Rescue (USAR) programs in many regions of the nation?

11 Public and Private Support Organizations

Robert S. Fleming
Rowan University

Learning Objectives

After completing this chapter, the reader should be able to do the following:

- Synthesize the role of national, state, and local support organizations in fire protection and emergency service
- Identify the types of support organizations that assist fire and emergency operations
- Describe the specific purposes of these support organizations
- Identify the various organizations that represent the fire service
- List the various trade publications for the fire service
- Identify the principal sources of fire-related information on the Internet

A fire service professional is responsible for ensuring that the fire department is prepared to respond effectively, efficiently, safely, and professionally to all incidents to which it is dispatched. Many resources are available to fire departments to ensure readiness and the level of fire protection in the communities that those departments serve and protect. The success of fire service professionals and their fire departments are determined in large part by their awareness, understanding, and use of these resources.

This chapter introduces the resources available. Fire service professionals reading this material will discover many valuable resources that can be instrumental for planning and delivering fire department services. The public and private support organizations that can be utilized throughout a fire service career are introduced in this chapter.

These organizations are designed to fill support roles within our overall system. While they do not respond with apparatus, equipment, and personnel, the information, expertise, and other resources that they can provide to fire departments and to fire service professionals will enhance a fire department's preparedness to fulfill its mission and meet and exceed the expectations of its stakeholders. It is therefore in a professional's best interest to learn about the resources available through these support organizations and to develop professional contacts with appropriate individuals within these organizations.

The support organizations introduced in this chapter are diverse in their individual organizational missions but collectively represent a wealth of information and resources for the fire service professional. There are both public and private organizations and these organizations exist at the local, county, regional, state, and national levels. As fire service professionals become familiar with each of these organizations, it is incumbent on them to reach out to the appropriate organizations and establish working relationships with them. These relationships will benefit both fire departments and professional careers.

Although the national organizations are referenced by name in this chapter, those support organizations that may exist at the state, regional, county, or local levels are discussed generically, and it is up to fire service professionals to investigate and learn about appropriate public and private support organizations. The knowledge developed regarding the mission, services, and resources of each relevant support organization and the working relationships developed with these organizations are mission-critical skills as professionals serve their fire department and advance in their career.

Continued success in the fire service is, in large part, based on the ability to remain current in the field and to engage in continuous professional development. Most of the public and private support organizations discussed in this chapter offer many opportunities for professional development.

▶ WORKING WITH SUPPORT ORGANIZATIONS

The public and private support organizations profiled in this chapter are valuable resources that contemporary fire departments must fully understand and utilize. These organizations include governmental and nongovernmental organizations and exist at the local, county, regional, state, and national levels.

Understanding Available Resources

The success of a fire service professional and that of a fire department requires a thorough understanding of the support organizations discussed in this chapter. Unless fire service professionals are aware of these organizations and their activities, they will not be able to access the benefits of affiliating or working with them. Fire service professionals must constantly seek new information that will enhance their decision-making capabilities. In so doing, fire departments and fire service careers are naturally advanced.

Determining Support Needs

It is not enough simply to be aware of these support organizations. Fire service professionals must learn enough about each of them to determine their relevance to them and their fire department. If a fire department currently offers Basic Life Support services and plans to enhance its service level by initiating Advanced Life Support services, its professionals need to be working closely with organizations such as its county or regional emergency medical services council and its state counterpart or state health department.

If a fire department plans to start its own hazardous materials response team, it would need to work closely with the appropriate training and certifying agencies within its state and its Local Emergency Planning Committee. Some states have hazardous materials technician associations that might provide useful assistance or guidance to the fire department and its members.

An important theme of this chapter is that there are many support organizations available to assist fire departments. It is up to the department to determine what assistance is needed and from which particular sources.

Accessing and Securing Needed Support

Once what is needed in the way of support has been identified, an individual or representative of the fire department must reach out to secure that needed assistance or guidance. As professionals reach out to the various public and private support organizations, they must recognize the importance of getting to know people within those organizations. The affiliations developed through such networking will stand the professional and the fire department in good stead in both the present and the future.

▶ LOCAL ORGANIZATIONS

Fire departments in the United States and in most countries around the world are based at the local level. They are chartered and organized to respond to fires and other emergencies within a predefined primary response territory. When needed, they fulfill mutual aid agreements and assist other emergency response agencies. Most departments are designed to function at the municipal level, covering one or more boroughs, townships, towns, parishes, or cities. Occasionally, fire departments are established to protect counties or regions.

Just as there are primary and mutual aid response organizations at the local level, there are support organizations within these geographical areas. These support organizations fall into two categories: (1) public or governmental agencies and (2) private or nongovernmental organizations. This distinction is important for a number of reasons, including understanding the mandate, authority, and enabling legislation of governmental agencies as well as the mission and purpose of nongovernmental agencies.

Governmental Agencies

A number of local governmental agencies or other public organizational entities exist, and the successful fire service professional will be knowledgeable about these agencies and their roles and responsibilities and will also have developed a working relationship with appropriate individuals within these organizations. The existence of these agencies varies with location. Their relevance to a particular fire department is based on where that department operates and the nature and level of the services that it provides.

The following four types of organizations represent typical governmental organizations at the local level. While their names may vary, their functions remain relatively consistent. The successful fire service professional will research these organizations and others that may exist within his or her jurisdiction. These organizations can provide tremendous assistance to a fire department, assuming that the department is thoroughly familiar with their roles and responsibilities, services, and capabilities and utilizes them when and as appropriate.

911 Communications Centers. One essential component of any system of emergency response is the capability to receive calls for emergency assistance, process those calls, dispatch needed emergency response resources, and support incident operations. These operational support entities are referenced by various names, including communications centers, dispatch centers, and radio rooms. They may be operated by various agencies, including local or county government, the fire department, or law enforcement. They may provide services to one community or to a region of communities. It is imperative that the fire department have a thorough understanding of the full range of support services available through these organizations, as well as their operational policies and procedures. See Chapter 6 for further information.

Emergency Management Agencies. Emergency management agencies (often referred to as EMAs) may exist at various levels of local and county government. These agencies have roles and responsibilities that are mandated by laws and regulations and are an important resource for the fire department in terms of planning for response to emergency incidents. They can also be key resources to the fire department in that they can support training, exercises, and response in a variety of ways, including making available important site-specific information and community plans that will enable the fire department to make more informed decisions both in preincident planning and during incident management.

Local Emergency Planning Committees.   In the United States, Local Emergency Planning Committees (LEPCs) are mandated under Title III of the Superfund Amendments and Reauthorization Act (SARA) of 1986 [1] and have responsibilities for ensuring community preparedness to various types of potential emergencies, particularly those involving the production, transportation, or use of hazardous chemicals. The members of the LEPC are drawn from various disciplines within the community, with representation from the various stakeholder groups. The fire department should be represented on the LEPC. The LEPCs have a mandated role in the development and approval of *Community Response Plans* for SARA sites. These plans reference the response agencies that have been identified to respond to a chemical emergency at a given facility. It is imperative that all involved fire departments and emergency response agencies secure and familiarize themselves with the plans that involve their response territories. These plans should serve as the basis for fire department training and exercises.

Emergency Medical Services Councils.   Emergency medical services councils have responsibility for ensuring the effectiveness, efficiency, and quality of emergency medical services provided within their jurisdiction, which is typically a county or region. Fire departments that provide emergency medical services must work closely with their respective emergency medical services council or other appropriate emergency medical services agency within their geographic area. Even if a fire department does not provide emergency medical services directly, it will still need to be thoroughly familiar with the policies, procedures, and protocols of its particular county or region.

Nongovernmental Organizations

Nongovernmental organizations are related to the provision of fire and emergency services to a community but are not direct arms or agencies of a governmental entity. Their mission and the services they provide are based on their mission statement and organizational charter. In addition to supporting the local fire department, these organizations also work collaboratively with local governmental agencies. These organizations include associations comprised of fire chiefs, fire fighters, fire marshals, fire inspectors, emergency medical services personnel, and law enforcement personnel.

Once again, the following organizations are those that are typical within most geographic areas. Successful fire service professionals should learn about all of the support organizations related to their fire department and should join and actively participate in those that are appropriate.

Fire Chiefs Associations.   Fire chiefs associations are common, particularly at the county level. These organizations provide forums for fire chiefs and other chief officers to meet and discuss issues of interest to the fire departments within the geographic region covered by the organization. Through the coordinated and professional activities of such an organization, the fire chiefs in a county can make sure that the necessary dispatch, response, and operational procedures are established and utilized to ensure fire department effectiveness, efficiency, and safety. The success of these organizations is based on the active participation of incumbent chief officers and their professionalism in focusing on an agenda of enhancing fire department response capabilities.

Police Chiefs Associations.   Law enforcement is a closely related discipline to the fire service. Fire department officers and members are called on to work cooperatively in many situations with members of the law enforcement community. Just as there are fire chiefs associations, there are also police chiefs associations. Although many of the issues and initiatives of concern to the members of the police chiefs association have no relevance to the fire department, issues such as operating safely on the highway, securing incident scenes, cooperating in mass evaluations, and conducting fire investigations and securing evidence require collaboration between the respective fire chiefs and police chiefs organizations.

Emergency Medical Services/Ambulance Associations.   In addition to the emergency medical services (EMS) councils that function under the auspices of a governmental agency, the operational or administrative chiefs of fire departments and other response agencies providing emergency medical services may form an organization that functions in similar manner to a fire chiefs association, but focuses on EMS issues.

▶ REGIONAL ORGANIZATIONS

In addition to public and private organizations that exist at the local level, public officials and emergency service leaders in many jurisdictions have formed regional organizations specific to the needs of that region. These organizations, which may be public or private, differ vastly based on the challenges found in each region.

In addition to the two following examples of regional organizations, many other regional groups have been established to identify and address regional issues that cross normal jurisdictional boundaries and require resources that no one geographic jurisdiction would be prepared to provide.

Mutual Aid Organizations

Mutual aid organizations are formed in recognition of the fact that the resource planning and response needs of many communities may require more extensive resources or expertise than a particular fire department or emergency response agency may have on its own. Examples of mutual aid organizations include those designed to prepare for and respond to major incidents such as refinery explosions, shipboard fires, or wildland fires. These organizations often span the geographic boundaries of a number of counties and may span the boundaries of more than one state, as does the Cumberland Valley Fireman's Association. A timely illustration of the activities of the Cumberland Valley Fireman's Association would be its Responder Highway Safety initiative designed to address the safety of responders while operating on highways.

Counterterrorism Task Forces

An integral component of homeland preparedness initiatives has been the establishment of regional terrorism task forces within each state. Each task force is comprised of representatives from a number of counties or geographic subdivisions that would likely be required to work together in the event of a terrorist event. These task forces have a major role in planning, equipment purchase and deployment, training, and exercises within that geographic region and are receiving federal funding to enhance and ensure the level of regional preparedness for terrorist events.

▶ STATE ORGANIZATIONS

Each of the 50 states has a number of state organizations and agencies that relate to the work of the fire department. Those aspiring to serve their fire departments well and experience progressive career growth are encouraged to learn about the state organizations within their state. As was the case with local organizations, these organizations are either public or private and are working under a legislative mandate, mission statement, or charter. It is extremely important to get to know the support that these organizations can provide to a fire department.

Governmental Agencies

Although the various state governments differ significantly in their organizational structure and stance with respect to the establishment and operation of state agencies whose charge and responsibilities relate to the fire departments of that state, there are a number of departments, agencies, or functions that are typical.

Within any state there will likely be other governmental entities with whom the successful fire department must be acquainted and develop appropriate working relationships. These agencies may work through established agencies at the county level such as those offices being established to address homeland preparedness at the state level. In some cases, fire departments will work directly with these agencies as is the case in the limited number of states that have established low-interest loan programs from which fire departments are eligible to secure loans to acquire needed mission-critical apparatus and facilities.

State Fire Training Agency. The fire training model varies from state to state, as do the resources that each state is willing to invest in the training and preparation of its fire fighters. Some states have centralized systems with most fire training that takes place outside of the fire station being conducted through the state system. In other states, the training system is more decentralized. Each state training system partners with the National Fire Academy in that the state training systems are an integral component of the National Fire Academy's delivery system for off-campus courses.

State Emergency Management Agency. A state emergency management agency has responsibility for ensuring the preparedness of that state to prevent, prepare for, and mitigate, as necessary, emergency situations. The state emergency management agencies serve as a link between the Federal Emergency Management Agency (FEMA) and county emergency management agencies. With the establishment of the Department of Homeland Security, the role of state emergency management agencies is expected to be enhanced.

State Fire Marshal. The roles and responsibilities of the state fire marshal's office vary based on the statutes and regulations of that state. The office of the state fire marshal may be an independent office, or housed within a division of fire safety, or an organizational unit of the state police. Line and staff officers of the fire department often engage in cooperative initiatives, including fire investigations, with members of the state fire marshal's office.

State Health Department.   The various states have health departments that fulfill roles and responsibilities that often impact the contemporary fire department. The role of these state agencies in the prevention and control of disease and illness has relevance to our first responders. State health departments typically have staff members with extensive expertise that will be of great value during these times of increased concern regarding the potential for biological or chemical terrorist attacks.

State Emergency Medical Services Agency.   The responsibilities for emergency medical services may fall under the state health department or may be under a separate agency within state government. Regardless of positioning within the state organizational structure, the functions of the state in regulating and supporting emergency medical services have increased significantly in recent years. These agencies work closely with the emergency medical services councils, or comparable agencies, that exist at the county or regional level.

Nongovernmental Organizations

The governmental agencies that have been established at the state level are often complemented by a number of nongovernmental or private organizations established to address statewide emergency services issues, similar to the comparable organizations that exist at the county level. These organizations often work on issues that are of interest to their counterpart organizations at the county and national levels. In some cases, they may be addressing issues that are state issues and unique in comparison to the national agenda.

There may be additional statewide organizations within any state that fire service professionals should be acquainted with. These, once again, will vary by state. As an example, approximately 10 states make provision for the position of fire police within the fire department organizational structure. In states that utilize the position of fire police, there may be state fire police organizations.

State Fire Chiefs Associations.   Most states have a fire chiefs association. In fact, some have more than one such organization that represent related, but sometimes divergent, interests based on variables such as department size and type or rank of the member chief officers. These organizations are designed to provide a forum for fire department chief officers to take positions on important issues and use their collective strength to educate or lobby members of the legislative or executive branch of state government. These organizations typically provide timely information to their members regarding proposed laws, regulations, initiatives, or budgetary priorities that could potentially negatively impact their fire departments.

State Fire Fighters Associations.   These organizations are similar to the state fire chiefs associations in that they are established to ensure a two-way flow of communications between fire fighters throughout a state and their elected officials. These organizations are often interested in the same issues and support the same initiatives as does the state fire chiefs association. On some occasions, as would be expected, there will be differences in orientation with respect to a given issue based on the fact that the state fire fighters association has a more diverse membership base than does the state fire chiefs association.

State Fire Instructors Associations.   The training of fire and emergency services personnel throughout our nation is in large part accomplished through county and regional fire training academies and through the local fire departments. Fire instructors play an instrumental role in the development and delivery of the training courses and programs necessary to ensure that first responders are prepared for response to any and all emergencies. Many states have instructors organizations that provide a forum for instructors to identify and address issues of importance to fire service training within that state.

Fire Service Institutes.   Some states have developed or are in the process of developing fire service institutes. These organizations are established to monitor proposed legislation and regulation at the state level and ensure that the emergency response community is aware of it and responds appropriately through initiatives designed to educate elected officials. The Pennsylvania Fire and Emergency Services Institute is an example of such an organization that was established based on the successful model of the Congressional Fire Services Institute discussed in the next section.

State Chapters of National/International Organizations.   National fire service organizations are discussed in the following section of this chapter. A number of these organizations have established chapters at the state level. National fire service organizations with recognized state chapters include the International Association of Arson Investigators, the International Association of Fire Chiefs, the International Society of Fire Service Instructors, and the National Volunteer Fire Council.

▶ NATIONAL ORGANIZATIONS

As was the case with local and state organizations, fire-related national organizations fall into the categories of governmental agencies and nongovernmental organizations. These organizations serve as valuable resources to fire service professionals and their fire departments. The services of many of the nongovernmental agencies may have a cost attached to them, in terms of a charge for a specific product and/or service or a membership fee that guarantees the member access to a number of products and/or services.

The national organizations discussed in this chapter represent the organizations that are most useful to a fire service professional. Through a professional's personal endeavors he or she may become acquainted with additional useful organizations.

Governmental Agencies

The agencies charged with fire-related issues at the federal level represent a wide range of diverse governmental agencies. Agencies, such as the United States Fire Administration, have a central focus regarding addressing our nation's fire problem. These agencies address fire protection as an integral element of their mission and resulting program activities and initiatives.

The federal government has been involved in fire protection for a number of decades. Prior to the 1970s the federal role was primarily one of protecting federal property and individuals working on or visiting that property. Congress significantly increased the role of the federal government in fire protection in the 1970s. Responsibilities for fire research, fire training, fire prevention, detection and suppression, technical and financial assistance, fire investigation, criminal investigation and enforcement, and development and administration of safety policy were delegated under the jurisdictions of 12 executive branch departments and 10 independent agencies.

Although a number of these departments and agencies have over the years provided support and assistance to local fire departments in a variety of ways, the funding of the recent Firefighter Assistance Grant Program represents a new era of federal support for the fire service. This grant program has been extremely successful in providing federal support to local fire departments that have significant documented resource needs.

Federal Emergency Management Agency. The Federal Emergency Management Agency (FEMA) was established to reduce loss of life and property and protect our nation's critical infrastructure from all types of hazards through a comprehensive, risk-based, emergency management program of mitigation, preparedness, response, and recovery. The Federal Emergency Management Agency has been a highly visible governmental agency in the preparation for and response to major emergency incidents. Under the governmental reorganization after September 11, 2001, FEMA now falls under the Emergency Preparedness and Response Directorate within the Department of Homeland Security. Fire protection professionals recognize the United States Fire Administration as a highly capable and valuable entity of Department of Homeland Security and FEMA. http://www.fema.gov

United States Fire Administration.   The United States Fire Administration (USFA) has a mission of reducing life and economic losses due to fire and related emergencies, through leadership, advocacy, coordination, and support. Through a variety of initiatives, the USFA is focusing on meeting a series of strategic objectives related to the reduction of fire fighter and civilian fire-related deaths. The National Fire Academy and the Emergency Management Institute deliver a variety of all-risks training programs under the auspices of the USFA. Figure 11.1 shows an organizational chart for the USFA. The fire service professional will benefit greatly from becoming aware of the many resources that are available through the USFA, includ-ing a comprehensive collection of available publications. http://www.usfa.fema.gov

National Fire Academy.   The National Fire Academy (NFA) exists to enhance the ability of fire and emergency services and allied professionals to deal more effectively with fire and related emergencies. The courses and programs are delivered through a diverse course delivery system that includes resident courses at the Emmitsburg, Maryland, campus of the National Emergency Training Center (NETC), shown in Figure 11.2, and courses delivered throughout the nation through cooperative working relationships with state and local fire training systems. A growing number of courses are available through distance delivery. http://usfa.fema.gov/fire-service/nfa.cfm

Emergency Management Institute.   The Emergency Management Institute (EMI) provides a nationwide program of resident and nonresident courses designed to enhance emergency management practices throughout the United States. These courses and programs are also offered through a diverse delivery system designed to be responsive to the needs of the local and state emergency management personnel who enroll in these educational offerings. http://www.training.fema.gov/emiweb

Department of Homeland Security.   As you learn about the various agencies of the U.S. government involved in some aspect of fire protection, it is important to consider the responsibilities of each agency in light of the recent

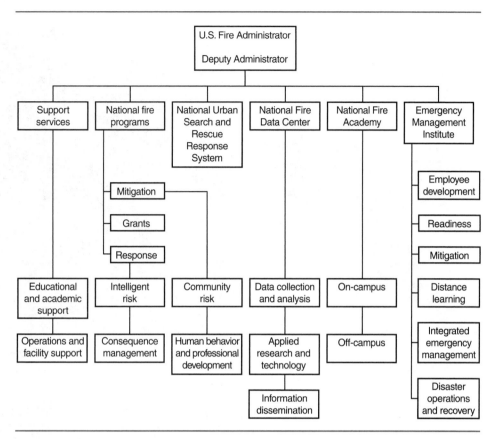

▶ **Figure 11.1**   USFA Organizational Chart.

major governmental reorganization that has taken place. The *Homeland Security Act of 2002* [2] created the Department of Homeland Security with the purpose of realigning the previous confusing patchwork of governmental activities, which fell under multiple governmental agencies, into a single department whose mission it is to protect our homeland.
http://www.dhs.gov/dhspublic

Mission of Department.   The mission of the Department of Homeland Security (DHS) includes the following three components:

• Prevent terrorist attacks within the United States
• Reduce America's vulnerability to terrorism
• Minimize the damage from potential attacks and natural disasters

▶ **Figure 11.2** FEMA's National Emergency Training Center in Emmitsburg, Maryland. *(Source: Photo by Jocelyn Augustino/FEMA News Photo.)*

Challenges of Domestic Terrorism. As a nation, we face unprecedented challenges regarding domestic terrorism. The bombings at the World Trade Center in 1993 and Oklahoma City in 1995 served as wake-up calls regarding the potential for domestic terrorism. The events of September 11, 2001, in three geographically dispersed locations clearly demonstrated the ever-present reality of domestic terrorism.

Consolidation of Governmental Agencies. The transition plan for establishing the Department of Homeland Security includes bringing 22 governmental agencies under this new department. Figure 11.3 shows an organizational chart for the department. This reorganization of the U.S. government is the largest in over half a century. As the agencies join the Department of Homeland Security, it is possible that their former missions and strategic goals may change to align them properly within the Department of Homeland Security. The Department of Homeland Security will consist of five directorates, or divisions, which will include all or parts of the following agencies:

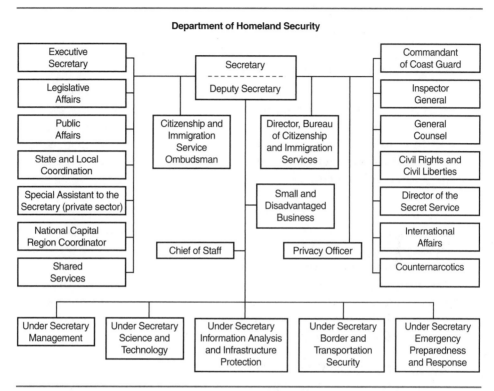

Department of Homeland Security

▶ **Figure 11.3** Organizational Chart for Department of Homeland Security.

- *Border and Transportation Security (BTS).* Customs Service, Immigration and Naturalization Service, Federal Protective Service, Transportation Security Administration, Federal Law Enforcement Training Center, Animal and Plant Health Inspection Service, and Office of Domestic Preparedness.

- *Emergency Preparedness and Response (EPR).* Federal Emergency Management Agency, Strategic National Stockpile and National Disaster Medical System, Nuclear Incident Response Team, Domestic Emergency Response Teams, National Domestic Preparedness Office.

- *Science and Technology (S & T).* Chemical, Biological, Radiological, and Nuclear (CBRN) Countermeasures Program, Environmental Measures Laboratory, National Bio-Weapons Defense Analysis Center, and Plum Island Animal Disease Center.

- *Information Analysis and Infrastructure Protection (IAIP).* Critical Infrastructure Assurance Office, Federal Computer Incident Response Center,

National Communications System, National Infrastructure Protection System, and Energy Security and Assurance Program.

- *Management.* This directorate will be responsible for departmental budget, management and personnel issues.

- *Other Critical Agencies*—In addition to the five Directorates, several other critical agencies are folding into the new Department or being created. These include: United States Coast Guard, United States Secret Service, Bureau of Citizenship and Immigration Services, Office of State and Local Government Coordination, Office of Private Sector Liaison, and Office of Inspector General.

Coordination of Local, State, and Federal Resources.   Local fire departments, emergency medical service organizations, law enforcement, and emergency management agencies will be the first to respond to future terrorist events and are thus considered our nation's "domestic defenders." Our system of domestic preparedness requires the coordination between these local resources and those that will respond to assist them from the state and federal levels. Effective preparedness involves planning, training, communications interoperability, and equipment acquisition.

Bureau of Alcohol, Tobacco, Firearms, and Explosives.   The Bureau of Alcohol, Tobacco, and Firearms (ATF) is housed within the Department of Justice. The ATF provides training and technical assistance to local fire officials and law enforcement authorities to assist in the prevention, detection, investigation, and prosecution of arson. http://www.atf.gov

Consumer Product Safety Commission.   The Consumer Product Safety Commission (CPSC) is engaged in a diverse range of initiatives to protect consumers from injuries and death resulting from consumer products. One such program is the Fire and Thermal Burn Program, which includes the investigation of injury patterns, data collection, research, and the promulgation and enforcement of mandatory standards regarding consumer products. http://www.cpsc.gov

Department of Housing and Urban Development.   The Department of Housing and Urban Development (HUD), through its Manufactured Housing and Construction Standards Division, promulgates and enforces rules regarding the safety and durability of manufactured housing, including the establishment of fire safety standards to qualify for eligibility for HUD loans and mortgage insurance policies. http://www.hud.gov

Department of Interior.   The Bureau of Land Management (BLM), of the Department of the Interior (DOI), provides protection against wildfires on millions of acres of public lands and natural resources. It also supports the Interagency Fire Center in Boise, Idaho, which provides logistical support for a number of agencies including the U.S. Forest Service and National Park Service. http://www.blm.gov

Department of Transportation.   While the Department of Transportation (DOT) has undergone a number of changes in conjunction with the establishment of the Department of Homeland Security, many of its traditional functions will not change. Various operational units within the department investigate, research, analyze, and provide for the safety of the various forms of transportation. Given the number of emergency incidents that involve some form of transportation, the work of first responders is directly related to the initiatives of the Department of Transportation. http://www.dot.gov

Environmental Protection Agency.   The Environmental Protection Agency (EPA) is charged with protecting human health and safeguarding the natural environment—air, water, and land—on which life depends. The EPA works closely with other federal agencies, states, and local governments to develop and enforce environmental regulations. The agency plays a leadership role in our nation's environmental science, research, education, and assessment efforts. http://www.epa.gov

National Institute for Occupational Safety and Health.   The National Institute for Occupational Safety and Health (NIOSH) is responsible for conducting research and making recommendations for the prevention of work-related disease and injury. This research includes investigating potentially hazardous working conditions, making recommendations and disseminating information on preventing workplace disease, injury, and disability, and providing training to occupational safety and health professionals. Through its Fire Fighter Fatality Investigation and Prevention Program, NIOSH attempts to determine factors that cause or contribute to fire fighter line-of-duty deaths. http://www.cdc.gov/niosh

National Institute of Standards and Technology.   The Building and Fire Research Laboratory, within the National Institute of Standards and Technology (NIST), conducts extensive laboratory, field, and analytical research to predict and measure the performance of building materials, components, systems, and practices under various conditions including fire. http://www.nist.gov

Occupational Safety and Health Administration. The Occupational Safety and Health Administration (OSHA) was established to save lives, prevent injuries, and protect the health of America's workers. The regulations promulgated by this agency have an impact on response agencies across our nation. http://www.osha.gov

U.S. Forest Service. An organizational unit within the United States Department of Agriculture (USDA), the United States Forest Service (USFS) is charged with protecting and preserving the nation's forests and grasslands. A major challenge within this mission is protecting these valuable national resources from the threat of fire. The U.S. Forest Service works with other state and local resources in addressing this fire problem. http://www.fs.fed.us

U.S. Nongovernmental Organizations

There are many private organizations at the national level whose mission or initiatives address fire protection related issues. The successful fire service professional will benefit from becoming familiar with the following national organizations.

American Fire Sprinkler Association. The American Fire Sprinkler Association (AFSA) provides a full range of resources to those who work in the fire sprinkler industry. Examples of these services are training programs available to the fire service on topics related to the design, installation, and inspection of automatic sprinklers and information and materials available to assist local authorities and fire officials in the development of sprinkler ordinances. http://www.firesprinkler.org

Certified Fire Protection Specialist Board. Through a cooperative partnership with the National Fire Protection Association (NFPA), the Certified Fire Protection Specialist (CFPS) Board recognizes qualified individuals dedicated to curtailing fire loss. CFPS certification documents competency of and offers professional recognition to individuals with responsibilities dealing with the application of fire safety, fire protection, and fire suppression technologies. http://www.nfpa.org/ProfessionalDev/CertificationPrograms/CFPS/CFPS.asp

Congressional Fire Services Institute. The Congressional Fire Services Institute (CFSI) is a nonprofit, nonpartisan policy institute charged with educating Members of Congress on fire and life safety issues. The CFSI has achieved great success in building consensus both on Capitol Hill and between fire service organizations. The CFSI works closely with the Congressional Fire Services Caucus, the largest caucus within Congress, to keep fire service issues in the forum of discussion and legislative action. http://www.cfsi.org

Fire and Emergency Manufacturers and Services Association.   The Fire and Emergency Manufacturers and Services Association (FEMSA) is the trade association for the fire and emergency services industry, representing manufacturers and service providers. This organization offers a number of marketing, educational, legislative, representation, and communication services on behalf of its member organizations.
http://www.femsa.org

Fire Department Safety Officers Association.   The Fire Department Safety Officers Association (FDSOA) is a nonprofit association dedicated to promoting safety standards and practices in the fire and emergency services. Its purpose is to provide safety officers and their fire departments with the necessary tools to perform their duties through education.
http://www.fdsoa.org

Fire Service Section of NFPA.   The Fire Service Section of the NFPA was organized in 1973 and is open to all NFPA members who are active or retired members of fire departments providing public fire prevention, fire suppression services, and/or EMS services or who are involved in training and/or education of fire department members. The objectives of this Section relate to advancing the professionalism and recognition of fire service professionals.
http://www.nfpa.org/MemberSections/Fire-Service/fire-service.asp

FM Global Research.   FM Global Research is an entity within FM Global, an insurance organization committed to state-of-the-art loss prevention engineering and research. Representative research initiatives of this organization that provide useful information and guidance to the fire service professional include projects related to developing state-of-the-art active protection systems, identifying facility hazards, analyzing potential for materials damage, and evaluating flammability of products and materials.
http://www.fmglobal.com

International Association of Arson Investigators.   The International Association of Arson Investigators (IAAI) is a professional membership organization dedicated to the control of arson and related crimes. The diverse membership of this organization is representative of all disciplines, organizations, and entities involved in fire and arson prevention and investigation. The organization offers a full range of educational programs and technical services, coupled with its highly recognized and respected *Certified Fire Investigator (CFI)* designation and strategic partnerships with other organizations. http://www.firearson.com

International Association of Black Professional Fire Fighters. The International Association of Black Professional Fire Fighters (IABPFF) was organized to create a liaison between black fire fighters across the nation. The purpose of the organization includes promoting interracial progress within the fire service, the recruitment and employment of competent blacks throughout the fire service, and the motivation of black fire fighters to seek advancement within the ranks of the fire service. http://www.iabpff.org

International Association of Fire Chiefs. The International Association of Fire Chiefs (IAFC) is a professional organization open to all fire and emergency services administrators and managers, career and volunteer, interested in improving the delivery of fire and emergency medical services. The mission of this association is to provide vision, information, education, services, and representation to enhance the professionalism and capabilities of its members. http://www.iafc.org

International Association of Fire Fighters. The International Association of Fire Fighters (IAFF) is an affiliate of the AFL-CIO, which represents individuals engaged as permanent and paid employees of a fire department. Services offered to members include technical assistance in occupational safety and health issues, labor relations expertise, educational programs and seminars, legislative representation, and public relations guidance. http://www.iaff.org

International Code Council. The International Code Council (ICC) was established in 1994 as a nonprofit organization whose purpose was to develop a set of comprehensive and coordinated national model construction codes. The founders of ICC are the three regional model code organizations: the Building Officials and Code Administrators International, Inc. (BOCA), the International Council of Building Officials (ICBO), and the Southern Building Code Congress International, Inc. (SBCCI). These three organizations have merged to form the ICC. http://www.iccsafe.org

International Fire Marshals Association. The International Fire Marshals Association (IFMA) is one of the sections of the NFPA. Membership in this section is open to fire marshals, fire prevention officers, and other officials with legal responsibility for fire prevention or investigation. The objectives of the IFMA involve enhancing the preparation and effectiveness of its members. http://www.nfpa.org/MemberSections/IFMA/IFMA.asp

International Fire Service Accreditation Congress. The International Fire Service Accreditation Congress (IFSAC) is a nonprofit peer-driven association that accredits programs that issue certificates for emergency service training or academic degrees in fire-related fields. This organization was established to ensure the quality of training received by individuals in the fire protection field. http://www.ifsac.org

International Fire Service Training Association.   The International Fire Service Training Association (IFSTA) was established as a nonprofit association of firefighting personnel dedicated to upgrading firefighting techniques and safety through training. It fulfills this mission by validating training materials for publication, developing training materials for publication, and incorporating new techniques and developments while deleting outdated and obsolete methods. Fire Protection Publications, an entity of Oklahoma State University, was established to publish and distribute the training texts developed and validated by IFSTA. http://www.ifsta.org

International Society of Fire Service Instructors.   The International Society of Fire Service Instructors (ISFSI) is a nonprofit professional membership organization consisting of fire and emergency service training professionals. The members of this organization represent all ranks within the fire and emergency services and are involved in the development and/or delivery of training programs. The organization exists to provide its members with the tools necessary to enhance their ability to ensure the effectiveness, efficiency, and safety of those fire and emergency service personnel that they train. http://www.isfsi.org

Metropolitan Fire Chiefs Section of the NFPA and IAFC.   The purpose of the Metropolitan Fire Chiefs Section is to serve as a forum for fire chiefs from large, metropolitan fire departments to share information and affect policy changes. Members of this Section are fire chiefs from cities or jurisdictions having a minimum staffing strength of 400 fully paid career fire fighters. http://www.nfpa.org/MemberSections/Metro/Metro.asp

National Association of State Fire Marshals.   The National Association of State Fire Marshals (NASFM) represents the most senior fire official of each of the 50 states and the District of Columbia. The mission of this association is to protect human life, property, and the environment from fire and to improve the efficiency and effectiveness of state fire marshals' operations. http://www.firemarshals.org

National Association of State Foresters.   The National Association of State Foresters (NASF) is a nonprofit organization that represents the directors of the state forestry agencies of the 50 states, 8 U.S. territories, and the District of Columbia. The organization provides services to its members to enable them to provide management assistance and protection services of our nation's forests. http://www.stateforesters.org

National Board on Fire Service Professional Qualifications. The National Board on Fire Service Professional Qualifications (NBFSPQ) has the primary mission of accrediting fire service training agencies that use the NFPA's fire service professional qualification standards. This mission is accomplished through the establishment and oversight of a national system of fire service professional qualifications, certification, and accreditation. http://www.nfpa.org/ProfessionalDev/CertificationPrograms/NBFSPQ/NBFSPQ.asp

National Fallen Firefighters Foundation. The National Fallen Firefighters Foundation (NFFF) was created in 1992 by the U.S. Congress to honor our nation's fallen fire fighters and to provide the necessary resources to assist their survivors in rebuilding their lives. This nonprofit organization delivers a variety of much-needed and well-received services to the surviving families and departments of fire fighters who die in the line of duty. Support of this organization will ensure that it will always have the necessary resources to fulfill its mission—critical services. http://www.firehero.org

National Fire Protection Association. The NFPA was founded in 1896 to reduce the burden of fire on the quality of life by advocating scientifically based consensus codes and standards and research and education for fire and related safety issues. The NFPA offers a number of membership categories designed to meet the needs of the fire protection professional. Since its inception, the NFPA has demonstrated its responsiveness to meeting the needs of the contemporary fire service professional. Products and services available through NFPA include codes and standards, member publications, technical publications, public education, engineering, fire analysis and research, public fire protection, and public affairs. http://www.nfpa.org

National Fire Sprinkler Association. The National Fire Sprinkler Association (NFSA) is a trade association comprised of manufacturers and installers of fire sprinklers and related equipment and services. This association provides publications, seminars, market development, and representation in codes and standards making to its members. http://www.nfsa.org

National Volunteer Fire Council. The National Volunteer Fire Council (NVFC) is a nonprofit membership association that represents the interests of volunteer fire fighters, EMS, and rescue services in the national policy arena and fire fighters on numerous national and international committees. The organization plays a crucial role in keeping its members apprised of legislative developments and provides educational offerings that address contemporary issues that its members are facing. http://www.nvfc.org

North American Fire Training Directors.   The North American Fire Training Directors (NAFTD) was established to serve as a forum for the enhancement and enrichment of state, provincial, and territorial fire training and education programs and their managers. The mission of the organization is to focus the combined, diverse talents and resources of its members to achieve a safe environment through enhancing and supporting the role of training and education for the fire and rescue services. Membership is limited to the single individual from each state, province, or territory responsible for that entity-wide fire training and education program. http://www.naftd.org

Underwriters Laboratories.   Underwriters Laboratories Inc. (UL) is a not-for-profit corporation established to promote public safety by conducting scientific investigation, study, experiments, and tests of materials, devices, products, equipment, constructions, methods, and systems. Through its work, Underwriters Laboratories seeks to protect the public by reducing or preventing injuries, loss of life, or property damage. http://www.ul.com

Women in the Fire Service.   Women in the Fire Service is a nonprofit organization established to provide networking, advocacy, and peer support for women and information on women's issues in the fire service to all. The organization offers a variety of publications, consulting, workshops, and conferences on issues related to gender integration in the fire service. http://www.wfsi.org

See the *Fire Protection Handbook* for a list of organizations outside the United States [3].

▶ PUBLICATIONS AND OTHER RESOURCES

It is imperative that fire service professionals become familiar with the various fire service publications available and conscientiously consider the merits of each with respect to individual informational and research needs. The publications used the most fall into several categories.

Reference Books and Manuals

Reference books and manuals serve as a valuable resource to fire departments. Representative publishers of these reference and training sources include Delmar Publishing, Fire Engineering, Fire Protection Publications, the National Fire Protection Association, PennWell Publishing, and the United States Fire Administration. It is essential that fire service professionals develop a thorough understanding of these tools of the trade and ensure that there is ready access to them through the fire department, fire training academy, or personal library.

Trade Publications

While the reference books and manuals just referenced will serve as an informational foundation for work in the fire service, trade publications will prove to be an essential supplement. Using trade publications, published on an established schedule in accordance with the publication's editorial calendar, will give professionals the opportunity to keep current in the field. These publications may provide generic coverage of topics of interest to the fire service or may specialize in a particular fire service discipline. Fire service professionals should seek to become an informed consumer of those publications that best serve their personal and professional needs and ensure that they have the ability to access the appropriate publications in a timely fashion.

Among the useful publications are the following representative fire service publications:

- *Fire and Arson Investigator* http://www.firearson.com
- *Fire Chief* http://www.firechief.com
- *Fire Engineering* http://www.fe.pennnet.com
- *Fire Technology* http://www.nfpa.org
- *Firehouse* http://www.firehouse.com
- *Industrial Fire World* http://www.fireworld.com
- *National Fire and Rescue* http://www.nfrmag.com
- *NFPA Journal* http://www.nfpa.org
- *NFPA News* http://www.nfpa.org
- *The Instructor* http://www.isfsi.org

Internet Resources

The information age in which we live has provided access to many resources through the use of the Internet. The Internet allows fire service professionals to conduct research and stay current in their profession from the comfort of their home or fire station. The Web sites provided throughout this chapter are for learning more about the wealth of available resources.

Government Agency Web Sites. All of the federal governmental entities referenced in this chapter have robust Web sites that will be extremely useful in personal and professional endeavors. Fire service professionals will find the Web sites FEMA, the USFA, the NFA, and the EMI extremely useful. The resources of the Learning Resource Center, housed at the National Emergency Training Center, can be readily accessed through the Internet and will prove crucial for researching issues. Many of the governmental agencies at the local and state levels also have informative Web sites. The conscientious fire professional will make it a mission to discover these Web sites and visit them frequently.

Fire Service Organization Web Sites.   Each of the national fire service organizations discussed in this chapter has a useful Web site available for use that can be explored using the Web address provided. It is equally important to discover the local and state organizations that have Web sites related to professional responsibilities.

Fire Service Publication Web Sites.   Each of the fire service publications noted in this chapter, as well as others not mentioned, has a Web site that often provides information even more timely than that found in their publications. These resources will provide knowledge of important developments within the fire service. These available information sources should be researched and one or more should be selected that meets ongoing informational needs.

▶ SUMMARY

This chapter, by design, has introduced many tools of the fire service professional's trade. While these tools do not replace the fire protection systems, fire apparatus and equipment, and fire department personnel discussed in previous chapters, they are useful tools in enabling fire service professionals to prepare their fire department to respond effectively, efficiently, safely, and professionally to and return from each and every emergency.

In this chapter we have examined the roles of various local, state, and national support organizations. We have discussed the purpose of these organizations and how they can assist and support the contemporary fire department and fire protection professionals. The need for becoming thoroughly familiar with appropriate public and private support organizations has been a common theme throughout the chapter. As well, a variety of additional informational sources in the forms of fire service publications and Internet Web sites was introduced.

For those who are new to the fire service, most if not all of the information contained in this chapter was likely new. For those with some time in the fire service, we trust that we have expanded the reader's knowledge regarding resources available to personnel and fire departments.

The contemporary world in which we live and work contains numerous unprecedented and many continuing challenges for the fire service. The organizations profiled in this chapter are available and are prepared to assist those seeking knowledge or help. They represent tools of the trade that professionals must be intimately familiar with and use as the need arises. We challenge fire service professionals to become familiar with and utilize these support organizations.

Group Activity

Imagine that you are a fire fighter interested in promotion within your department. What sources or resources that are listed in this chapter would you use? Why? Discuss your individual answers in your small group.

Review Questions

1. If a local fire department plans to start a hazardous materials response team, who would they need to work closely with?

2. What are four types of organizations representing typical governmental organizations at the local level?

3. What are two government organizations and two nongovernment organizations at the state level that are a valuable resource to fire service professionals?

4. What are two government organizations and two nongovernment organizations at the national level that are a valuable resource to fire service professionals?

5. Name three fire service trade publications.

6. Which government agencies have Web sites that may be useful to fire fighters?

12 Careers in Fire Protection

Douglas Forsman
Union Colony Fire/Rescue Authority

▲ Learning Objectives

After completing this chapter, the reader should be able to do the following:

- Identify public and private sector jobs in the areas of fire protection and fire science
- Give examples of fire protection careers other than fire fighter/paramedic
- Explain the importance of education and training in fire service careers
- State the duties and responsibilities of fire fighter/paramedic
- Explain the hiring process of at least two public sector jobs
- Explain the hiring process of at least two private sector jobs
- Explain the role of career development for those already in the fire service
- Describe the testing and promotion process in the fire service
- Identify the instruments/processes used to rank fire service personnel for suitability for promotion
- Describe the value of diversity within the fire service
- Describe the role and value of professional certification in the fire service

---

As is the case in many broad fields of study, there are a number of specialty areas in which opportunities for careers exist in the field of fire protection. This chapter discusses a number of those opportunities and some of the routes to various careers. It also provides the reader with some guidance to confront employee selection criteria and issues.

In the next few pages the reader will find a strong relationship between training, education, and a successful career in fire protection, at literally any level. This emphasis on technical knowledge and skills are noted as the key

417

factors to landing and maintaining challenging positions in a diverse and changing world.

Perhaps the easiest way to examine fire protection career opportunities is to discuss them in terms of two broad categories: Public Sector and Private Sector. In addition, although it cannot be classified as a career, opportunities for volunteer service are also discussed.

► PUBLIC SECTOR FIRE-FIGHTING CAREERS

The fire service stands out as the most common fire protection career field in the public sector. According to NFPA estimates, there were about 1,078,300 fire fighters (all ranks) in the United States in 2001. Approximately 293,500 of them are full-time (career) fire fighter positions with the remainder being active part-time (call or volunteer). Although a few of these positions may be in the private sector, the vast majority are employees of local governments charged with providing for the health, safety, and welfare of their local residents. Unlike many nations around the world, local governments that are wholly separate entities from each other primarily operate the fire service in North America. In the United Kingdom, most European nations, and many South American countries, the fire service is operated at a national level.

Not all public sector fire departments operate in the traditional role of protecting a defined local community. In fact, the federal government, state government, and institutions operate many fire departments. The United States Department of Defense employs a large number of both civilian and military fire fighters to protect its bases around the world. Many federal and state prisons have their own fire departments as do places such as Department of Energy laboratories and some Veterans Administration hospitals.

Career Ladder System

Most fire departments hire their personnel into a career ladder system that includes the following rungs, in rank order. Those ranks listed below up to and including shift officer typically work one of several variations on a 24-hour schedule.

Recruit Fire Fighter.   This rank is the basic entry level. Most incumbents in this position spend a good bit of their first year in intensive training and are generally regarded as on probation for a period of time. They learn the general skills and knowledge associated with firefighting and usually emergency medical treatment. In addition, the recruit fire fighter will spend time learning the specifics of a particular organization and the community that it serves.

Fire Fighter.   This rank is generally a permanent rank in an organization, although it may have a number of pay steps within it. There is extensive training for this position as well. The skills and knowledge required to become certified as a Fire Fighter I and Fire Fighter II are described in NFPA 1001, *Standard for Fire Fighter Professional Qualifications* [1], which is shown in Figure 12.1. In addition to fire fighter duties, this position may well be assigned to special teams that require unique skills in areas such as hazardous materials, technical rescue, public education, fire inspection, and several others.

With a majority of fire departments in the United States now delivering emergency medical services, most fire fighters are required to become and maintain through much of their entire career, a certification as an emergency medical first responder of an emergency medical technician (EMT). This latter certification has three widely accepted steps that are know as EMT-Basic, EMT-Intermediate, and EMT-Paramedic. These three steps, in

▶ **Figure 12.1**   Cover of NFPA 1001, *Standard for Fire Fighter Professional Qualifications, 2002.*

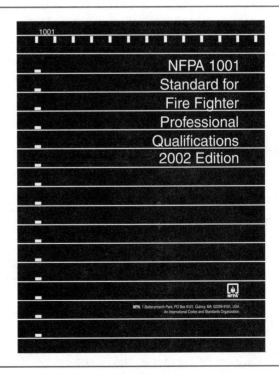

NFPA 1001
Standard for
Fire Fighter
Professional
Qualifications
2002 Edition

NFPA, 1 Batterymarch Park, PO Box 9101, Quincy, MA 02269-9101, USA.
An International Codes and Standards Organization.

themselves, require a great deal of training and education that lead up to certification or licensure.

Driver/Operator. Although not all fire services include this step in their career ladders, many do. Some fire departments may have the Driver/Operator rank (often called Engineer, Chauffer, Equipment Operator, etc.), but do not require its attainment as a prerequisite for further advancement. The Driver/Operator is most often responsible to safely and effectively drive and operate the sophisticated fire apparatus in use by fire departments today. The incumbent spends a good bit of training time learning how to operate fire pumps, hydraulic ladders, elevating platforms, and various specialized power tools. In addition, the Driver/Operator is typically responsible for the inspection and operability of a piece of fire apparatus and its assigned equipment. The specific qualifications for certification as a Fire Apparatus Driver/Operator can be found in NFPA 1002, *Standard for Fire Apparatus Driver/Operator Professional Qualifications* [2].

Company Officer. The company officer position is generally associated with the first line of supervision in the fire service. It is often titled as lieutenant or captain, or there may be both levels within a fire department. Company officers train for and have the responsibility of supervising the basic work group of the fire service, a fire company, which generally consists of two or three other employees who are fire fighters or driver/operators. When an organization has both captains and lieutenants, it generally implies that a captain will oversee a multiple-company fire station and have one or more lieutenants to supervise in addition to the captain's own fire company. The qualifications to be certified as a fire officer are contained in NFPA 1021, *Standard for Fire Officer Professional Qualifications,* where Levels I and possibly II would apply to the company officers [3]. Company officers are often required to maintain some level of EMT certification.

Shift Officer. This position is responsible for the supervision of all or a sizeable portion of the on-duty fire companies in a fire department. Several company officers report to the shift officer. This position, sometimes designated as battalion chief or shift officer, is generally assigned to work the same hours as fire company personnel, but is also expected to perform a number of administrative and management tasks that may well require extra time. The qualifications for certification at this level of fire officer are again found in NFPA 1021, where Level II or possibly Level III would apply [3].

Division Officer. The division officer in a fire department, who is also sometimes known as the assistant chief or deputy chief, would typically be assigned to manage a large function within the department such as fire prevention, field operations, emergency medical services, emergency management services, support services, or administration. However, in many medium and small fire departments, the division officer may be required to manage several of these areas or the responsibilities may even be assigned to a shift officer. Officers assigned at the division level generally work standard business hours and may well be on call for large emergency events after normal hours. Qualifications for this level of fire officer are again found in NFPA 1021, where Level III would apply [3].

Fire Chief.   This position is the senior management official in most fire departments. The fire chief usually reports to a city manager, mayor, or a special district board of directors. The position has the ultimate responsibility for the management of the fire department and in that role supervises whatever management officers are in place. The extensive qualifications required for this position are described in NFPA 1021, under Level IV [3].

Job Requirements

Entry-level career fire service positions are highly coveted by many young men and women. The pay, benefits, and stability are regarded as attractive in most communities. However, many people are attracted to the positions because of the level of personal reward that comes from helping people in need. Whatever the reason, those wishing to enter the fire service need to be aware of the process necessary to obtain employment in their desired location. Most communities have some combination of the following elements in their requirements and selection process. However, the order in which the steps are accomplished varies widely.

Minimum Age.   For reasons of physical development and mental maturity, most fire departments require that a candidate for fire fighter employment must have attained a certain age before that candidate either applies or is hired. That minimum age may vary from community to community, but generally is in the range of 18 to 21 years. Some communities have excellent cadet programs that allow younger people to participate in limited roles that are not usually compensated. These cadet or explorer programs are often a great way to get a head start on the training and education required for a fire service career.

Education.   Fire departments typically desire employees who have demonstrated achievement in academic pursuits. Accordingly, there generally is a requirement for a minimum attained level of education such as a high school diploma, a general education diploma, or some level of college work. Applicants with advanced and/or strong educational credentials will likely have some measure of advantage.

Certifications. Depending upon the breadth of the candidate pool, some communities may require certain certifications in order to be considered for employment. The most commonly required certifications are Fire Fighter I and/or EMT-Basic. In a few cases, communities that offer advanced life support services may well require that employees hold an EMT-Intermediate or an EMT-Paramedic certification. Regardless of the minimum requirements, the candidate that brings previously attained certifications to the application process is attractive to potential fire service employers.

Criminal Background Check. The fire service is clearly a position of public trust and thus requires people of good character. In order to ensure that candidates for fire fighter positions can meet community expectations for trust, a thorough criminal background check is usually accomplished during the hiring process. Community standards for the results of background checks differ and a candidate should seek out answers to any questions in this area.

Written Test. Most communities require candidates for fire fighter positions to complete a written test. Although the type and purpose of the tests vary depending on their source, most communities are interested in testing a candidate's general knowledge as well as personality traits to see how well the candidate matches the profile of a successful fire fighter. Many communities offer some sample questions or other information regarding their particular test prior to a candidate taking the test.

Oral Interview. Nearly all communities use some form of oral interview for screening employment candidates. The interviews may be one-on-one with a chief officer or may be done by a group of people. Oral interviews typically are intended to measure several skills and traits of the candidate. Some of those skills and traits include but are not limited to judgment, motivation, communication skills, knowledge, social appropriateness, and leadership. It is difficult to prepare or study for an oral interview. However, most candidates will do as well as possible if they are properly rested and have a positive attitude. Some communities may choose to conduct more than one oral interview. In those cases, the later oral interview may well be the deciding factor in which individuals are hired.

Agility/Fitness Test. Firefighting is a difficult and physically taxing job. Figure 12.2 shows a typical drill tower used for fire training at the University of

▶ **Figure 12.2** Drill Tower at the University of Kansas. (*Source: Fire Protection Handbook, NFPA, 2003, Fig. 16.9.*)

Kansas. Communities have an obligation to sort out candidates who are not capable of doing the specific job related tasks. This separation is usually accomplished by a series of simulations in a timed test environment.

Although a number of tests and test scenarios exist, many communities have adopted the Candidate Physical Ability Test or CPAT. The CPAT test was developed in the mid 1990s as a joint project among the International Association of Fire Fighters (the labor union that represents most career fire fighters in North America), the International Association of Fire Chiefs, and several metropolitan fire departments. This well-designed and validated test involves the timed completion of several workstations that simulate common physical tasks in firefighting. It is a test not only of an individual's strength, but also his/her agility and endurance. Licensees who use the test must agree to a strict set of conditions that include informing candidates well in advance as to every aspect of the test and how they might prepare for it.

Physical Examination.   Almost without exception, a comprehensive physical examination is required before an individual is hired. This examination not only helps to determine if an individual's health is sufficient to allow firefighting duties, but it also establishes a baseline set of data that can be used to detect any health problems that may be encountered over the length of a career. Typically, annual physical examinations are required throughout a fire service career. For more information, see NFPA 1582, *Standard on Comprehensive Occupational Medical Program for Fire Departments* [4].

Other.   Two items worthy of mention are the psychological profile and the polygraph test. The psychological profile is a written test, an interview by a mental health professional, or both. Its purpose is to determine the general suitability of an individual for the employment he or she is seeking. It is based on what is perceived to be the profile of a successful employee. The polygraph or lie detector test is a simple test using a machine that will indicate whether an individual is telling the truth about questions that pertain to past conduct or declarations made in the pre-employment process. Both of these screening tools are used by some agencies. However, they are both quite controversial with respect to their accuracy and the ethics of their applicability.

In many communities, the recruitment and selection process for new employees is a function of a Civil Service Board or similar entity. These boards were developed to insulate the selection process from any political forces that might be in play in a given community. Accordingly, candidates should not be surprised if the fire department is not responsible for the selection process in some cases. Civil Service Boards typically work very hard to ensure that the selection process is relevant and completely fair.

Nepotism policies also need to be noted by candidates. *Nepotism* is roughly defined as the practice of showing favoritism to a relative in the scope of employment. Policies designed to preclude nepotism are common in both the public and private sector. Nepotism has long been an issue in some fire departments as the desire to be a fire fighter is passed down through generations. Candidates should be aware of nepotism policies of any potential employer. Many policies are well designed not to discriminate, but to preclude relatives supervising relatives. These policies are ultimately good for the employee and the employer.

Hiring Process

Fire service employers have learned over the years just how important it is to recruit and hire quality personnel. According to the United States Bureau of Labor Statistics, fire-fighting jobs are among the top 10 career fields in employee longevity. Most employers understand that the relationship between a community and its fire fighters is, for the most part, a long-term investment in each other.

Even more important to the employers is the desire to find those employees who will offer the best chance of a successful performance in providing quality customer service. Written tests and oral interviews often focus on attempting to measure a candidate's ability to show compassion to people in need as well as to be extremely competent technically.

Screening.   Communities have an obligation to conduct a screening process that is open, fair, and compliant with applicable laws. Most employers spend many dollars advertising for candidates and then screening them to find the very best. They constantly work to find testing instruments and a process that achieves community objectives and complies with the law.

Candidates should understand that the process is complex and may vary from place to place. Usually, there are certain criteria that are purely pass or fail, and these are typically the minimum qualifications such as age, education, criminal background, physical agility/fitness, and physical examination. However, the employer has an obligation to provide the standard to which a candidate is being measured in each of these areas.

Scoring.   The written test and oral interview(s) are often evaluated using a numerical scoring system that establishes a rank order eligibility list of candidates. When this is the case, candidates must understand that there may be a final selection process that allows the fire department to choose from the top 5 or 10 (or more) names on the list. Often this final screening may be accomplished with the aid of another oral interview that is not part of the numerical ranking process for establishing the eligibility list.

Preference Points.   One further point to consider is the matter of preference points. Candidates need to be aware of any preference points that may be given in establishing an eligibility list. These points are typically given for honorable service in the military or for being the descendant of a fire fighter killed in the line of duty. Although there is a trend away from using preference points, candidates should be aware that using any applicable preference points for which they might be eligible can make a significant difference in the numerical scores.

Residency. Finally, the matter of residency within the political boundaries of a potential employer can be an issue. Some communities require that candidates must establish residence prior to making application or that they must live in the jurisdiction prior to being hired. Others may allow several months or a year after hiring to move into the political area. Others require that employees must live within so many minutes of travel time or so many miles. Still others do not have any residency requirements. Understanding these restrictions, if any, is important prior to application.

In general, landing a career position in the fire service requires perseverance, preparation, hard work, and a strong desire to be successful. The reward for success is a highly respected and valued career in a personally rewarding field.

Examples of Hiring Process

City Level. A medium-size city in Illinois maintains a fully career fire department with several stations. A board of fire and police commissioners on which the fire chief sits as an ex-officio member governs its hiring process for entry-level employees. The screening and hiring process is conducted as a cooperative effort of the fire department and the city's personnel department, all under the auspices of the board.

Candidates must be 21 years of age at the time of hiring, but may apply earlier. Because there are typically large numbers of applicants for a few fire fighter positions, the entire pool of candidates is given a written test that is purchased from a private firm that has done extensive validation studies. The test is a basic knowledge test and has no questions relating to the fire service. Those who pass the test (score 70 percent or more) go on to a locally designed physical agility test that is scored on the basis of pass/fail. Those candidates that pass the physical agility test then participate in a group oral interview with several other candidates. Members of the board are trained as assessors and accomplish the weighted scoring.

These numerical scores from the written and oral tests are combined and provide the basis for the rank order list. Up to 5 points can be added for service with the United States military.

When the final list is published and a vacancy occurs, the fire chief may select from the top five people at his/her discretion considering their education, training, and experience. After being selected, candidates must then pass an extensive physical examination, a psychological screening, and a criminal background check. Successful candidates are then sent on to a fire academy for training.

District Level.   A fire protection district in Colorado has an extensive hiring process that it coordinates along with representatives from its collective bargaining unit and a city's human resources department. There are usually many candidates for a few positions and the test is given every 2 years.

Candidates must be 21 years of age to apply and must have a certification as an Emergency Medical Technician-Basic prior to employment. They must also live within a stated distance of the district boundaries within 1 year after being employed.

The district's testing process starts with all candidates taking the CPAT as it is administered locally. Candidates pay a small fee to take this test. Those who pass the test take a written test the next day. The written test is based on a basic fire fighter text. In both the physical ability test and the written test, candidates are informed well in advance as to how they should prepare.

The cutoff score for the written test is established by the ratio with the number of positions currently available. Those candidates above the cutoff score move on to a set of individual oral interviews with members of the fire department leadership. They are assigned a score from the interview process and that score combined with the written score forms the rank order listing.

Hiring is accomplished from the list in which the fire chief (or his/her designees) can pick from the top 10 people. A second round of interviews may be conducted among the candidates to make this final selection. The final choice can also take into consideration the candidates' credentials and experience. Finally, candidates must pass a physical examination and a criminal background investigation.

▶ PUBLIC SECTOR NONFIRE-FIGHTING CAREERS

There are a number of fire protection career fields in federal, state, provincial, or local government that do not involve actual fire-fighting duties. Although individuals who were reassigned from firefighting may accomplish these jobs, there are many cases in which employees have been hired specifically for these duties. Examples of these careers are as follows.

Career Opportunities

Fire and Life Safety Educator.   This position plays an important role in the fight against injury and death from all accidental causes. The fire and life safety educator often works with a broad spectrum of a community to define the fire and life safety issues in a community and then to address those issues in a multifaceted approach. While early education efforts on the part of the fire service were focused on the fire problem, most programs now attack the entire injury problem using tools such as NFPA's *Risk Watch*® program. Qualifications for certification as a fire and life safety educator can be found in NFPA 1035, *Standard for Professional Qualifications for Public Fire and Life Safety Educator* [5].

Fire Inspector/Code Enforcement Officer. Fire inspectors are usually responsible for the inspection of new or existing buildings for compliance with highly technical fire protection and safety-related codes and standards. They are part of the team that ensures a safe environment for the public, not only from fire but from other perils as well. Most fire inspectors attempt to achieve compliance on the part of the public through persuasion and the desire of most building owners to provide a safe facility. However, they typically do have enforcement powers and thus need to be aware of proper legal procedures. In smaller organizations, the fire inspector may well have roles in fire and life safety education and/or fire investigations. Most fire inspectors work normal business hours. The qualifications for certification as a fire inspector can be found in NFPA 1031, *Standard for Professional Qualifications for Fire Inspector and Plan Examiner* [6].

Plans Reviewer. This job generally entails the detailed review of architectural and/or engineering drawings for compliance with applicable fire and life safety codes, standards, and laws. The plans may apply to new or renovation projects and may encompass very simple to complex problems. Plans reviewers must have a strong understanding of fire protection and building technology as well as codes and standards, all gained either through formal education or extensive experience or both. The qualifications for certification as a plans reviewer may be found in NFPA 1031, *Standard for Professional Qualifications for Fire Inspector and Plan Examiner*, specifically in the section for Plans Reviewers [6].

Fire Investigator. Establishing the cause and origin of fires is the primary function of fire investigators. They use techniques learned in extensive training to establish forensic evidence for determining what caused a fire. In a large fire, the fire investigator may be on-site for many days, gathering information from witnesses and physical evidence in an attempt to rule out causes and eventually establish a cause and origin. In some states, provinces, or communities, fire investigators may have full police powers. However, in most cases, they will work with the police department when that is deemed necessary. The qualifications for certification as a fire investigator are found in NFPA 1033, *Standard for Professional Qualifications for Fire Investigator* [7].

Fire Protection Engineer.   This highly technical position is becoming more common in the public sector fire services at both the local and state/provincial level. Fire protection engineers work with architects, engineers, developers, builders, and others to solve complex fire protection and safety issues. They may also work with fire department management in addressing planning for resource deployment and facilities. Only those individuals who have attained academic degrees in engineering or engineering technology are usually considered for these positions. In most jurisdictions, a professional engineer's license is required. For further information, see *The SFPE Handbook of Fire Protection Engineering* [8] (or the Society of Fire Protection Engineers Web site at http://www.sfpe.org).

Training Officer.   Positions in the field of fire service training exist at several levels and in both state/provincial and local settings. Most fire departments have one or more persons designated as the training officer. Their responsibilities likely include the planning, scheduling, coordinating, delivering, and evaluating of training for all departmental personnel. Because the fire service is necessarily training intensive, there is considerable work in this area, even in small organizations. If the training division for a given entity has several people assigned, it is likely that there will be a training hierarchy and thus a career ladder in this specific area. Most states in the United States and provinces and territories in Canada operate supplemental training programs for constituent fire departments. These agencies also offer public sector careers in fire service training. Qualifications for fire service trainers and positions within a training division hierarchy are described in NFPA 1041, *Standard for Fire Service Instructor Professional Qualifications* [9].

Wildland Fire Fighter.   Some states, provinces, and federal governments hire either seasonal or full-time employees to fight wildland fires. This labor-intensive activity requires a set of skills separate from those of structural firefighting. These positions require a high level of physical fitness and may result in extended periods of long shifts at large fires. Qualifications for certification in several levels of wildland fire fighter certification can be found in publications produced by the National Wildfire Coordinating Group and to a more limited degree in NFPA 1051, *Standard for Wildland Fire Fighter Professional Qualifications* [10].

Hiring Process

In any of the preceding areas in which large enough organizations exist, there may well be a career ladder potential. With the exception of the fire investigator standard where only one level exists, the standards referenced provide qualifications for increasing levels of competence and responsibility. The process to access upper level positions in most of these public sector fields usually involves a well-defined promotional procedure.

The entry-level requirements for the career fields discussed in the preceding section may encompass many of the same elements as those described for firefighting. Typically, however, there is not the rigorous physical agility/fitness or physical health standards that apply to fire fighters. Candidates should be careful to examine all of the requirements and the elements of the selection process to make sure that there are no surprises or misunderstandings.

In a number of cases, the nonfire-fighting jobs may be filled through promotions or transfers from fire-fighting positions. And, in some cases, candidates from both fire-fighting positions and from outside an organization may be competing for these positions. In either case, there is usually a defined process of testing, interviews, and selection that must be understood by all involved.

▶ PRIVATE SECTOR FIRE PROTECTION CAREERS

Fire protection careers in the private sector are many and varied. For the purposes of this chapter, the private sector refers to both for-profit and not-for-profit entities other than those discussed in the public sector portion of this chapter. As is the case with public sector positions, fire protection personnel are typically engaged in saving life and property from fire and related causes. The rewards, both personal and monetary, make the private sector an excellent career option. The following are a few of the fields with fire protection career opportunities.

Fields with Career Opportunities

Industry.   Industry is a very broad term and in the context of this discussion, refers to manufacturing, distribution, and service sectors. The fire protection career options in industry are focused primarily in the loss prevention area. Many businesses, small or large, have one or more individuals responsible for the following activities:

- Fire prevention inspections
- Fire prevention education
- Fire protection system inspection and maintenance
- Process hazard evaluation and protection
- Fire investigation and hazard mitigation

- Contractor monitoring
- Fire insurance carrier liaison
- Fire brigade participation
- Fire brigade leadership

Particularly in smaller industries, these and other safety related functions are often combined in a few or even one designated position. Examples of other job duties that may be combined with these duties as well are an insurance manager or a human relations manager.

The qualifications for fire protection positions in industry are not always as clear or consistent as they are in the fire service. However, there are some standards that will help to understand some of the specific skills and duties. Fire brigade member functions are described in NFPA 1081, *Standard for Industrial Fire Brigade Member Professional Qualifications* [11]. Additional information regarding job functions can be found in a number of other NFPA standards and in documents available from the U.S. Department of Labor, and the Occupational Safety and Health Administration.

It is also noteworthy to mention the difference between the qualifications for industrial/institutional fire brigades and fire departments. As a point of explanation, public fire service in many countries outside of North America are often called *fire brigades*. However, for the purposes of this discussion, the term *fire brigades* refers to those operations within an industrial or institutional complex that provide emergency response primarily to the complex. Fire brigades, as defined in NFPA 600, *Standard on Industrial Fire Brigades* are divided into incipient, interior structural, and exterior structural categories [12]. The incipient level requires only minimal training and assumes that only small fires will be attacked using personnel with fire extinguishers or small hoses, without the benefit or need for personal protective equipment. The latter two categories require more extensive training and assume that brigade members will make substantial fire attacks. Although the training in many ways resembles that given to municipal fire fighters, there are subtle differences. The certifications for industrial and standard fire fighter positions are not necessarily interchangeable.

Health Care.   This often highly regulated industry is truly a field unto itself. Although many of the job categories mentioned in the previous section for industry apply here, the regulations differ and are generally very stringent. Fire protection positions in the health care industry are usually associated with a large medical facility or with the corporate office of a group of medical facilities.

Insurance.   The fire insurance industry has long recognized the importance of good fire protection practices. They have for many years hired and even trained fire protection engineers, consultants, and inspectors to assist in loss control efforts. These positions typically work with a company's clients and underwriters to establish fire insurance rates and to assist clients in achieving the best possible fire loss experience. Many of these positions involve extensive knowledge of NFPA and insurance industry standards. While no specific standard exists for job qualifications, many of these positions require either licensure or certification as a fire protection engineer, certified fire protection specialist, or a Certified Safety Professional.

Fire Protection Systems.   Fire protection system designers and fire protection engineers and technicians are often required by architectural and engineering firms. These companies along with some design/build firms are responsible for the design of most sprinkler and other fire protection systems in modern buildings. In addition, many fire protection system installation and service companies hire qualified technicians. All of these companies are responsible to make sure that the proper fire protections systems are installed and maintained in relation to the given building and its intended use. The highly technical field of system design offers excellent rewards for work that involves the application of engineering principles in designing just the right system for a particular fire protection problem. The installation and service positions also play an important role in ensuring the proper performance of fire protection and life safety systems. Qualifications for these positions require strong skills and knowledge in the physical sciences including hydraulics, fire behavior, heat transfer, and others. The installation and maintenance positions also require a good sense of mechanics. Certification or licensure as a fire protection engineer or certified fire protections specialist and/or fire protection system installer is required in most cases.

Building.   Many large design and construction firms employ their own fire protection staff. Individuals in these positions often coordinate the many aspects of projects that deal with fire and life safety, checking for compliance on the part of contractors and the correctness of design. These positions offer a good mix of field and office work. They are usually financially rewarding as well as offering the opportunity to see projects carried through to completion. While licensure or certification may not be required, they are usually desirable and will likely make a candidate more desirable. A good sense of fire protection principles and the construction process are essential.

Hiring Process

Most private sector companies recruit and hire fire protection and life safety positions in the same manner as they do for nearly all positions. Interested individuals should regularly review professional journals, employment journals, newspapers, Web sites, and other sources of position-available postings. In addition, companies often use recruiting firms for engineering and other professional level openings. Finally, larger companies may regularly recruit individuals by visiting with graduating seniors through college placement offices on a particular campus.

Candidates for these private sector positions should make sure they have the education, training, and skills necessary for the position in which they are interested. In addition, it is wise to prepare further by staying abreast of technology through memberships in associations such as the National Fire Protection Association and the Society for Fire Protection Engineers and reviewing the literature in their specific fields. Internships or other positions that provide a logical career experience ladder to a more desirable position make a great deal of sense. Most candidates will find that a quality resume and a set of interviews will be the beginning of the hiring process. Most employers will check references before offering a job.

▶ VOLUNTEER OPPORTUNITIES

This chapter is intended to discuss career opportunities in fire protection. Although volunteer opportunities probably do not fit the definition of a career, the discussion would be remiss if volunteerism were not discussed.

The public fire service represents an excellent chance for motivated people to volunteer in the cause of fire protection and life safety, no matter what their vocation might be. Arguably, nearly 80 percent of the fire fighters in North America serve in that role as volunteer or paid-on-call personnel. Volunteer fire fighters have always taken a great amount of pride in the quality of service that they provide to their friends and neighbors throughout North America's smaller communities. However, today's volunteers have a great deal of pressure to maintain specialized skills and training and the various certifications required by a myriad of governments. The pressure of the time required to effectively train and function as a volunteer fire fighter has created difficulty for some communities in finding volunteers. That being said, there are many excellent volunteer fire organizations that offer opportunities to willing and qualified people in this personally rewarding endeavor.

Many young people take advantage of these opportunities to obtain training and develop skills that will make them more qualified for career fire service positions, a win-win scenario for the volunteer, the community, and his/her eventual employer.

▶ CERTIFICATION AND LICENSURE

The need for certification and licensure in certain fields of practice has been mentioned in several areas. Understanding the process and requirements for these professional benchmarks is important to prospective candidates. While the details of each category are too cumbersome for inclusion in this text, the basic differences are worthy of explanation.

Certification for fire service job levels, such as fire fighter, fire officer, or fire instructor, and emergency medical service levels, such as EMT-Basic, EMT-Intermediate, or EMT-Paramedic, are often accomplished through a process of testing sponsored either by a local government or a state or provincial entity. Accredited testing processes require the successful completion of written and skills examinations, many of which are based on the job performance requirements set forth in NFPA Professional Qualification Standards or nationally recognized emergency medical curricula.

Certification as a Certified Fire Protection Specialist is available through NFPA. This comprehensive program requires the completion of a study regimen and a comprehensive examination that is given periodically throughout the year.

The National Institute for Certification in Engineering Technologies (NICET) offers a number of certifications in the fire protection system design, installation, and maintenance fields. NICET certifications require testing, documentation of work history, and recommendations. A similar program is available from the Board of Certified Safety Professionals. This Certified Safety Professional™ program is more focused on the broad spectrum of safety and requires a comprehensive program of study, experience, and examination.

Licensure as a fire protection engineer is available in a number of states, provinces, and countries. As a field of practice in engineering, the licensing process and agency for fire protection engineers is consistent with other engineering fields in a given jurisdiction (usually a state). Reciprocity among states, provinces, and countries may exist, but is not always the case.

EDUCATION AND TRAINING

Throughout the previous discussion of career opportunities, the concepts of training and education are mentioned as important preparation. Across the spectrum of fire protection, emergency medical, and life safety positions, there are extensive education and training requirements, not only for entry-level jobs, but also on a continuing basis during a career. In addition to programs offered through traditional college, community college, and vocational school sources, some specialized agencies offer focused training programs. In nearly every state and province and in many countries there is an agency charged with fire service training. These agencies may be housed in a fire marshal's office, a university, an insurance regulatory agency, or other departments of government. The United States National Fire Academy and its counterparts in other countries offer excellent programs of training and education primarily focused on career advancement for existing fire service employees and volunteers.

Education

There is a subtle difference between training and education. For the purposes of this discussion, *education* can be considered an organized program of study that focuses on concepts, theory, and applicable sciences. Education may be in a traditional mode, such as classroom lecture, laboratory work, etc., or nontraditional mode, such as distance learning, Web-based programs, etc. However, education should be from an accredited school, college, or institute and result in a diploma, certificate, or degree.

Having an appropriate level of education is attractive to many potential employers. It demonstrates a level of interest in a specific technical area. In addition, the successful completion of a program of study typically signals to an employer that an individual has the interest and initiative to undertake and complete a challenge. Accordingly, some employers may ask for a college degree in literally any field of study. However, a strong education program in fire protection, emergency medical services, or a related field will provide the scientific, social, and skills base for the positions addressed in this chapter.

Education is a life-long process. Many people decide to seek a college degree or an advanced degree after they have been employed in their selected field, usually to advance their understanding of concepts and/or prepare themselves for upper level positions either in an organization or a specific field. In some cases, continuing education credits are required to maintain specific certifications or licenses.

Training

Training, again for the purposes of this discussion, can be considered a focused and specialized program associated with a field of practice, skill, or task. Training usually includes the learning of specific skills and the practical application of those skills as practice. It may also be delivered in either traditional or nontraditional modes and may result in a certification.

Training for a specific job classification or even for a specific position is a good strategy for employment candidates. As mentioned earlier, some fire departments or other employers require that applicants complete certain training prior to applying for a position. For example, it is not uncommon for a fire department to require candidates to be credentialed as a Fire Fighter I and/or Emergency Medical Technician-Basic prior to hiring.

Many positions will require specific training and refresher training throughout an employee's tenure. Maintenance of skills associated with high-risk, low-frequency functions is a particularly important aspect of continuing training. Specific activities such as hazardous materials abatement and technical rescue fall into this category. Maintaining special and basic skills and knowledge through training is an important investment for the employer's role of providing quality service and the employee's ability to operate effectively and safely.

Particularly in relation to training, it is very important for both the individual and the employer to maintain thorough records. Employers may well be obligated by law or by their insurance carriers to retain proof of certain levels of competence for its employees. And, it is to the advantage of employees to maintain records of training achievements in order to apply for employment pay steps or promotions. Both training and education are about safety, effectiveness, and customer service. These three concepts are important to the success of both the employer and the employee.

Fire and Emergency Services Professional Development Model

The U.S. Fire Administration (USFA) hosts the annual Fire and Emergency Services Higher Education (FESHE) conference on its campus in Emmitsburg, Maryland. These conferences are a combination of presentations, problem solving, and consensus-building sessions that result in higher education-related products or recommendations for national adoption.

The professional development model is one product finalized at the 2002 FESHE IV conference. See Figure 12.3. It is not a promotion model ad-

National Professional Development Model

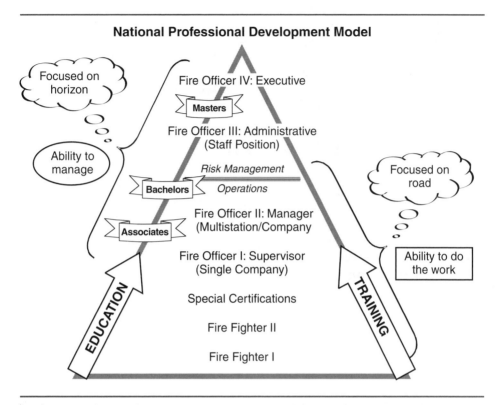

▶ **Figure 12.3** Professional Development Model

dressing credentials; rather, it is an experience-based model that recommends an efficient path for fire service professional development supported by collaboration between fire-related training, higher education, and certification providers.

In 2001, the National Fire Science Curriculum Committee (NFSCC) was formed to develop standard titles, descriptions, outcomes, and outlines for each of the six core courses. In 2002, the FESHE IV conference attendees approved the model courses and outlines. The major publishers of fire-related textbooks are committed to providing texts for some or all of these courses. *Fundamentals of Fire Protection* is based on the FESHE objectives for the first of those core courses, also called Fundamentals of Fire Protection.

It was recommended that all fire science associate degree programs require these courses as the theoretical core on which their major is based. The course outlines address the need for a uniformity of curriculum and content among the fire science courses within the United States' 2-year programs. Many schools already offer these courses in their programs, while others are in the process of adopting them. Once adopted, these model courses address the need for problem-free student transfers between schools. It also promotes crosswalks for those who apply their academic coursework toward satisfaction of the national qualification standards necessary for fire-fighter certifications and degrees.

► PROMOTIONAL OPPORTUNITIES

Looking ahead at the promotional opportunities with an employer is a good idea, even at the time of initial hiring. Most people want to move into more responsible and financially rewarding positions as they mature in a career field and/or organization.

Factors in Promotions

Many fire departments will offer five to seven ranks in an organization, all requiring a different level of preparation and competitive testing. The private sector also offers many challenging opportunities for advancement, however, there is usually less structure to the promotional process. In both the public and private sector, there are certain career areas that are reasonably "flat," that is, there may only be one or two supervisory levels that require a given set of technical skills. However, if employees are willing to expand their horizons in terms of skills and knowledge, there will likely be many opportunities. The following are some key things to think about in relation to moving up the ladder.

Initiative.   People who take the initiative to obtain additional training and education may have an advantage. Doing so demonstrates to employers a serious desire to better one's self and to explore interests that could be important to the organization. Doing so without being forced into the effort sends great signals to decision makers.

Work Ethic.   Having and demonstrating a strong work ethic is important. If an employee willingly works hard for the organization and goes the extra mile, that person is likely to be fondly viewed in the subjective portions of any promotional process. Taking on a task that will result in a project being completed for the department and some valuable new knowledge being gained for the employee is a win-win situation.

Certifications.   Working toward and obtaining advanced certifications shows initiative, provides advanced knowledge, and may save money for an organization. Many fire departments are willing to let employees certify at a level above their current positions, just to help them prepare for promotion. The NFPA Professional Qualification Standards are, for the most part, divided into benchmark levels that describe the job performance requirements for various ranks or positions. For instance, NFPA 1021, *Standard on Fire Officer Professional Qualifications*, has four certification levels [3]. Level 1 or 2 is designed for company and station officers. Level 3 is designed for middle management and divisional positions. Finally, Level 4 is designed for the highest-ranking chief officers. Gaining advanced certifications is a clear indicator that a person is preparing for a promotion.

Reading and Studying. There is no substitute for reading and studying. Advanced positions require advanced knowledge and skills that are undoubtedly available in many good publications and other sources. Reading and studying for a promotional test would seem to be a requisite to do well for most people.

Willingness to Work Out of Grade. Accepting opportunities to work out of grade should strengthen an employee's knowledge of the next level up in an organization. Accomplishing the boss's work and feeling a different level of pressure if just for a limited time are great preparation for assuming the job permanently.

Experience with the Selection Process. Employees who would like to be promoted should experience the selection process whenever possible. Some organizations have prerequisites that must be met in order to compete for a promotion. As soon as they are met, employees should begin experiencing the tests, interviews, and other elements of the promotional selection process. Although it may be frustrating to fail a few times, the experience is invaluable.

Promotion Process Elements

Understanding the promotional process used by a given employer is important to the potential promotee. In the private sector and some public sector situations, the process may be as simple as a selection being made by a senior manager based solely on past work experience and/or an interview. However, in the fire service, there is usually a much more involved process that may be operated by the fire department, a human relations agency, a consulting firm, or a civil service board. Some of the following elements may be included:

Tenure in Rank.   Many organizations require that a candidate for promotion must have attained a certain lower rank and been in that position for a period of months or even years, prior to taking a promotional test. For instance, a department may require an employee to be a lieutenant for one year before taking a captain's exam.

Certifications.   Certification or licensure at a certain level might be required prior to taking a test.

Past Performance.   Some testing processes specifically require the evaluating authority to take into consideration past performance evaluations. Conversely, others specifically prohibit such consideration.

Written Tests.   It is not uncommon for a promotional process to include a written test. Hopefully, the test has been scientifically validated for the position and will be a good measurement of the skills and knowledge needed to perform the job. When a written test is used, it is common to have a reading list published so that candidates may study the materials used to construct the exam.

Assessment Centers.   Some employers use a process called an assessment center. It presents a number of scenarios and settings that produce joblike simulations under reasonably stressful time constraints. Candidates are observed by a group of trained assessors who assign weighted scores for such things as judgment, leadership, and time management. Typically the elements of an assessment center would include some role playing, an in-basket exercise, and an opportunity to demonstrate leadership.

Oral Interviews.   Either as a function of the testing process or as a final element of selection, many promotions require an oral interview. It is usually conducted with a group of evaluators either from within or outside of the organization or with the individual who will ultimately make the promotional decision.

Research/Position Paper.   Some organizations may require that candidates submit a research paper, position paper, or other written exercise. This paper is a valuable tool in assessing a person's ability to write and communicate well.

Particularly in the fire service, the results of promotional testing will likely be a rank order list of candidates who successfully complete the testing process. The promoting authority may be compelled to take the first person on the list, but more often would be given the choice of the top three or more.

Workplace Diversity

Particularly in the United States, diversity in the workplace is an important issue. Since the mid 1970s, federal rules and guidelines have been in place to either encourage or force organizations to diversify their workforce on the basis of race and gender. While the courts continue to render decisions on such programs as Affirmative Action, there remains in most communities an honest desire to diversify the workforce with quality people.

Candidates for both entry-level and promotional opportunities should be keenly aware of the impact of any policy regarding diversity. They are typically not intended to preclude anyone from an opportunity to compete and to be successful, but to make sure that everyone has a chance to compete and be selected.

 SUMMARY

There are some great opportunities in both the public and private sector for fire protection careers and/or volunteer work. The key issues in successfully preparing for and securing a good position are the following:

- Attaining proper education and training
- Seeking the job that fits the career objectives
- Understanding the scope of the various positions
- Understanding and participating in the selection process
- Obtaining and maintaining appropriate credentials
- Preparing for and seeking out advancement opportunities

Fire protection and related fields offer rewards beyond those of just salary and benefits. Protecting life and property is a very noble cause. Compassionate people will do well in these careers and they will be rewarded in their level of achievement.

 Group Activity

Each group is to do an electronic search for fire protection related jobs information. Lists of jobs should include the qualifications for each job and the procedures necessary to apply for the position. The group that comes up with the most descriptions/positions is the winner. (The instructor can set the time limit. This assignment may be done outside of class and the results reported in class.)

Review Questions

1. What is the most common fire protection career field in the public sector?

2. What are some career opportunities in the public sector for nonfire-fighting careers?

3. What are some fields in the private sector with career opportunities for fire protection personnel?

4. Which federal department employs a large number of both civilian and military fire fighters to protect its bases around the world?

5. Which officer in a fire department would typically be assigned to manage a large function within the department such as fire prevention, field operations, emergency medical services, emergency management services, support services, or administration?

6. Who usually reports to a city manager, mayor, or a special district board of directors?

7. Most communities will require candidates for fire fighter positions to complete what?

8. Almost without exception, what type of examination will be required before an individual is hired?

9. Who in the department often works with a broad spectrum of the community to define the fire and life safety issues in a community and then to address those issues in a multifaceted approach?

10. Who has the primary function of establishing the cause and origin of a fire?

11. Fire protection _____ _____ and fire protection engineers and technicians are often required by architectural and engineering firms.

12. What are some factors in promotions for fire department personnel?

Suggested Readings

Crane, Bennie L. and Julian Williams. *Personal Empowerment: Achieving Individual and Departmental Excellence.* Tulsa, OK: Fire Engineering Books, 2002.

International Fire Service Training Association. *Fire Service Orientation and Terminology,* 3rd ed. Stillwater, OK: Fire Protection Publications, 1993.

Stein, Paul H. "Mastering Fire Department Entrance Interviews," Video. Tulsa, OK: Fire Engineering Books.

References

Chapter 1

1. Information gathered from www.angliacampus.com.

2. Information gathered from www.portglasgow4u.co.uk/socialhis/.

3. Lyons, Paul. *Fire in America!* Boston: National Fire Protection Association, 1976.

4. Ditzel, Paul. *Fire Engines, Firefighters.* New York: Bonanza Books, 1984.

5. Cote, A. E., ed. *Fire Protection Handbook*, 19th ed. Quincy, MA: National Fire Protection Association, 2003.

6. Harlow, Alvin F. *The Serene Cincinnatians.* New York: E. P. Dutton, 1950.

7. Capron, Walter. *The U.S. Coast Guard.* New York: Franklin Watts, 1965.

8. Smith, Dennis. *History of Firefighting in America.* New York: The Dial Press, 1978.

9. National Fire Protection Association. *Reconstruction of Tragedy: The Beverly Hills Supper Club Fire.* Quincy, MA.

10. Tremblay, Kenneth J. "Catastrophic Fire Deaths." *NFPA Journal.* September/October, 1994.

11. *The Standard Grading Schedule for Grading Cities and Towns of the United States with Reference to Their Fire Defenses and Physical Conditions.* National Board of Fire Underwriters, 1916.

12. NFPA 1710, *Standard for the Organization and Deployment of Fire Suppression Operations, Emergency Medical Operations, and Special Operations to the Public by Career Fire Departments*, 2001 edition.

13. NFPA 1720, *Standard for the Organization and Deployment of Fire Suppression Operations, Emergency Medical Operations, and Special Operations to the Public by Volunteer Fire Departments*, 2001 edition.

14. Harry E. Hickey. *Fire Suppression Rating Schedule Handbook.* Louisville, KY: Chicago Spectrum Press, 2002.

15. NFPA 472, *Standard for Professional Competence of Responders to Hazardous Materials Incidents*, 2002 edition.

16. NFPA 1001, *Standard for Fire Fighter Professional Qualifications*, 2002 edition.

17. NFPA 1002, *Standard for Fire Apparatus Driver/Operator Professional Qualifications*, 2003 edition.

18. NFPA 1003, *Standard for Airport Fire Fighter Professional Qualifications*, 2000 edition.

19. NFPA 1006, *Standard for Rescue Technician Professional Qualifications*, 2003 edition.
20. NFPA 1021, *Standard for Fire Officer Professional Qualifications*, 2003 edition.
21. NFPA 1031, *Standard for Professional Qualifications for Fire Inspector and Plan Examiner*, 2003 edition.
22. NFPA 1033, *Standard for Professional Qualifications for Fire Investigator*, 2003 edition.
23. NFPA 1035, *Standard for Professional Qualifications for Public Fire and Life Safety Educator*, 2000 edition.
24. NFPA 1041, *Standard for Fire Service Instructor Fighter Professional Qualifications*, 2002 edition.
25. NFPA 1051, *Standard for Wildland Fire Fighter Professional Qualifications*, 2002 edition.
26. NFPA 1061, *Standard for Professional Qualifications for Public Safety Telecommunicator*, 2002 edition.
27. NFPA 1071, *Standard for Emergency Vehicle Technician Professional Qualifications*, 2000 edition.
28. NFPA 1081, *Standard for Industrial Fire Brigade Professional Qualifications*, 2001 edition.

Chapter 2

1. Karter, M. "Fire Loss in the United States." Quincy, MA: National Fire Protection Association, 2002 and earlier years.
2. LeBlanc, P. and Fahy, R. "Full Report: Firefighter Fatalities in the United States—2001." Quincy, MA: National Fire Protection Association, 2002.
3. Ahrens, M. "U.S. Fire Problem Overview Report." Quincy, MA: National Fire Protection Association, 2003.
4. NFPA 901, *Uniform Coding for Fire Protection*, 1976 edition.
5. Hall, J. "Intentional Fires and Arson." Quincy, MA: National Fire Protection Association, 2003.
6. Bureau of Justice Statistics summarized in J. Hall's "Intentional Fires and Arson." Quincy, MA: National Fire Protection Association, 2003.
7. FBI's *Crime in the U.S.* series, summarized in J. Hall's "Intentional Fires and Arson." Quincy, MA: National Fire Protection Association, 2003.
8. Karter, M. "U.S. Fire Experience by Region." Quincy, MA: National Fire Protection Association, 2003.
9. Hall, J. "Patterns of Fire Casualties in Home Fires by Age and Sex." Quincy, MA: National Fire Protection Association, 2001.
10. Hall, J. "U.S. Fire Death Patterns by State." Quincy, MA: National Fire Protection Association, 2002.

11. Hall, J. *International Comparison Reports: U.S.A. vs. Canada, U.S.A. vs. Japan, U.S.A. vs. Sweden, U.S.A. vs. U.K.* Quincy, MA: National Fire Protection Association.

12. Tri-Data Corporation for the USFA. *Fire Death Rate Trends: An International Perspective,* May 1997, contract EMW-95-C-4717, downloaded from http://www.usfa.fema.gov/downloads/pdf/publications/internat.pdf on April 23, 2003.

13. Hall, J. R., Jr., and B. Harwood. "The National Estimates Approach to U.S. Fire Statistics." *Fire Technology*, Vol. 25, no. 2, May 1989, pp. 99–113.

14. Rohr, K. "Structure Fires in Dormitory Properties: Statistical Analysis." Quincy, MA: National Fire Protection Association, 2002.

Chapter 3

1. NFPA 921, *Guide for Fire and Explosion Investigations*, 2004 edition.

2. NFPA 30, *Flammable and Combustible Liquids Code*, 2003 edition.

3. NFPA 10, *Standard for Portable Fire Extinguishers*, 2002 edition.

4. NFPA 1, *Uniform Fire Code*™, 2003 edition.

5. *NFPA 5000*™, *Building Construction and Safety Code*™, 2003 edition.

Chapter 4

1. Karter, Michael. "Fire Loss in the United States During 2001." Quincy, MA: National Fire Protection Association, 2002.

2. NFPA 101®, *Life Safety Code*®, 2003 edition.

3. NFPA 221, *Standard for Fire Walls and Fire Barrier Walls*, 2000 edition.

4. NFPA 220, *Standard for Types of Building Construction*, 1999 edition.

5. NFPA 801, *Standard for Fire Protection Facilities Handling Radioactive Materials*, 2003 edition.

6. NFPA 1977, *Standard for Fire Protection Clothing and Equipment for Wildland Fire Fighting*, 1998 edition.

7. NFPA 79, *Electrical Standard for Industrial Machinery*, 2002 edition.

8. Fleischer, H.O. "The Performance of Wood in Fire." Report No. 2202. Madison, WI: Forest Products Laboratory, 1960.

9. Fitzgerald, Robert W. "Structural Integrity During Fire." Cote, A.E., ed. *Fire Protection Handbook,* 19th ed. Quincy, MA: National Fire Protection Association 2002.

10. Brannigan, Francis L. *Building Construction for the Fire Service.* Quincy, MA: National Fire Protection Association, 1992.

11. Fleischman, C. "Analytical Methods of Determining Fire Resistance of Concrete Members." DiNenno, P.J., ed. *The SFPE Handbook of Fire Protection*

Engineering, 3rd ed. Quincy, MA: National Fire Protection Association and Bethesda, MD: Society of Fire Protection Engineers, 2002.

12. ASTM E-119-00a, *Standard Test Methods for Fire Tests of Building Construction and Materials.*
13. NFPA 251, *Standard Methods of Tests of Fire Endurance of Building Construction and Materials,* 1999 edition.
14. *NFPA 5000^{TM}, Building Construction and Safety Code^{TM},* 2003 edition.
15. NFPA 80, *Standard for Fire Doors and Fire Windows,* 1999 edition.
16. NFPA 255, *Standard Method of Test of Surface Burning Characteristics of Building Materials,* 2000 edition.
17. ASTM E-84-03b, *Standard Test Method for Surface Burning Characteristics of Building Materials.*

Chapter 5

1. Brunacini, Alan. *Essentials of Fire Department Customer Service.* Stillwater, OK: International Fire Service Training Association, 1996.

Chapter 6

1. NFPA 1710, *Standard for the Organization and Deployment of Fire Suppression Operations, Emergency Medical Operations, and Special Operations to the Public by Career Fire Departments,* 2001 edition.
2. NFPA 1720, *Standard for the Organization and Deployment of Fire Suppression Operations, Emergency Medical Operations, and Special Operations to the Public by Volunteer Fire Departments,* 2001 edition.
3. NFPA 1971, *Standard on Protective Ensemble for Structural Fire Fighting,* 2002 edition.
4. NFPA 1981, *Standard on Open-Circuit Self-Contained Breathing Apparatus for Fire and Emergency Services,* 2002 edition.
5. NFPA 1961, *Standard on Fire Hose,* 2002 edition.
6. NFPA 1931, *Standard on Design of and Design Verification Tests for Fire Department Ground Ladders,* 1999 edition.
7. NFPA 1932, *Standard on Use, Maintenance, and Service Testing of Fire Department Ground Ladders,* 1999 edition.
8. NFPA 10, *Standard for Portable Fire Extinguishers,* 2002 edition.
9. *NFPA Guide to Portable Fire Extinguishers,* 2003 edition.
10. NFPA 70E, *Standard for Electrical Safety in the Workplace,* 2004 edition.
11. NFPA 1983, *Standard on Fire Service Life Safety Rope and System Components,* 2001 edition.

12. NFPA 1901, *Standard for Automotive Fire Apparatus*, 2003 edition.

13. NFPA 1906, *Standard for Wildland Fire Apparatus*, 2001 edition.

14. NFPA 72®, *National Fire Alarm Code®*, 2002 edition.

15. NFPA 1221, *Standard for the Installation, Maintenance, and Use of Emergency Services Communications Systems*, 2002 edition.

Chapter 7

1. Crosby, E. U., H. A. Fiske, and H. W. Forester. *Crosby-Fiske-Forester Handbook of Fire Protection*, 7th ed. New York: Lancaster Press, 1924.

2. "The President's Conference on Fire Prevention." Federal Works Agency, Washington D.C., 1947.

3. National Commission on Fire Prevention and Control. *America Burning.* Washington D.C.: U.S. Government Printing Office, 1973.

4. International Association of Fire Chiefs. *Wingspread IV.* Dothan, AL: IAFC Foundation, 1996.

5. NFPA 13, *Standard for the Installation of Sprinkler Systems*, 2002 edition.

6. *SFPE Engineering Guide to Performance-Based Fire Protection Analysis and Design of Buildings.* Quincy, MA: National Fire Protection Association, 2000.

7. NFPA 1, *Uniform Fire Code™*, 2003 edition.

8. *NFPA 5000™, Building Construction and Safety Code™*, 2003 edition.

9. *NFPA 101®, Life Safety Code®*, 2003 edition.

10. FBI's *Crime in the U.S.* series, summarized in J. Hall's *Intentional Fires and Arson.* Quincy, MA: National Fire Protection Association, 2003.

Chapter 8

1. *NFPA 5000™, Building Construction and Safety Code™*, 2003 edition.

2. NFPA *101®, Life Safety Code®*, 2003 edition.

3. *NFPA 72®, National Fire Alarm Code®*, 2002 edition.

4. NFPA 1150, *Standard on Fire-Fighting Foam Chemicals for Class A Fuels in Rural, Suburban, and Vegetated Areas*, 1999 edition (formerly NFPA 298).

5. NFPA 1145, *Guide for the Use of Class A Foams in Manual Structural Fire Fighting*, 2000 edition.

6. NFPA 13, *Standard for the Installation of Sprinkler Systems*, 2002 edition.

7. NFPA 22, *Standard for Water Tanks for Private Fire Protection*, 1998 edition.

8. NFPA 20, *Standard for the Installation of Stationary Pumps for Fire Protection*, 2003 edition.

9. NFPA 291, *Recommended Practice for Fire Flow Testing and Marking of Hydrants*, 2002 edition.

10. Milosh T. Puchovsky and Kenneth E. Isman, eds. *Fire Pump Handbook*, Quincy, MA: National Fire Protection Association, 1998.

11. NFPA 13R, *Standard for the Installation of Sprinkler Systems in Residential Occupancies up to and Including Four Stories in Height*, 2002 edition.

12. NFPA 13D, *Standard for the Installation of Sprinkler Systems in One- and Two-Family Dwellings and Manufactured Homes*, 2002 edition.

13. NFPA 15, *Standard for Water Spray Fixed Systems for Fire Protection*, 2001 edition.

14. NFPA 69, *Standard on Explosion Prevention Systems*, 2002 edition.

15. NFPA 13E, *Recommended Practice for Fire Department Operations in Properties Protected by Sprinkler and Standpipe Systems*, 2000 edition.

16. NFPA 14, *Standard for the Installation of Standpipe and Hose Systems*, 2003 edition.

Chapter 9

1. NFPA 921, *Guide for Fire and Explosion Investigations*, 2004 edition.

Chapter 10

1. NFPA 1561, *Standard on Emergency Services Incident Management System*, 2002 edition.

2. NFPA 1600, *Standard on Disaster/Emergency Management and Business Continuity Programs*, 2004 edition.

3. NFPA 1221, *Standard for the Installation, Maintenance, and Use of Emergency Services Communications Systems*, 2002 edition.

4. NFPA 1620, *Recommended Practice for Pre-Incident Planning*, 2003 edition.

5. NFPA 1901, *Standard for Automotive Fire Apparatus*, 2003 edition.

6. NFPA 1906, *Standard for Wildland Fire Apparatus*, 2001 edition.

7. NFPA 1925, *Standard on Marine Fire-Fighting Vessels*, 2004 edition.

8. NFPA 1405, *Guide for Land-Based Fire Fighters Who Respond to Marine Vessel Fires*, 2001 edition.

9. NFPA 414, *Standard for Aircraft Rescue and Fire-Fighting Vehicles*, 2001 edition.

10. NFPA 1710, *Standard for the Organization and Deployment of Fire Suppression Operations, Emergency Medical Operations, and Special Operations to the Public by Career Fire Departments*, 2001 edition.

11. NFPA 402, *Guide for Aircraft Rescue and Fire Fighting Operations*, 2002 edition.

12. NFPA 403, *Standard for Aircraft Rescue and Fire-Fighting Services at Airports*, 2003 edition.

13. NFPA 405, *Recommended Practice for the Recurring Proficiency Training of Aircraft Rescue and Fire-Fighting Services*, 1999 edition.

14. NFPA 424, *Guide for Airport/Community Emergency Planning*, 2002 edition.

15. NFPA 1003, *Standard for Airport Fire Fighter Professional Qualification*, 2000 edition.

16. NFPA 600, *Standard on Industrial Fire Brigades*, 2000 edition.

17. NFPA 1081, *Standard for Industrial Fire Brigade Member Professional Qualifications*, 2001 edition.

18. Title 29 CFR Part 1910.134, "Respiratory Protection."

19. NFPA 1500, *Standard on Fire Department Occupational Safety and Health Program*, 2002 edition.

20. NFPA 1720, *Standard for the Organization and Deployment of Fire Suppression Operations, Emergency Medical Operations, and Special Operations to the Public by Volunteer Fire Departments*, 2001 edition.

21. NFPA 1006, *Standard for Rescue Technician Professional Qualifications*, 2003 edition.

22. NFPA 1670 *Standard on Operations and Training for Technical Search and Rescue Incidents*, 2004 edition.

23. *Superfund Amendments and Reauthorization Act (SARA) of 1986*, Title III, "The Emergency Planning and Community Right-to-Know Act," Public Law 99-499, Oct. 17, 1986.

24. NFPA 472, *Standard for Professional Competence of Responders to Hazardous Materials Incidents*, 2002 edition.

25. NFPA 473, *Standard for Competencies for EMS Personnel Responding to Hazardous Materials Incidents*, 2002 edition.

26. NFPA 704, *Standard System for the Identification of the Hazards of Materials for Emergency Response*, 2001 edition.

27. NFPA 1991, *Standard on Vapor-Protective Ensembles for Hazardous Materials Emergencies*, 2000 edition.

28. NFPA 1992, *Standard on Liquid Splash-Protective Ensembles and Clothing for Hazardous Materials Emergencies*, 2000 edition.

29. NFPA 1994, *Standard on Protective Ensembles for Chemical/Biological Terrorism Incidents*, 2001 edition.

Chapter 11

1. *Superfund Amendments and Reauthorization Act (SARA) of 1986*, Title III, "The Emergency Planning and Community Right-to-Know Act," Public Law 99-499, Oct. 17, 1986.

2. *Homeland Security Act of 2002*, Public Law 107-296, Nov. 25, 2002.

3. Cote, A. E., ed. *Fire Protection Handbook,* 19th ed. Quincy, MA: National Fire Protection Association, 2003.

Chapter 12

1. NFPA 1001, *Standard for Fire Fighter Professional Qualifications,* 2002 edition.

2. NFPA 1002, *Standard for Fire Apparatus Driver/Operator Professional Qualifications,* 1998 edition.

3. NFPA 1021, *Standard for Fire Officer Professional Qualifications,* 2003 edition.

4. NFPA 1582, *Standard on Comprehensive Occupational Medical Program for Fire Departments,* 2003 edition.

5. NFPA 1035, *Standard for Professional Qualifications for Public Fire and Life Safety Educator,* 2000 edition.

6. NFPA 1031, *Standard for Professional Qualifications for Fire Inspector and Plan Examiner,* 2003 edition.

7. NFPA 1033, *Standard for Professional Qualifications for Fire Investigator,* 2003 edition.

8. DiNenno, P. J., ed. *The SFPE Handbook for Fire Protection Engineering,* 3rd edition. Quincy, MA: National Fire Protection Association and Bethesda, MD: Society of Fire Protection Engineers, 2002.

9. NFPA 1041, *Standard for Fire Service Instructor Professional Qualifications, 2002* edition.

10. NFPA 1051, *Standard for Wildland Fire Fighter Professional Qualifications,* 2002 edition.

11. NFPA 1081, *Standard for Industrial Fire Brigade Member Professional Qualifications,* 2001 edition.

12. NFPA 600, *Standard on Industrial Fire Brigades,* 2000 edition.

Answer Key for Chapter Review Questions

Chapter 1, Fire in History and Contemporary Life

1. The London of 1666 was a city of <u>half-timbered</u> and <u>pitch</u> covered medieval buildings, mostly with thatched roofs.
2. Housing materials and fire insurance
3. They could make money by offering fire cover policies to owners.
4. Employing men to fight fires
5. With a metal badge or mark that was fixed to the outside of a building
6. Spirits and gunpowder
7. To enforce fire laws
8. To purchase firefighting equipment
9. Leather hose
10. They bored a hole in the log.
11. Automatic sprinkler
12. National Fire Protection Association
13. Gas lines and water mains were broken.
14. A bin of rags
15. Accreditation, certification, and customer service

Chapter 2, Understanding America's Fire Problem

1. Residential properties
2. Radiated or conducted heat from operating equipment
3. That a crime was committed, and then to prove guilt
4. Only 15 percent to 20 percent resulted in arrest
5. Children under six and people over 65 years of age
6. High state of death rates
7. The United States and Canada
8. National Fire Incident Reporting System (NFIRS)
9. Cooking equipment

Chapter 3, Understanding Fire Behavior

1. Fuel, oxygen, and heat
2. The fuel and the oxidizer
3. They must be intimately mixed.
4. This is ignition that occurs in the presence of an intense heat source such as an arc, spark, or a flame.
5. Conduction, convection, and radiation
6. The most common classification system depends on the type of fuel burning and gives a general indication of the hazards involved. This system uses letters to designate the fuel type: A, B, C, D, and K.

Chapter 4, Building Design and Construction

1. Significant improvements include automatic fire suppression systems, automatic smoke detection, and improved egress systems.
2. Decisions need to be made early in the design or planning process for a new building to provide an effective and safe design from a fire safety standpoint.
3. Life safety
4. Compression, tension, and shear
5. The proper proportions of cement and water, size, type and amount of aggregates, and thorough mixing
6. Glass is not very fire-resistive but does resist the passage of smoke.
7. Many plastic products give off toxic fire gases during thermal decomposition.
8. Elevators and escalators

Chapter 5, Fire Department Structure and Management

1. Suppression includes the major goal or mission of saving lives and protecting property. To accomplish this goal, fire departments are organized into smaller <u>tactical</u> units or <u>companies</u>.
2. In many areas, the number of EMS runs greatly exceeds the number of fire calls companies respond to in any given year.
3. Good planning
4. Capital budgets detail large expenditures that generally are expected to last more than 1 year, such as apparatus, fire stations, and fire hose. Operating budgets are the day-to-day expenditures such as the salary and benefits of personnel.

5. Chain of Command, Unity of Command, Span of Control, Division of Labor, Delegation of Authority, and Discipline

6. There are several steps in the cycle of discipline. These steps are often referred to as the progressive system of discipline.

7. Maslow suggested that the primary motivator of people is their needs.

8. Assessment Centers are sometimes used to determine how an individual may respond in a management position.

9. Mitigation, preparedness, response, and recovery

10. The basic core of abilities that aid individuals in getting along with others in their environment whether at home or at work

Chapter 6, Fire Department Facilities and Equipment

1. The first criterion is time and distance, based on what the response times are and what the distance is. The second criterion is call volume, based on how many responses one facility can handle at a time.

2. Training facilities are typically located in areas that are remote from residential areas because of noise and smoke. Large metro fire departments will have additional support facilities. These can include: administrative, vehicle maintenance shop, communications center, fire prevention, fire investigation facility, logistics (warehouse), reserve apparatus storage, museums, wellness centers, and building maintenance.

3. Coats and trousers (turnouts or turnout gear) are constructed of three layers: outer shell, moisture barrier, and thermal barrier.

4. Forcible entry tools

5. Life safety rope and utility rope

6. The pumper

7. An elevating platform has an area at the end of the aerial device for firefighters to stand and operate.

8. The Communications Center

Chapter 7, Preventing Fire Loss

1. President Harry S. Truman convened the Conference on Fire Prevention on May 6–8, 1947 in Washington D.C.

2. Better building designs, the use of technology (including automatic fire sprinklers and detection systems), and the revision and strengthening of fire and building codes

3. Education, engineering, and enforcement

4. A focus on customer service, a reiteration of the need to commit to fire prevention and public education, and the need to support adoption of codes and standards

5. The United States Fire Administration and the National Fire Academy

6. Life safety programs

7. Working with the community on non-enforcement <u>fire</u> <u>prevention</u> activities often builds the confidence in the community that the fire department needs to manage an effective code enforcement program.

8. Performance-based designs

9. Planning and zoning ordinances regulate what type of occupancies can be built, where they can be built, how they can be built, and in some cases how they can be designed.

10. A regulatory agency

11. Standards are different than codes and regulations, in that they specify in great detail how something should be installed, what materials should be used, where the standard is <u>applicable</u>, and sometimes more importantly where the standard is <u>not</u> <u>applicable</u>.

Chapter 8, Controlling Fire Loss through Engineered Fire Protection Systems

1. Automatic fire detection and alarm systems, automatic fire suppression systems, manual fire suppression systems, and automatic smoke control/exhaust

2. Notification time, reaction time, and egress time

3. Recognize that there is a fire—*detection*; sound the alarm to let occupants know that there is a problem—*notification*; and carry out preplanned actions such as notifying the fire department, closing doors or activating smoke control equipment.

4. The ionization principle

5. <u>Heat</u> <u>detectors</u> placed near the ceiling are in an ideal location to intercept the hot gas stream and detect the temperature change (increase) characteristic of fires

6. By sensing the rate of change in a room's temperature

7. Flame detection systems

8. Audible notification appliances

9. Trouble signals

10. Remote supervising stations

11. Water

12. Dry chemicals
13. A wet pipe sprinkler system
14. Exhaust it to a safe location outside the building

Chapter 9, Fire Investigation

1. Fire investigation is the process of determining cause of the fire. If the fire investigation concludes that cause of the fire is incendiary, then the crime of arson has been committed and an arson investigation is undertaken.
2. "Scientific method"
3. The exterior of the building
4. Within the area of origin
5. When carrying out overhaul operations, potential <u>evidence</u> such as business and personal records, equipment manuals, operational documents and files may be encountered.
6. The first responders
7. Reduction in <u>fire losses</u> is achieved through building codes, fire protection standards, and fire prevention codes and regulations that directly address the causes of fires and the factors that control fire growth and spread.

Chapter 10, Planning for Emergency Response

1. The purpose of <u>incident command systems (incident commanders)</u> is to provide a flexible, adaptable structure for the management and coordination of emergency incident operations.
2. Unity of Command
3. The number of persons directly reporting to a supervisor
4. At a single company incident (e.g., light brush fire adjacent to a dwelling), the <u>Company Officer</u> is the Incident Commander who personally fulfills all the basic management functions.
5. Unified Command
6. Commercial or industrial occupancies
7. The Federal Emergency Management Agency (FEMA)

Chapter 11, Public and Private Support Organizations

1. Appropriate training and certifying agencies within the state and Local Emergency Planning Committee

2. Emergency management agencies, 911 communications centers, Local Emergency Planning Committees, and Emergency Medical Services Councils

3. Government: state fire training agency and the state fire marshal; Non-government: state fire chiefs association and state firefighters association

4. Government: Federal Emergency Management Agency and United States Fire Administration; Non-government: International Association of Fire Chiefs and American Fire Sprinkler Association

5. *Fire Engineering, NFPA Journal®*, and *National Fire and Rescue* are three examples but many others exist.

6. Federal Emergency Management Agency, United States Fire Administration, National Fire Academy, and Emergency Management Institute

Chapter 12, Careers in Fire Protection

1. The fire service

2. Code enforcement officer, fire investigator, and training officer

3. Health care, insurance, and construction

4. U.S. Department of Defense

5. The division officer

6. The fire chief

7. A written test

8. A comprehensive physical examination

9. The fire and life safety educator

10. A fire investigator

11. Fire protection <u>system designers</u> and fire protection engineers and technicians are often required by the architect and engineering firms.

12. Initiative, work ethic, and certifications

Index